Applied Mathematical Sciences
Volume 78

Applied Mathematical Sciences

(continued following index)

Bernard Dacorogna

Direct Methods in the Calculus of Variations

With 10 Figures

Springer-Verlag

Berlin Heidelberg New York
London Paris Tokyo

Bernard Dacorogna
Département de Mathématiques
École Polytechnique Fédérale de Lausanne
CH-1015 Lausanne, Switzerland

Editors

F. John
Courant Institute of
Mathematical Sciences
New York University
New York, NY 10012
USA

J. E. Marsden
Department of
Mathematics
University of California
Berkeley, CA 94720
USA

L. Sirovich
Division of
Applied Mathematics
Brown University
Providence, RI 02912
USA

Mathematics Subject Classification (1980): Primary number: 49; secondary numbers: 35, 73

ISBN 3-540-50491-5 Springer-Verlag Berlin Heidelberg New York
ISBN 0-387-50491-5 Springer-Verlag New York Berlin Heidelberg

Library of Congress Cataloging-in-Publication Data
Dacorogna, Bernard, 1953–
Direct methods in the calculus of variations / Bernard Dacorogna.
p. cm. – (Applied mathematical sciences; v. 78)
Bibliography: p. Includes index.
ISBN 0-387-50491-5 (U.S.)
1. Calculus of variations. I. Title. II. Series: Applied mathematical sciences (Springer-
Verlag New York Inc.); v. 78. QA1.A647   vol. 78   [QA315]   510 s–dc 19   [515'.64]
88-31552 CIP

Media conversion: EDV-Beratung Mattes, Heidelberg
2141/3140-5 4 3 2 1 0   Printed on acid-free paper

Preface

In recent years there has been a considerable renewal of interest in the classical problems of the calculus of variations, both from the point of view of mathematics and of applications. Some of the most powerful tools for proving existence of minima for such problems are known as direct methods. They are often the only available ones, particularly for vectorial problems. It is the aim of this book to present them. These methods were introduced by Tonelli, following earlier work of Hilbert and Lebesgue.

Although there are excellent books on calculus of variations and on direct methods, there are recent important developments which cannot be found in these books; in particular, those dealing with vector valued functions and relaxation of non convex problems. These two last ones are important in applications to nonlinear elasticity, optimal design.... . In these fields the variational methods are particularly effective. Part of the mathematical developments and of the renewal of interest in these methods finds its motivations in nonlinear elasticity. Moreover, one of the recent important contributions to nonlinear analysis has been the study of the behaviour of nonlinear functionals under various types of convergence, particularly the weak convergence. Two well studied theories have now been developed, namely Γ-convergence and compensated compactness. They both include as a particular case the direct methods of the calculus of variations, but they are also, both, inspired and have as main examples these direct methods.

This monograph is addressed to readers having some elementary notions of functional analysis and Sobolev spaces; however, most of the facts which we use, concerning these notions, are listed and sometimes proved in Chapter 2. Chapter 3 is concerned with minimization problems involving only scalar functions, while Chapter 4 deals with vector valued functions. Finally, in Chapter 5 we study the relaxation of non-convex problems in the scalar as well as the vectorial case. In an appendix we give some applications to nonlinear elasticity and optimal design of the theory developed earlier.

This book is part of a more extended project that originally arose jointly with L. Boccardo. We finally decided to split it into two parts. This monograph corresponds to the first of these and essentially deals with the calculus of variations. The second part will be concerned with nonlinear elliptic partial differential equations and will appear later.

A large part of the present book has been influenced by long discussions with L. Boccardo, particularly as concerns Chapter 3 and the plan of the monograph. Without his collaboration this book would never have been written.

I would like to thank I. Ekeland who, from the beginning, showed enthusiasm for this project and encouraged me strongly to go ahead with it. It was while giving a graduate course in Paris-Dauphine that the project of writing a book originated.

I want to thank J.C. Evard who helped me in writing and in dealing with complicated notations in the Appendix of Chapter 4.

My thanks also go to B. Kawohl who pointed out several mistakes in the manuscript; to P. Ciarlet who communicated to me the manuscript of his recent book which helped me in the writing of the Appendix and of the Bibliography; to J.M. Ball, P. Marcellini, J. Moser, E. Zehnder; to my colleagues in Lausanne C.A. Stuart, B. Zwahlen and to many others with whom I had interesting and helpful discussions.

Particular thanks go to Mrs G. Rime who not only skilfully and rapidly typed the manuscript, but also was patient with all changes and constraints I imposed on her.

Springer-Verlag and the editors of the Applied Mathematical Sciences series have been encouraging and efficient during the process of reviewing and editing this book.

Finally I would like to thank the institutions that supported me during the writing of this book, the Ecole Polytechnique Fédérale de Lausanne, the Université de Paris-Dauphine and the Eidgenössische Technische Hochschule in Zürich.

Lausanne, October 1988   B. DACOROGNA

Table of Contents

CHAPTER 1

Introduction

1.1 General Considerations and Some Examples

1.1.1 Statement of the Problem and Some Examples

In this book we shall be concerned with one of the central problems of the calculus of variations which is to find among all functions with prescribed boundary condition, those which minimize a given functional. More precisely let

$$(P) \qquad \min\left\{I(u) : u \in X \text{ and } u = u_0 \text{ on } \partial\Omega\right\}$$

where

$$I(u) = \int_\Omega f(x, u(x), \nabla u(x))\, dx \tag{1.1}$$

and $\Omega \subset \mathbf{R}^n$ is a bounded open set, $u : \Omega \subset \mathbf{R}^n \to \mathbf{R}^m$, $\nabla u \in \mathbf{R}^{nm}$, $f : \Omega \times \mathbf{R}^m \times \mathbf{R}^{nm} \to \mathbf{R}$ is a continuous function, u_0 is a given function and X is a Banach space.

The story of this problem is as old as the calculus of variations itself and we shall give some brief historical comments below, but we first give some examples.

EXAMPLES.

i) One of the first problems studied by methods from the calculus of variations is the so called brachistochrone problem, where $n = m = 1$ and

$$f(x, u, u') = \frac{\sqrt{1 + u'^2}}{\sqrt{2g(u - a)}} \tag{1.2}$$

where g and a are constants and $u > a$.

ii) Another classical example is the Fermat principle in geometrical optics, where $n = m = 1$ and

$$f(x, u, u') = g(x, u)\sqrt{1 + u'^2} . \tag{1.3}$$

iii) If we turn our attention to multiple integrals, i.e. $n > 1$, then the most celebrated example is the so called Dirichlet integral where $n > 1, m = 1$ and

$$f(x, u, \nabla u) = |\nabla u|^2 . \tag{1.4}$$

iv) Another extensively studied example is the minimal surface problem in non parametric form where $n > 1, m = 1$ and

$$f(x, u, \nabla u) = \sqrt{1 + |\nabla u|^2} . \tag{1.5}$$

In this case, the integral I represents the area of the surface defined by the function u over Ω.

v) Related to the previous case is the minimal surface problem in parametric form where $m = n + 1$ and

$$f(x, u, \nabla u) = \left[\sum_{\nu=1}^{n+1} \left(\frac{\partial(u_1, \ldots, u_{\nu-1}, u_{\nu+1}, \ldots, u_{n+1})}{\partial(x_1, \ldots, x_n)} \right)^2 \right]^{1/2} \tag{1.6}$$

where $\frac{\partial(u_1, \ldots, u_n)}{\partial(x_1, \ldots, x_n)}$ denotes the determinant of the matrix $\left(\frac{\partial u_i}{\partial x_j} \right)_{1 \leq i,j \leq n}$. In particular if $n = 2$, then

$$f(x, u, \nabla u) = \left[\left(\frac{\partial(u_2, u_3)}{\partial(x_1, x_2)} \right)^2 + \left(\frac{\partial(u_3, u_1)}{\partial(x_1, x_2)} \right)^2 + \left(\frac{\partial(u_1, u_2)}{\partial(x_1, x_2)} \right)^2 \right]^{1/2} .$$

As in the preceding example, the integral I represents the area of the surface defined by u.

vi) The last example we shall mention is the case where $m = n$ and

$$f(x, u, \nabla u) = g(\det \nabla u) \tag{1.7}$$

where $g : \mathbb{R} \to \mathbb{R}$. In particular if $g(x) = |x|$, then (1.7) represents the Jacobian of the change of coordinates $u : \mathbb{R}^n \to \mathbb{R}^n$.

The calculus of variations has generated a great deal of mathematical work since the pioneering time of Bernoulli and Euler. During the 19th century, new methods were developed and important contributions were made by Lagrange, Riemann, Weierstrass, Jacobi, Hamilton ... These methods are called classical. With the beginning of the 20th century, different techniques were introduced by Hilbert and Lebesgue in connection with the study of the Dirichlet integral. These methods were then generalized by Tonelli and are now known as the direct methods; since then these methods have been extensively used and generalized and we shall mention in the next chapters some of the more recent developments. Of course different approaches were also introduced (e.g. Morse theory), but we shall not be concerned with them in this book. We shall close these very brief historical comments by mentioning that among the celebrated problems of Hilbert, the 19th and 20th are devoted to the study of problem (P). For an extensive history of the calculus of variations, one can refer to Goldstine, Hildebrandt ...

It is clear that the problems of the calculus of variations have very wide applications in other fields of mathematics and in many areas of physics, economics, biology .... The early problems, such as the brachistochrone, were related to classical mechanics. In recent years, part of the renewal of interest in variational methods finds its origins in nonlinear elasticity. We shall, at the end of the book, give some applications to elasticity.

Let us now return to the minimization problem (P). We see that three natural questions arise in connection with (P).

 i) Existence of solutions in a "reasonable" space X
 ii) Uniqueness of solutions
iii) Properties of solutions (for instance, regularity).

In this monograph, we shall be concerned essentially with the first question. In fact, we shall concentrate our attention to two different approaches, the classical and the direct.

1.1.2 The Classical Approach

As in the finite dimensional case, one way of studying the problem is by finding the zeroes of the "derivative" of I, $I'(u) = 0$ known as the *Euler equation*, and then studying the positivity of the second variation around solutions of the Euler equations. To do so, there are numerous necessary or sufficient conditions to be studied, namely Euler equations, Weierstrass, Legendre or Jacobi conditions ... In this book, we shall mostly study the Euler equations associated to I, which are, if $u \in C^2$,

$$(E) \quad I'(u) = \left(-\sum_{i=1}^{n} \frac{\partial}{\partial x_i} \left(\frac{\partial f}{\partial \xi_{i\alpha}}(x, u, \nabla u) \right) + \frac{\partial f}{\partial u_\alpha}(x, u, \nabla u) \right)_{1 \leq \alpha \leq m} = 0$$

where $f = f(x, u, \xi)$.

We now give two examples of such equations

EXAMPLES.

i) Let $m = 1$, $n \geq 1$ and consider the problem

$$(P) \qquad \min\left\{ I(u) = \int_\Omega |\nabla u|^2 \, dx \; : \; u = u_0 \text{ on } \partial\Omega \right\} .$$

Then the associated Euler equation is

$$(E) \qquad \begin{cases} -\Delta u = 0 & \text{in } \Omega \\ u = u_0 & \text{on } \partial\Omega \end{cases} . \qquad (1.8)$$

ii) Similarly for the minimal surface case in non parametric form ($m = 1, n \geq 1$) and

$$(P) \qquad \min\left\{ I(u) = \int_\Omega \sqrt{1 + |\nabla u|^2} \, dx \; : \; u = u_0 \text{ on } \partial\Omega \right\}$$

$$(E) \qquad \begin{cases} -\sum_{i=1}^n \dfrac{\partial}{\partial x_i}\left[(1 + |\nabla u|^2)^{-1/2} \dfrac{\partial u}{\partial x_i} \right] = 0 & \text{in } \Omega \\ u = u_0 & \text{on } \partial\Omega \end{cases} . \qquad (1.9)$$

1.1.3 Direct Methods

The direct methods, which are the main concern of this monograph, consist in dealing directly with the functional I. In order to understand more clearly the essence of these techniques, we start with the finite dimensional case. One way of proving existence of minima is to find minimizing sequences belonging to a bounded and closed set (since we are in a finite dimensional space this means that the set is compact) and one therefore may extract a convergent subsequence, which by continuity of the function converges necessarily to a minimum. In fact, it is not necessary that the function I be continuous

(i.e. $\lim_{\nu \to \infty} I(u_\nu) = I(u)$ whenever $u_\nu \to u$) but only lower semicontinuous

(i.e. $\lim_{\nu \to \infty} I(u_\nu) \geq I(u)$), since we are interested in minima.

The idea of the direct methods is to reproduce this analysis in the infinite dimensional case and to find i) minimizing sequences belonging to a bounded and closed set and to ensure that ii) I is lower semicontinuous. The problem is more delicate than before in the sense that in infinite dimensional space the first hypothesis is in general not sufficient to allow extraction of a convergent subsequence. This is possible only in a weaker topology than the usual one. The second requirement is then that the functional I should be lower semicontinuous with respect to this weak topology, i.e. the one which ensures the compactness of the minimizing sequence. We shall show that this last requirement is fulfilled by a sufficiently large class of functionals I, namely those with convex integrands f.

In order to be more precise, one needs to introduce the space of functions X under consideration and define what we mean by weak topology.

A reasonable space of functions X must be complete under integral type norms and therefore can only be a Sobolev type space $W^{1,p}$, and not a space of smooth functions, e.g. C^1 functions; the question of knowing that if the solution found which is in X is then sufficiently regular (say C^1) is a problem of regularity theory.

Having chosen X in this way we see that in order to get compactness of the minimizing sequences the easiest way is to endow X with its weak topology. Recall that

$$u_\nu \rightharpoonup u \ (\text{weakly}) \ \Leftrightarrow \ \langle u_\nu; u^* \rangle \to \langle u; u^* \rangle \ \text{for every} \ u^* \in X^* \,,$$

where X^* is the dual of X and $\langle \bullet; \bullet \rangle$ is the dual pairing. One of the remarkable features of these topologies is that uniformly bounded sequences are precompact (provided X is reflexive), i.e.

$$\|u_\nu\|_X \leq K \ \Leftrightarrow \ u_\nu \rightharpoonup u \ (\text{up to a subsequence}) \,. \tag{1.10}$$

Returning to the problem (P)

$$(P) \quad \min \left\{ I(u) = \int_\Omega f(x, u(x), \nabla u(x)) \, dx \ : \ u = u_0 \ \text{on} \ \partial\Omega, \ u \in X \right\}$$

we see therefore, in view of property (1.10), that in order to ensure compactness in the weak topology, it is sufficient to ensure boundedness of minimizing sequences and this can be easily obtained by imposing an appropriate behaviour on the function f at infinity (as in the finite dimensional case). The second part, i.e. to verify the lower semicontinuity of I with respect to weak convergence, is however much more delicate and will be investigated thoroughly in this monograph. Let us just recall that this property is intimately related to the convexity of f, with respect to the variable ∇u. (Note already that this is the case if f is as in the examples one through four above.)

1.2 The Direct Methods

We now describe the content of this monograph and we shall only mention
the most important theorems; but we first introduce some notations. Recall
that

$$I(u) = \int_\Omega f(x, u(x), \nabla u(x))\, dx$$

and $u : \Omega \subset \mathbb{R}^n \to \mathbb{R}^m (\nabla u \in \mathbb{R}^{nm})$.

  i) We shall speak of the *scalar case* if either $n = 1$ or $m = 1$. Note that
     the first one, $n = 1$, corresponds to the case of single integrals, while the
     case $m = 1$ and $n > 1$ corresponds to the case of multiple integrals. Note
     also that the Euler equation $I'(u) = 0$ is in the case $n = 1$ an *ordinary*
     differential equation (or a system if $m > 1$) and in the case $m = 1, n > 1$
     a *single partial* differential equation.
 ii) We shall speak of the *vectorial case* whenever $m, n > 1$. In this case the
     Euler equations are *systems of partial* differential equations.

Both cases present different features and we shall concentrate on the scalar
case in Chapter 3 and on the vectorial case in Chapter 4 and 5.

1.2.1 Preliminaries

In Chapter 2 we introduce the necessary background for the succeeding chapters. We shall usually state theorems and prove only some of them. In the
first section we study, in particular the notion of weak concergence in L^p
and we prove an important and well known theorem, which generalizes the
Riemann-Lebesgue lemma, on weak limits of periodic functions. We also give
very briefly, and only to set up the notations, the definitions of Sobolev spaces
as well as the Sobolev and Rellich imbedding theorems.

In the second section of Chapter 2, we give the basic results of convex
analysis that we shall need. In particular, we state different versions of the
Hahn-Banach theorem, which allows us to define the notion of duality for
convex functions. Finally, we close this chapter by proving the Carathéodory
theorem for sets $M \subset \mathbb{R}^n$, asserting that the closed convex hull of M is
obtained through $(n + 1)$ convex combinations of elements of M.

1.2.2 Abstract Results and the Scalar Case

Chapter 3 is divided into four sections. The first one deals with abstract
functionals $I : X \to \mathbb{R}$, where X is a Banach space, the second and the third
study the particular case where I is of the form

$$I(u) = \int_\Omega f(x, u(x), \nabla u(x)) \, dx$$

and X the Sobolev space $W^{1,p}$. In the fourth section we derive the Euler equations associated to I.

In the first section we establish the following well known result

Theorem 1. *Let X be a reflexive Banach space, $I : X \to \mathbb{R}$ lower semicontinuous, then*
Part 1: If I is convex over X then I is (sequentially) weakly lower semicontinuous i.e.,

$$\liminf_{\nu \to \infty} I(u_\nu) \geq I(u)$$

whenever $u_\nu \rightharpoonup u$ in X (weakly).
Part 2: Let I be weakly lower semicontinuous and coercive, i.e.,

$$\lim_{\|u\| \to \infty} \frac{I(u)}{|u|} \geq \alpha > 0 \ ,$$

then

$$(P) \qquad\qquad\qquad \inf\{I(u) : u \in X\}$$

admits at least one solution.
Part 3:
i) Let $\bar{u}$ be a minimum for (P) and let I be Gâteaux differentiable, then

$$I'(\bar{u}) = 0$$

ii) If $I'(\bar{u}) = 0$ and I is convex, then $\bar{u}$ is a minimum for (P).

We then apply the above abstract results to the case of the calculus of variations, i.e.,

$$I(u) = \int_\Omega f(x, u(x), \nabla u(x)) \, dx$$

where $\Omega \subset \mathbb{R}^n$ is a bounded open set, $u : \Omega \to \mathbb{R}^m$, $\nabla u \in \mathbb{R}^{nm}$, $f : \Omega \times \mathbb{R}^m \times \mathbb{R}^{nm} \to \mathbb{R}$ is continuous and $X = \{u \in W^{1,p}(\Omega) : u = u_0 \text{ on } \partial\Omega\}$.

In Chapter 3 we restrict our attention to the scalar case i.e., $m = 1$ (or $n = 1$). We study what conditions on f ensure first the convexity, then the weak lower semicontinuity of the functional I.

In Section 2, we discuss the first problem and we obtain a necessary condition, namely

Theorem 2. *If $I : W_0^{1,p}(\Omega) \to \mathbb{R}$ is convex, then $f(x, u, \bullet)$ is convex.*

We then study sufficient conditions on several examples, the simplest and most important case being the one with no explicit dependence on u, i.e.,

$$f(x, u, \xi) \equiv f(x, \xi)$$

where the convexity of f turns out to be, trivially, sufficient for the convexity of I. Therefore in view of the abstract result (Theorem 1) the case of the Dirichlet integral is solved since then

$$f(x, u, \xi) = |\xi|^2$$

is obviously convex and coercive and therefore

$$(P) \qquad \inf \left\{ \int_\Omega |\nabla u|^2 \, dx \, : \, u = u_0 \text{ on } \partial\Omega, \ u \in W^{1,2}(\Omega) \right\}$$

has a minimum.

We then restrict ourselves to the case $m = n = 1$ and we study some examples such as $f(x, u, \xi) = a(u)\xi^2$ or $f(x, u, \xi) = g(u) + h(\xi)$ and we obtain necessary and sufficient conditions on the function f such that the functional I is convex over $W_0^{1,p}$.

In Section 3, we turn our attention to the weak lower semicontinuity of the functional I. The main result of this chapter is

Theorem 3. *I is (sequentially) weakly lower semicontinuous if and only if $f(x, u, \bullet)$ is convex.*

This result is of fundamental importance and has a long history. This theorem originated with Hilbert and Lebesgue with their study of the Dirichlet integral. However, the first one to obtain a general result was Tonelli. His result was then generalized in many ways by Serrin, Cacciopoli-Scorza-Dragoni, Morrey, Mac Shane, De Giorgi, Berkowitz, Cesari, Ioffe, Olech, Ekeland-Témam, Marcus-Mizel, Rockafellar, Marcellini-Sbordone . . . .

Note that the four first examples given at the beginning of the introduction have convex integrands while the two last do not.

From Theorem 1 and 3 one sees that if f is coercive (i.e., $f(x, u, \xi) \geq a|\xi|^p$ with $p > 1$) and convex, then the existence of a minimum follows immediately.

At the end of this section we present the well known theorem of Weyl and Carathéodory on invariant integrals,

Theorem 4. *I is invariant, i.e.,*

$$I(u) = \int_\Omega f(x, u(x), \nabla u(x)) \, dx \equiv \text{constant}$$

for every $u \in W^{1,\infty}(\Omega)$ *with* $u = u_0$ *on* $\partial\Omega$ *if and only if there exists* $\Phi :$ $\Omega \times \mathbb{R} \to \mathbb{R}^n$ *and a function* $\beta : \mathbb{R}^n \to \mathbb{R}$ *such that*

$$f(x, u, \nabla u) = \mathrm{div}[\Phi(x, u(x))] + \beta(x) .$$

The above result has important applications in the field theories of the calculus of variations.

Finally, in Section 4 of this chapter, we show how to derive the Euler equations associated to the minimization problem, under various types of conditions on f and on the minimum $\bar{u}$. We close this section with the presentation of the so called Lavrentiev phenomenon, which exhibits some pathological behaviour when the usual hypotheses of growth and coercivity conditions are violated.

1.2.3 The Vectorial Case

In Chapter 4 we turn our attention to the vectorial case, i.e.,

$$I(u) = \int_\Omega f(x, u(x), \nabla u(x)) \, dx \tag{2.1}$$

where $u : \Omega \subset \mathbb{R}^n \to \mathbb{R}^m$, with $n, m > 1$.

We study, as in Chapter 3, under what conditions on f the functional I is weakly lower semicontinuous. The results are very different from those of the scalar case. Although the convexity of $f(x, u, \bullet)$ is still sufficient to ensure the weak lower semicontinuity of I, it is far from being necessary and furthermore it is not satisfied in many important examples such as

EXAMPLES.

i) Minimal surfaces in parametric form where $m = n + 1$ and

$$f(x, u, \nabla u) = \left[\sum_{\nu=1}^{n+1} \left(\frac{\partial(u_1, \ldots, u_{\nu-1}, u_{\nu+1}, \ldots, u_{n+1})}{\partial(x_1, \ldots, x_n)} \right)^2 \right]^{1/2} . \tag{2.2}$$

ii) If $m = n$ and

$$f(x, u, \nabla u) = g(\det \nabla u) \tag{2.3}$$

where $g : \mathbb{R} \to \mathbb{R}$ is convex.

Nevertheless, the two examples above lead to weakly lower semicontinuous functionals and one therefore needs to introduce, following Morrey, a necessary and sufficient condition known as *quasiconvexity*. In practice, however, this condition is hard to verify and we are led to introduce (c.f. Morrey and Ball) a stronger and a weaker condition called *polyconvexity* and *rank one convexity* respectively.

We now define these notions and give some examples (for simplicity we drop the dependence on x and u in the function f).

Definitions.

i) $f : \mathbf{R}^{nm} \to \mathbf{R}$ is said to be *quasiconvex* if

$$\frac{1}{\operatorname{meas} D} \int_D f(\xi + \nabla\varphi(x))\,dx \geq f(\xi)$$

for every bounded domain $D \subset \mathbf{R}^n$, for every $\xi \in \mathbf{R}^{nm}$ and for every $\varphi \in C_0^\infty(D; \mathbf{R}^m)$, i.e. $\varphi = 0$ on ∂D.

ii) $f : \mathbf{R}^{nm} \to R$ is said to be *rank one convex* if

$$\lambda f(\xi) + (1 - \lambda)f(\eta) \geq f(\lambda\xi + (1 - \lambda)\eta)$$

for every $\lambda \in [0,1]$, $\xi, \eta \in \mathbf{R}^{nm}$ with $\operatorname{rank}\{\xi - \eta\} \leq 1$.

iii) $f : \mathbf{R}^{nm} \to \mathbf{R}$ is said to be *polyconvex* if there exists a convex function g such that

$$f(\xi) = g(T(\xi))$$

where $T(\xi)$ is a vector composed of all $s \times s$, $1 \leq s \leq \inf\{n, m\}$, minors of the matrix $\xi \in \mathbf{R}^{nm}$.

EXAMPLES.

i) If $m = n = 2$, then the last definition can be read as $T(\xi) = (\xi, \det \xi)$ and

$$f(\xi) = g(\xi, \det \xi)$$

so the function

$$f(\xi, \det \xi) = |\xi|^2 + (\det \xi)^2$$

where $|\xi|^2 = \sum_{i,j=1}^2 (\xi_{ij})^2$ is polyconvex (but not convex).

ii) The two examples in (2.2) and (2.3) are polyconvex (but not convex).

One of the main theorems that we shall obtain is

Theorem 5.

i) f convex $\overset{\Leftarrow}{\nRightarrow}$ f polyconvex $\overset{\Leftarrow}{\nRightarrow}$ f quasiconvex $\overset{\overset{?}{\Leftarrow}}{\Rightarrow}$ f rank one convex

$$\Updownarrow \qquad\qquad\qquad \Updownarrow$$

I is weakly lower   The Euler equations
semicontinuous $I'(u) = 0$ are elliptic.

ii) In the scalar case, i.e. $m = 1$ or $n = 1$, all these notions are equivalent.

The above theorem is essentially due to Morrey and has been refined by Meyers, Acerbi-Fusco, Marcellini … .

We then give numerous examples and counter-examples illustrating these notions.

As usual, it is then easy to deduce, from Theorem 1 and 5, the existence of minima.

We conclude this chapter with a typical phenomenon of the vectorial case, which is that there are *nonlinear* weakly continuous functions f and we just give here an example.

Theorem 6. *Let $m = n = 2$, $p > 2$ and*

$$u_\nu \rightharpoonup u \ in \ W^{1,p}(\Omega; \mathbb{R}^2)$$

then

$$\det \nabla u_\nu \rightharpoonup \det \nabla u \ in \ L^{p/2}(\Omega).$$

Following Ball and Reshetnyak, we shall give a complete characterization of such functions for general m and n, namely that they are linear combinations of minors of the matrix $\nabla u \in \mathbb{R}^{nm}$.

1.2.4 Nonconvex Integrands

In the last chapter, we study the problem (P) in the case when f is not convex (not quasiconvex, in the vectorial case). In general, such problems have no classical solution (i.e. in a Sobolev space) as the following simple examples indicate.

EXAMPLES.

i) (Bolza) Let $m = n = 1$, $\Omega = (0,1)$ and

$$(P) \qquad \inf\left\{ I(u) = \int_0^1 (u^2 + (u'^2 - 1)^2)\, dx \ : \ u(0) = u(1) = 0, \right.$$

$$\left. u \in W^{1,4}(0,1)\right\},$$

then (P) has no solution.

ii) Let $m = 1$, $n = 2$, $\Omega = (0,1)^2$ and

$$(P) \qquad \inf\left\{ I(u) = \int_0^1 \int_0^1 (u_x^4 + (u_y^2 - 1)^2)\, dx dy \ : \ u = 0 \ \text{on} \ \partial\Omega, \right.$$

$$\left. u \in W^{1,4}(\Omega)\right\}$$

where $u_x = \frac{\partial u}{\partial x}$ and $u_y = \frac{\partial u}{\partial y}$, then (P) has no solution.

Therefore, one is lead to introduce the so called *relaxed* problem (QP) which is a regular problem and then study the relationship between (P) and (QP).

The chapter is divided into two sections. The first one introduces the notion of convex, polyconvex, quasiconvex and rank one convex envelopes which are defined as

$$Cf = \sup\{g \leq f : g \text{ convex}\}$$

$$Pf = \sup\{g \leq f : g \text{ polyconvex}\}$$

$$Qf = \sup\{g \leq f : g \text{ quasiconvex}\}$$

$$Rf = \sup\{g \leq f : g \text{ rank one convex}\}$$

These envelopes, and particularly Qf, play an important role in the relaxation theorem below. The convex envelope Cf has been intensively studied in convex analysis, c.f. Fenchel, Moreau, Rockafellar ... .

In view of Theorem 5, we always have

$$Cf \leq Pf \leq Qf \leq Rf \leq f \ ,$$

and in the scalar case the different envelopes are all equal.

Following Kohn-Strang and Dacorogna, we characterize these envelopes using the Hahn-Banach and Carathéodory theorem. We then give some examples where one can compute explicitly all the different envelopes.

In the second section, we return to the study of

$$(P) \qquad \inf\left\{ I(u) = \int_\Omega f(x, u(x), \nabla u(x))\, dx \ : \ u = u_0 \text{ on } \partial\Omega \right\}$$

where $f(x, u, \bullet)$ is not convex (not quasiconvex, in the vectorial case). We then introduce the *relaxed* problem

$$(QP) \qquad \inf\left\{ \bar{I}(u) = \int_\Omega Qf(x, u(x), \nabla u(x))\, dx \ : \ u = u_0 \text{ on } \partial\Omega \right\}$$

where Qf is the quasiconvex envelope of f, with respect to the last variable, and we shall show that

Theorem 7.

i) $\inf(P) = \inf(QP)$.

ii) More precisely, if $\bar{u} \in W^{1,p}$ is a solution of (QP), then there exists $u_\nu \in W^{1,p}$, a minimizing sequence of (P), such that

$$u_\nu = u_0 \ \text{on} \ \partial\Omega$$

$$u_\nu \rightharpoonup \bar{u} \ \text{in} \ W^{1,p}$$

$$I(u_\nu) \to \bar{I}(\bar{u}) \ .$$

This approach of relaxation is also used intensively in control theory and it goes back to L.C. Young and Mac Shane. The above theorem is due to L.C. Young, Ekeland, Ekeland-Témam, Ioffe-Tihomirov, in the scalar case, and to Dacorogna in the vectorial case.

1.2.5 Applications to Nonlinear Elasticity

The above theory can be applied to various contexts in physics, economics, biology .... We shall here discuss some applications to nonlinear elasticity and optimal design. As we shall see in the Appendix, one can write the energy function for a hyperelastic material as

$$I(u) = \int_\Omega f(x, u(x), \nabla u(x)) \, dx$$

where $\Omega \subset \mathbb{R}^n$ $(n = 1, 2, 3)$ is a reference configuration, $u : \Omega \to \mathbb{R}^n$ represents a deformation of the elastic body and $f(x, u, \nabla u)$ is usually given by $W(x, \nabla u) + \Psi(x, u)$, where W is the stored energy function and Ψ is a body force potential. One also prescribes some boundary conditions such as Dirichlet or mixed boundary conditions. The problem is then to minimize I subject to these conditions.

We shall decompose the Appendix into two parts. The first one will present applications of Chapters 3 and 4 to existence and uniqueness results. We shall first state and prove the celebrated theorem of Ball on existence of absolute minimizers for such problems. We also give some uniqueness results due to Knops and Stuart.

The second part of the Appendix deals with non convex, in fact non quasiconvex, problems. It will show how the relaxation theorem and the actual computation of the different envelopes Pf, Qf and Rf are used. These non convex problems are important in modelling, for example, phase transitions in different contexts, such as equilibrium of gases, changes of crystals due to variation of the temperature. They can also be used in optimal design theory.

CHAPTER 2

Preliminaries

2.1 L^p and Sobolev Spaces

In this section we only give the definitions and main theorems that we shall need in the next chapters. Most of the theorems are standard and their proofs as well as a deeper analysis are available in several classical textbooks.

2.1.1 Weak Convergence in L^p

We give here some details on the notion of weak convergence and we refer for a longer development to Dunford-Schwartz [1] or to Yosida [1]. We first define abstractly the notion of weak convergence and apply then these results to L^p spaces.

2.1.1.1 Abstract Results

We start with the definition.

Definition. Let X be a Banach space, X^* its dual and $\langle \bullet; \bullet \rangle$ the bilinear canonical pairing over $X \times X^*$.

i) We say that, $x_n,\, x \subset X$, x_n *converges weakly* to x and we denote

$$x_n \rightharpoonup x \quad \text{in } X$$

if

$$\langle x_n; x^* \rangle \to \langle x; x^* \rangle$$

for every $x^* \in X^*$.

ii) We say that, x_n^*, $x^* \in X^*$, x_n^* *converges weak* $*$ *to* x^* *and we denote*

$$x_n^* \xrightarrow{\;*\;} x^* \quad \text{in } X^*$$

if

$$\langle x; x_n^* \rangle \rightarrow \langle x; x^* \rangle$$

for every $x \in X$.

We then have the following theorem.

Theorem 1.1.

I) Let X be a Banach space.

* i) Let $x_n \rightharpoonup x$, then there exists $K > 0$ such that*

$$\|x_n\| \leq K \;,$$

* furthermore*

$$\|x\| \leq \liminf_{n\to\infty} \|x_n\| \;.$$

* ii) Let $x_n^* \xrightarrow{\;*\;} x^*$, then there exists $K > 0$ such that*

$$\|x_n^*\|_{X^*} \leq K$$

* furthermore*

$$\|x^*\|_{X^*} \leq \liminf_{n\to\infty} \|x_n^*\|_{X^*} \;.$$

* iii) If $x_n \rightarrow x$ (strongly), then $x_n \rightharpoonup x$ (weakly).*

* iv) If $x_n^* \rightarrow x^*$ (strongly), then $x_n^* \xrightarrow{\;*\;} x^*$ (weak $*$).*

II) Let X be a reflexive Banach space, let $K > 0$ and let

$$\|x_n\| \leq K$$

then there exists $x \in X$ and a subsequence $\{x_{n_j}\}$ of $\{x_n\}$ such that

$$x_{n_j} \rightharpoonup x \quad \text{in } X \;.$$

III) Let X be a separable Banach space, let $K > 0$ and let

$$\|x_n^*\|_{X^*} \leq K$$

then there exists $x^* \in X^*$ and a subsequence $\{x_{n_j}^*\}$ of $\{x_n^*\}$ such that

$$x_{n_j}^* \xrightarrow{\;*\;} x^* \quad \text{in } X^* \;.$$

We next mention an important result which hints on the importance of the notion of convexity when one deals with weak convergence.

Theorem 1.2 (Mazur lemma). *Let X be a Banach space and let*

$$x_\nu \rightharpoonup x \quad \text{in } X .$$

Let $\varepsilon > 0$, then there exist $n = n(\varepsilon)$ an integer, $\alpha_i = \alpha_i(\varepsilon) \geq 0$ with $\sum_{i=1}^{n} \alpha_i = 1$ such that

$$\left\| x - \sum_{i=1}^{n} \alpha_i x_i \right\| \leq \varepsilon .$$

2.1.1.2 Weak Convergence in L^p

We now see how the above results can be applied to L^p spaces, but before that we first introduce the following notations.

Definitions.

i) Let $1 \leq p < \infty$ and let $\Omega \subset \mathbb{R}^n$ be an open set. A measurable function $f : \Omega \to \mathbb{R}$ is said to be in $L^p(\Omega)$ if

$$\|f\|_{L^p} \equiv \left(\int_\Omega |f(x)|^p dx \right)^{1/p} < \infty .$$

ii) Let $p = \infty$ and $\Omega \subset \mathbb{R}^n$ be an open set. A measurable function $f : \Omega \to \mathbb{R}$ is said to be in $L^\infty(\Omega)$ if

$$\|f\|_{L^\infty} \equiv \inf\{\alpha : |f(x)| \leq \alpha \text{ a.e. in } \Omega\} < \infty .$$

REMARKS.

i) Let $1 \leq p \leq \infty$, we denote by p' the conjugate exponent of p, i.e., $1/p + 1/p' = 1$, where it is understood that if $p = 1$ then $p' = \infty$ and reciprocally.

ii) Let $1 \leq p < \infty$, then the dual space of $L^p(\Omega)$ is $L^{p'}(\Omega)$. Note also that the dual space of $L^\infty(\Omega)$ contains strictly $L^1(\Omega)$.

iii) The notion of weak convergence in $L^p(\Omega)$ becomes:
   Case 1: Let $1 \leq p < \infty$, then $f_n \rightharpoonup f$ in $L^p(\Omega)$ if

$$\int_\Omega f_n(x) g(x) \, dx \to \int_\Omega f(x) g(x) \, dx$$

for every $g \in L^{p'}(\Omega)$.

Case 2: Let $p = \infty$, then $f_n \overset{*}{\rightharpoonup} f$ in $L^\infty(\Omega)$ if

$$\int_\Omega f_n(x)g(x)\,dx \;\rightarrow\; \int_\Omega f(x)g(x)\,dx$$

for every $g \in L^1(\Omega)$.

With the help of the above definitions and of Theorem 1.1 we have

Theorem 1.3. *Let $\Omega \subset \mathbb{R}^n$ be an open set.*

Case 1: *Let $p = 1$ and let*

$$f_n \rightharpoonup f \quad \text{in } L^1(\Omega)$$

then there exists $K > 0$ such that

$$\|f_n\|_{L^1} \leq K \ .$$

Case 2: *Let $1 < p < \infty$.*
i) Let

$$f_n \rightharpoonup f \quad \text{in } L^p(\Omega)$$

then there exists $K > 0$ such that

$$\|f_n\|_{L^p} \leq K \ .$$

ii) Conversely if $\|f_n\|_{L^p} \leq K$, then there exists $f \in L^p(\Omega)$ and a subsequence $\{f_{n_j}\}$ of $\{f_n\}$ such that

$$f_{n_j} \rightharpoonup f \quad \text{in } L^p(\Omega) \ .$$

Case 3: *Let $p = \infty$.*
i) Let

$$f_n \overset{*}{\rightharpoonup} f \quad \text{in } L^\infty(\Omega)$$

then there exists $K > 0$ such that

$$\|f_n\|_{L^\infty} \leq K \ .$$

ii) Conversely if $\|f_n\|_{L^\infty} \leq K$, then there exists $f \in L^\infty(\Omega)$ and a subsequence $\{f_{n_j}\}$ of $\{f_n\}$ such that

$$f_{n_j} \overset{*}{\rightharpoonup} f \quad \text{in } L^\infty(\Omega) \ .$$

REMARK. The above theorem follows directly from Theorem 1.1 and from the fact that $L^p(\Omega)$ is separable if $1 \leq p < \infty$ and reflexive only if $1 < p < \infty$. We shall see below that in order to get the converse of Case 1 in the above theorem, one needs to impose an additional hypothesis to the sequence $\{f_n\}$ besides the uniform boundedness.

We now give a criteria to check if a given sequence converges weakly to a limit and we shall then give some examples which will be intensively used in the next chapters.

Lemma 1.4. *Let $\Omega \subset \mathbb{R}^n$ be a bounded open set, then*

Case 1: let $p = 1$

$$f_n \rightharpoonup f \quad \text{in } L^1(\Omega)$$

if and only if
(1) $\|f_n\|_{L^1} \leq K$
(2) $\lim_{n\to\infty} \int_D [f_n(x) - f(x)]\, dx = 0$ for every cube $D \subset \Omega$
(3) for every $\varepsilon > 0$, there exists $\lambda = \lambda(\varepsilon) > 0$ such that if E is a measurable subset of Ω with $\operatorname{meas} E < \lambda(\varepsilon)$, then

$$\int_E |f_n(x)|\, dx \;<\; \varepsilon \ \text{for every } n \ .$$

Case 2: Let $1 < p < \infty$,

$$f_n \rightharpoonup f \quad \text{in } L^p(\Omega)$$

if and only if
(1) $\|f_n\|_{L^p} \leq K$
(2) $\lim_{n\to\infty} \int_D [f_n(x) - f(x)]\, dx = 0$ for every cube $D \subset \Omega$.
Case 3: Let $p = \infty$,

$$f_n \stackrel{*}{\rightharpoonup} f \quad \text{in } L^\infty(\Omega)$$

if and only if
(1) $\|f_n\|_{L^\infty} \leq K$
(2) $\lim_{n\to\infty} \int_D [f_n(x) - f(x)]\, dx = 0$ for every cube $D \subset \Omega$.

REMARKS.

i) Case 2 and 3 of the lemma are simple exercices. Case 1 is, however, a deeper result and is known as the Dunford-Pettis theorem and the property (3) as the property of equiintegrability.

ii) In Lemma 1.4 one can also take Ω to be unbounded, but we shall restrict ourselves to the case of bounded domains.

PROOF. The necessity of (1) and (2) in the three cases is obvious, (1) resulting from the Banach-Steinhaus theorem, (2) being a consequence of the fact that χ_D (the characteristic function of the set D) is in $L^\infty(\Omega)$.

The necessity of (3) in the case of L^1 weak convergence is more involved and we refer to Yosida [1] for a proof.

Case 1: We now turn our attention to the sufficiency of conditions (1), (2) and (3).

Observe that there is no loss of generality if we choose $f \equiv 0$. We wish to show that for every $g \in L^\infty(\Omega)$, then

$$\int_\Omega f_n(x)g(x)\,dx \rightarrow 0 \ .$$

Let $\varepsilon > 0$ be fixed, then, from condition (3), there exists $\lambda = \lambda(\varepsilon)$ such that

$$\int_{\Omega_\lambda} |f_n(x)|\,dx \leq \varepsilon$$

where $\Omega_\lambda = \{x \in \Omega : |f_n(x)| \geq \lambda \text{ for every } n\}$. Having fixed $\lambda(\varepsilon)$ in this way, we can find $a_i \in \mathbb{R}$ and cubes $D_i \subset \Omega$ such that

$$\|g - \sum_i a_i \chi_{D_i}\|_{L^1} \leq \frac{\varepsilon}{\lambda(\varepsilon)}$$

$$\sup\{|a_i|\} \leq K\|g\|_{L^\infty} \ ,$$

where $K > 0$ is a constant. Therefore

$$\left| \int_\Omega f_n(x)g(x)\,dx \right| \leq \sum_i |a_i| \left| \int_{D_i} f_n(x)\,dx \right|$$

$$+ \int_\Omega |f_n(x)| \, |g(x) - \sum_i a_i \chi_{D_i}(x)| \ dx$$

$$\leq \sum_i |a_i| \left| \int_{D_i} f_n(x)\,dx \right| + \lambda \int_{\Omega - \Omega_\lambda} |g(x) - \sum_i a_i \chi_{D_i}(x)| \ dx$$

$$+ \int_{\Omega_\lambda} |f_n(x)| \left(|g(x)| + \sum_i |a_i| \, |\chi_{D_i}(x)| \right) \ dx$$

$$\leq \sum_i |a_i| \left| \int_{D_i} f_n(x)\,dx \right| + \lambda \|g - \sum_i a_i \chi_{D_i}\|_{L^1}$$

$$+ K\|g\|_{L^\infty} \int_{\Omega_\lambda} |f_n(x)|\,dx \ .$$

Letting $n \to \infty$ and using the arbitrariness of ε we have indeed obtained the result.

Case 2 and Case 3 can be treated similarly using Hölder inequality instead of the equiintegrability. □

We now give an important example, which in some special cases is known as Riemann-Lebesgue lemma.

Theorem 1.5. *Let $\Omega = \prod_{i=1}^{n}(a_i, b_i)$ and let $f \in L^p(\Omega)$, $1 \le p \le \infty$. Extend f by periodicity from Ω to $\mathbb{R}^n$. Let*

$$f_\nu(x) = f(\nu x) \ ,$$

then

$$f_\nu \rightharpoonup \bar{f} \equiv \frac{1}{\operatorname{meas}\Omega} \int_\Omega f(x)\,dx \quad \text{in } L^p(\Omega), \text{ as } \nu \to \infty$$

if $1 \le p < \infty$ and

$$f_\nu \overset{*}{\rightharpoonup} \bar{f} \quad \text{in } L^\infty(\Omega), \text{ as } \nu \to \infty$$

if $p = \infty$.

REMARK. We shall see in the proof that the sequence $\{f_\nu\}$ is in fact equiintegrable in $L^p(\Omega)$.

EXAMPLES.

i) *Riemann-Lebesgue lemma.* Let $\Omega = (0, 2\pi)$ and $f(x) = \sin x$, then

$$\sin \nu x \overset{*}{\rightharpoonup} 0 \quad \text{in } L^\infty(0, 2\pi), \text{ as } \nu \to \infty \ .$$

ii) Let $\Omega = (0, 1)$, $0 < \lambda < 1$, $\alpha, \beta \in \mathbb{R}$ and

$$f(x) = \begin{cases} \alpha & \text{if } x \in (0, \lambda) \\ \beta & \text{if } x \in (\lambda, 1) \end{cases}$$

then

$$f_\nu \overset{*}{\rightharpoonup} \lambda\alpha + (1 - \lambda)\beta \quad \text{in } L^\infty(0, 1), \text{ as } \nu \to \infty \ .$$

Note that f_ν takes the values α and β on sets of measure λ and $(1 - \lambda)$ respectively, and its weak limit is a convex combination of α and β, namely $\lambda\alpha + (1 - \lambda)\beta$.

PROOF. The theorem is well known, but since the proof is not always easy to find in the literature we shall give it here.

In order to prove the theorem we shall show that if $1 \leq p < \infty$ then

(1) $\|f_\nu\|_{L^p} \leq K$

(2) $\lim_{\nu \to \infty} \int_D (f_\nu(x) - \bar{f}) \, dx = 0$ for every cube $D \subset \Omega$

(3) for every $\varepsilon > 0$, there exists $\lambda = \lambda(\varepsilon) > 0$ such that if E is a measurable subset of Ω with $\operatorname{meas} E < \lambda(\varepsilon)$, then

$$\int_E |f_\nu(x)| \, dx < \varepsilon, \quad \text{for every } \nu .$$

For the case $p = \infty$, we shall only show (1) and (2). We decompose the proof into three steps.

Step 1: Observe first that

$$\int_\Omega |f_\nu(x)|^p \, dx = \int_\Omega |f(\nu x)|^p \, dx = \frac{1}{\nu^n} \int_{\nu\Omega} |f(x)|^p \, dx = \int_\Omega |f(x)|^p \, dx$$

where we have used the periodicity of f; therefore

$$\|f_\nu\|_{L^p} = \|f\|_{L^p}$$

(in the case $p = \infty$, the above identity is trivial) and thus (1).

Step 2: In order to show (2), we first observe that if D is a cube contained in Ω then up to a rotation, there exist $\alpha, \beta \in \mathbf{R}^n$ such that

$$D = \alpha + \beta\Omega = \prod_{i=1}^{n} (\alpha_i + \beta_i a_i, \alpha_i + \beta_i b_i) .$$

We therefore have

$$\int_D (f_\nu(x) - \bar{f}) \, dx = \int_{\alpha+\beta\Omega} (f(\nu x) - \bar{f}) \, dx = \frac{1}{\nu^n} \int_{\nu\alpha+\nu\beta\Omega} (f(y) - \bar{f}) \, dy$$

$$= \frac{1}{\nu^n} \int_{\nu\alpha+[\nu\beta]\Omega} (f(y) - \bar{f}) \, dy$$

$$+ \frac{1}{\nu^n} \int_{\nu\alpha+(\nu\beta-[\nu\beta])\Omega} (f(y) - \bar{f}) \, dy ,$$

where $[x]$ denotes the integer part $x \in \mathbf{R}$. Using the periodicity of f we find that

$$\int_D (f_\nu(x) - \bar{f})\, dx = \left(\frac{[\nu\beta]}{\nu}\right)^n \int_\Omega (f(y) - \bar{f})\, dy$$

$$+ \frac{1}{\nu^n} \int_{\nu\alpha + (\nu\beta - [\nu\beta])\Omega} (f(y) - \bar{f})\, dy$$

$$= \frac{1}{\nu^n} \int_{\nu\alpha + (\nu\beta - [\nu\beta])\Omega} (f(y) - \bar{f})\, dy$$

i.e.,

$$\left| \int_D (f_\nu(x) - \bar{f})\, dx \right| \le \frac{1}{\nu^n} \int_\Omega |f(y) - \bar{f}|\, dy \ ,$$

where we have used again the periodicity of f. Letting $\nu \to \infty$ we have indeed obtained (2).

Step 3: It now remains to show (3), i.e., the equiintegrability of $\{f_\nu\}$. Let $\delta > 0$ and define for $x \in \Omega$

$$g_\delta(x) = \begin{cases} \min\{\delta, f(x)\} & \text{if } f(x) \ge 0 \\ -\min\{\delta, -f(x)\} & \text{if } f(x) \le 0 \end{cases}$$

and

$$h_\delta(x) = f(x) - g_\delta(x) \ .$$

Observe that for every $\varepsilon > 0$, there exists $\delta = \delta(\varepsilon)$ sufficiently large so that

$$\int_\Omega |h_\delta(x)|^p\, dx < \frac{\varepsilon}{2} \ .$$

Extend g_δ and h_δ by periodicity from Ω to $\mathbb{R}^n$, then as in Step 1 we have

$$\int_\Omega |h_\delta(\nu x)|^p\, dx = \int_\Omega |h_\delta(x)|^p\, dx < \frac{\varepsilon}{2} \ .$$

Therefore

$$\int_E |f(\nu x)|^p\, dx \le \int_E |g_\delta(\nu x)|^p\, dx + \int_E |h_\delta(\nu x)|^p\, dx < \delta^p\ \text{meas}\, E + \frac{\varepsilon}{2} \ .$$

Hence if we choose $\lambda(\varepsilon) = \varepsilon/2\delta^p$ we immediately obtain

$$\text{meas}\, E < \lambda(\varepsilon) \Rightarrow \int_E |f_\nu(x)|^p < \varepsilon \ . \qquad \square$$

We conclude this section with a related example.

EXAMPLE. Let $\Omega = \prod_{i=1}^{n}(a_i, b_i)$. Let $f_\nu \in L^p(\Omega)$, $1 < p < \infty$, be extended by periodicity from Ω to $\mathbb{R}^n$. Assume also that

i) $\|f_\nu\|_{L^p} \leq K$

ii) $\int_\Omega f_\nu(x)\, dx \to \bar{f}$ as $\nu \to \infty$.

Let

$$g_\nu(x) = f_\nu(\nu x)$$

then

$$g_\nu \rightharpoonup \bar{f} \quad \text{in } L^p(\Omega) \text{ as } \nu \to \infty \ .$$

Furthermore the result is in general false if $p = 1$.

PROOF. The proof is identical to Step 1 and 2 in the above theorem if $1 < p < \infty$. However if $p = 1$, we have in general no equiintegrability of the g_ν as the following example shows. Let $\Omega = (0, 1)$ and

$$f_\nu(x) = \begin{cases} \nu & \text{if } 0 < x < \frac{1}{\nu} \\ 0 & \text{if } \frac{1}{\nu} \leq x < 1 \ . \end{cases}$$

Then

$$\int_0^1 |f_\nu(x)|\, dx = \|f_\nu\|_{L^1} = \bar{f} = 1 \ .$$

However if

$$E_\lambda = \{x \in (0, 1) : g_\nu(x) = f_\nu(\nu x) \geq \lambda, \text{ for every } \nu \geq \nu_0\}$$

then for every $\varepsilon > 0$, there is no $\lambda = \lambda(\varepsilon)$ independent of ν such that

$$\int_{E_\lambda} |g_\nu(x)|\, dx = \int_{E_\lambda} |f_\nu(\nu x)|\, dx < \varepsilon, \text{ for every } \nu \geq \nu_0 \ .$$

2.1.2 Sobolev Spaces

We mention here some important results on Sobolev spaces that we shall use in the next chapters. For the proofs of the next theorems as well as for a more complete analysis we refer to Adams [1], Brézis [1], Lions-Magènes [1], Morrey [1].

2.1.2.1 Definitions and Properties

We first introduce the definition of Sobolev spaces.

Definitions.

i) Let $\Omega \subset \mathbf{R}^n$ be open, let $s \in \mathbf{N}$ (the set of integers) and $1 \le p \le \infty$, then

$$W^{s,p}(\Omega) \equiv \{u : u \in L^p(\Omega), \ \nabla^\alpha u \in L^p(\Omega), \ \alpha = 1, 2, \ldots, s\}$$

where $\nabla^\alpha u$ denotes the matrix of the αth derivative, in the sense of distributions, of the function u.

ii) To the space $W^{s,p}(\Omega)$ is associated the norm

$$\|u\|_{W^{s,p}} = \left(\sum_{\alpha=0}^{s} \|\nabla^\alpha u\|_{L^p}^p \right)^{1/p} \quad \text{if } 1 \le p < \infty$$

$$\|u\|_{W^{s,\infty}} = \max_{0 \le \alpha \le s} \{\|\nabla^\alpha u\|_{L^\infty}\} \text{ if } p = \infty \ .$$

iii) $W_0^{s,p}(\Omega)$ denotes the closure of $C_0^\infty(\Omega)$ in $W^{s,p}(\Omega)$.

iv) $W^{-s,p'}(\Omega)$ with $\frac{1}{p} + \frac{1}{p'} = 1$ denotes the dual space of $W_0^{s,p}(\Omega)$.

REMARKS.

i) If $u : \Omega \subset \mathbf{R}^n \to \mathbf{R}^m$, i.e. u is a vector valued function, we shall denote the Sobolev space by $W^{s,p}(\Omega; \mathbf{R}^m)$.

ii) $W^{s,p}(\Omega)$ is a Banach space, separable if $1 \le p < \infty$ and reflexive if $1 < p < \infty$.

iii) If $1 \le p < \infty$, the $C^\infty(\Omega)$ functions are dense in $W^{s,p}$ endowed with its norm.

iv) If Ω is bounded, then $W^{1,\infty}(\Omega)$ is the space of Lipschitz functions.

We now quote the Sobolev and Rellich imbedding theorems.

Theorem 1.6 (Sobolev and Rellich). *Let $\Omega \subset \mathbf{R}^n$ be a bounded open set with Lipschitz boundary and let $1 \le p \le \infty$.*

Case 1: If $1 \le p < n$, then

$$W^{1,p}(\Omega) \subset L^q(\Omega) \ for \ every \ 1 \le q \le \frac{np}{n-p}$$

and the imbedding is compact for every $1 \le q < \frac{np}{n-p}$.

Case 2: If $p = n$, then

$$W^{1,p}(\Omega) \subset L^q(\Omega) \ for \ every \ 1 \le q < \infty$$

and the imbedding is compact.

Case 3: If $p > n$, then

$$W^{1,p}(\Omega) \subset C(\bar{\Omega})$$

and the imbedding is compact.

REMARKS.

i) The regularity of the boundary $\partial\Omega$ in the theorem can be weakened, cf.
Adams [1]. Note that if the space $W^{1,p}$ is replaced by $W_0^{1,p}$, then no
regularity of the boundary is required.

ii) The last imbedding (Case 3) can be improved, in the sense that one can
replace $C(\bar\Omega)$ by spaces of Hölder continuous functions.

iii) The compact imbedding can be read in the following way. Let

$$u_\nu \rightharpoonup u \quad \text{in } W^{1,p}(\Omega)$$

Case 1: If $1 \leq p < n$, then

$$u_\nu \to u \quad \text{in } L^q(\Omega), \ 1 \leq q < \frac{np}{n-p}$$

Case 2: If $p = n$, then

$$u_\nu \to u \quad \text{in } L^q(\Omega), \ 1 \leq q < \infty$$

Case 3: If $p > n$, then

$$u_\nu \to u \quad \text{in } L^\infty(\Omega) \ .$$

We conclude this section with an important inequality (cf. Morrey [2],
Hardy-Littlewood-Polya [1], Brézis [1]).

Theorem 1.7 (Poincaré inequality).

*i) Let Ω be a bounded open set with Lipschitz boundary and let $1 \leq p < \infty$,
then*

$$\|u\|_{L^p} \leq K\|\nabla u\|_{L^p}$$

for every $u \in W_0^{1,p}(\Omega)$ and for some $K > 0$.

ii) If $n = 1$, $\Omega = (0,1)$ and $u \in W_0^{1,2}(0,1)$, then

$$\|u\|_{L^2} \leq \frac{1}{\pi}\|u'\|_{L^2} \ .$$

REMARK. Note that the constant $\frac{1}{\pi}$ is the best possible and it is attained
whenever $u(x) = \sin \pi x$.

2.1.2.2 Approximation by Piecewise Affine Fucntions

In the next chapters we shall often need to approximate functions in Sobolev
spaces by piecewise affine functions, i.e. functions whose gradients are piece-

wise constants. This is a standard procedure in numerical analysis where piecewise affine functions are called finite elements.

Definition. Let Ω be a bounded open set of $\mathbb{R}^n$, we define $\mathrm{Aff}(\Omega)$ to be the set of functions $u \in W^{1,\infty}(\Omega)$ such that there exists $I \in \mathbb{N}$ (an integer) and $\Omega_i \subset \Omega$ disjoint open sets such that

i) $\bar{\Omega} = \cup_{i=1}^{I} \bar{\Omega}_i$

ii) $\mathrm{grad}\, u$ is constant on each Ω_i, $1 \le i \le I$.

We shall denote by $\mathrm{Aff}_0(\Omega) = \{u \in \mathrm{Aff}(\Omega) : u = 0 \text{ on } \partial\Omega\}$.

We then have the following result (for a proof, see for example Ekeland-Témam [1]).

Theorem 1.8. *Let Ω be a bounded open set with Lipschitz boundary.*
Part 1: Let $u \in W_0^{1,\alpha}(\Omega)$ with $1 \le \alpha < \infty$, then there exists $u_\nu \in \mathrm{Aff}_0(\Omega)$ such that

$$u_\nu \to u \quad \text{in } W^{1,\alpha}(\Omega) \text{ as } \nu \to \infty \ .$$

Part 2: Let $u \in W_0^{1,\infty}(\Omega)$, then there exists $u_\nu \in \mathrm{Aff}_0(\Omega)$ such that

$$\begin{cases} u_\nu \to u & \text{in } W^{1,\alpha}(\Omega) \text{ for every } 1 \le \alpha < \infty \text{ as } \nu \to \infty \\ \|\nabla u_\nu\|_{L^\infty} \le K(\|\nabla u\|_{L^\infty}) \end{cases}$$

where K is a constant.

2.2 Convex Analysis

We now give a brief introduction on convex analysis. We refer to Rockafellar [1] and to Ekeland-Témam [1] for further references and for the proofs that we do not give.

2.2.1 Convex Functions

2.2.1.1 Basic Definitions

We start by recalling some definitions and notations.

Definitions. Let X be a Banach space and let

$$f : X \to \bar{\mathbb{R}} \equiv \mathbb{R} \cup \{+\infty\} \ .$$

i) f is said to be *lower semicontinuous* if

$$\liminf_{x_\nu \to \bar{x}} f(x_\nu) \ge f(\bar{x}) \ .$$

ii) f is said to be *convex* if

$$\lambda f(x) + (1 - \lambda)f(y) \geq f(\lambda x + (1 - \lambda)y)$$

for every $x, y \in X$, $\lambda \in [0, 1]$.
iii) The *domain* of f is defined as

$$\operatorname{dom} f \equiv \{x \in X : f(x) < +\infty\} \ .$$

iv) The *epigraph* of f is defined as

$$\operatorname{epi} f \equiv \{(x, \alpha) \in X \times \mathbb{R} : f(x) \leq \alpha\} \ .$$

v) The *level set* of height α, $\alpha \in \mathbb{R}$, of f is defined as

$$\operatorname{level}_\alpha f \equiv \{x \in X : f(x) \leq \alpha\} \ .$$

EXAMPLE. The *characteristic function*. We let for $E \subset X$

$$\chi_E(x) = \begin{cases} 0 & \text{if } x \in E \\ +\infty & \text{if } x \notin E \ , \end{cases}$$

then

$$\operatorname{dom} \chi_E = E$$

$$\operatorname{epi} \chi_E = E \times [0, +\infty) = E \times \mathbb{R}_+$$

$$\operatorname{level}_\alpha \chi_E = \begin{cases} \emptyset & \text{if } \alpha < 0 \\ E & \text{if } \alpha \geq 0 \ . \end{cases}$$

We now state without proof the following result.

Theorem 2.1.

Part 1: The following three conditions are equivalent

i) f *is lower semicontinuous*
ii) epi f *is closed*
iii) $\operatorname{level}_\alpha f$ *is closed for every* $\alpha \in \mathbb{R}$.

Part 2: The following two conditions are equivalent

i) f *is convex*
ii) epi f *is convex.*

Furthermore if f *is convex, then* $\operatorname{level}_\alpha$ *is convex for every* $\alpha \in \mathbb{R}$.

REMARK. Note that in general the convexity of $\text{level}_\alpha f$ for every $\alpha \in \mathbb{R}$ does not imply the convexity of f, as the following example indicates; let

$$f(x) = \begin{cases} 0 & \text{if } x \leq 0 \\ 1 & \text{if } x > 0 , \end{cases}$$

then

$$\text{level}_\alpha f = \begin{cases} \emptyset & \text{if } \alpha < 0 \\ (-\infty, 0] & \text{if } 0 \leq \alpha < 1 \\ \mathbb{R} & \text{if } \alpha \geq 1 \end{cases}$$

is convex for every $\alpha \in \mathbb{R}$, while f is not convex. In the context of optimization, functions, whose level sets are convex, are sometimes called quasiconvex; we shall not be concerned with such functions and when we shall use the notion of quasiconvexity in Chapters 4 and 5 it will have different meaning.

We close this very brief introduction by recalling Jensen inequality.

Theorem 2.2 (Jensen inequality). *Let $\Omega \subset \mathbb{R}^n$ be a bounded open set, $u \in L^1(\Omega)$ and $f : \mathbb{R} \to \mathbb{R}$ be convex, then*

$$f\left(\frac{1}{\text{meas}\,\Omega} \int_\Omega u(x)\,dx\right) \leq \frac{1}{\text{meas}\,\Omega} \int_\Omega f(u(x))\,dx \ .$$

2.2.1.2 Continuity of Convex Functions

We now turn our attention to the continuity of convex functions and we obtain the following classical result.

Theorem 2.3. *Let $X_1, \ldots, X_n$ be Banach spaces and let $f : X_1 \times \ldots \times X_n \to \bar{\mathbb{R}} = \mathbb{R} \cup \{+\infty\}$ be convex in each variable and $f \not\equiv +\infty$.*

Part 1: If f is bounded from above in a neighbourhood of a point x, then f is continuous at x.

Part 2: If $X_1 = \ldots = X_n = \mathbb{R}$, then f is locally Lipschitz on $\text{int}(\text{dom}\,f)$.

REMARKS.

i) Note that if $f : X_1 \times \ldots \times X_n \to \bar{\mathbb{R}}$ is convex, then f is convex in each variable, the converse being false as it is easily seen from the example $X_1 = X_2 = \mathbb{R}$ and $f(x,y) = xy$.

ii) The above theorem is usually stated for convex functions and not for convex functions in each variable, but in Chapters 4 and 5 we shall need this stronger version. The proof is however a straightforward adaptation of the classical result.

PROOF.

Part 1: There is no loss of generality if we assume that $x = 0$ and $f(0) = 0$. Since f is bounded in a neighbourhood of $x = 0$, there exists $\lambda > 0$ and $a \in \mathbb{R}$ such that

$$\|x\| = \max\{\|x_i\|; \ i = 1, \ldots, n\} \le \lambda \Rightarrow f(x) \le a \ . \tag{1}$$

Fix $\varepsilon > 0$ and without loss of generality assume that $\varepsilon \le an$. We now show that

$$\|x\| \le \frac{\varepsilon}{an}\lambda \Rightarrow |f(x)| \le \varepsilon \ . \tag{2}$$

Using the convexity of f we have

$$f(x) = f(x_1, \ldots, x_n) = f(\delta(\frac{x_1}{\delta}, x_2, \ldots, x_n) + (1 - \delta)(0, x_2, \ldots, x_n))$$

$$\le \delta f(\frac{x_1}{\delta}, x_2, \ldots, x_n) + (1 - \delta)f(0, x_2, \ldots, x_n)$$

where $\delta = \varepsilon/an$. Repeating the process with the second variable we have

$$f(x) \le \delta f(\frac{x_1}{\delta}, x_2, \ldots, x_n) + (1 - \delta)\delta f(0, \frac{x_2}{\delta}, \ldots, x_n)$$

$$+ (1 - \delta)^2 f(0, 0, x_3, \ldots, x_n) \ .$$

Iterating the process we obtain, using the fact that $(1 - \delta) < 1$,

$$f(x) \le \delta[f(\frac{x_1}{\delta}, x_2, \ldots, x_n) + f(0, \frac{x_2}{\delta}, \ldots, x_n) + f(0, 0, \frac{x_3}{\delta}, \ldots, x_n)$$

$$+ \ldots + f(0, \ldots, 0)] \ .$$

If we now assume that $\|x\| \le \delta\lambda = \frac{\varepsilon}{an}\lambda$, we deduce immediately that

$$\|x\| \le \frac{\varepsilon}{an}\lambda \Rightarrow f(x) \le \varepsilon \ . \tag{3}$$

In order to obtain (2), we need to show that $f(x) \ge -\varepsilon$ and this is done similarly. We have

$$0 = f(0, \ldots, 0) = f\left(\frac{1}{1 + \delta}(0, \ldots, 0, x_n) + \frac{\delta}{1 + \delta}\left(0, 0, \ldots, \frac{-x_n}{\delta}\right)\right)$$

$$\le \frac{1}{1 + \delta}[f(0, \ldots, 0, x_n) + \delta f(0, \ldots, 0, \frac{-x_n}{\delta})]$$

thus

$$f(0,\ldots,0,x_n) \geq -\delta f(0,\ldots,0,\frac{-x_n}{\delta}) \ .$$

Iterating the process as above we obtain that

$$\|x\| \leq \frac{\varepsilon}{an}\lambda \Rightarrow f(x) \geq -\varepsilon$$

and thus (2) and the continuity of f at x.

Part 2: We now prove that in the finite dimensional case, the condition that f be bounded in a neighbourhood of a point can be dropped.

Step 1: We first prove that if $x \in \text{int}(\text{dom}\, f)$, then f is continuous at x. There is no loss of generality, if we suppose as above that $x = 0$, therefore since $0 \in \text{int}(\text{dom}\, f)$, there exists $\varepsilon > 0$ such that

$$\{x = (x_1,\ldots,x_n) \in \mathbb{R} : \|x\| = \max\{|x_i| : i = 1,\ldots,n\} \leq 2\varepsilon\} \subset \text{dom}\, f \ . \tag{4}$$

Letting

$$a = \max\{f(\varepsilon_1,\varepsilon_2,\ldots,\varepsilon_n) : \varepsilon_i = -\varepsilon, 0, \varepsilon \text{ for every } i = 1,\ldots,n\} \tag{5}$$

we deduce from (4) that $a < +\infty$. We now claim that

$$\|x\| \leq \varepsilon \Rightarrow f(x) \leq a \ . \tag{6}$$

In order to prove (6), observe that if $0 \leq x_n \leq \varepsilon$ and $\varepsilon_i = -\varepsilon, 0, \varepsilon$, then the convexity of f with respect to the last variable implies that

$$\begin{aligned}
f(\varepsilon_1,\varepsilon_2,\ldots,\varepsilon_{n-1},x_n) \leq{}& \frac{x_n}{\varepsilon}f(\varepsilon_1,\ldots,\varepsilon_{n-1},\varepsilon) \\
&+ (1 - \frac{x_n}{\varepsilon})f(\varepsilon_1,\ldots,\varepsilon_{n-1},0) \\
\leq{}& \frac{x_n}{\varepsilon}a + \left(1 - \frac{x_n}{\varepsilon}\right)a = a \ .
\end{aligned} \tag{7}$$

Using (7) and the convexity of f with respect to x_{n-1} we have, if $0 \leq x_{n-1} \leq \varepsilon$, that

$$\begin{aligned}
f(\varepsilon_1,\ldots,\varepsilon_{n-2},\varepsilon_{n-1},x_n) \leq{}& \frac{x_{n-1}}{\varepsilon}f(\varepsilon_1,\ldots,\varepsilon_{n-2},\varepsilon,x_n) \\
&+ \left(1 - \frac{x_{n-1}}{\varepsilon}\right) f(\varepsilon_1,\ldots,\varepsilon_{n-2},0,x_n) \leq a \ .
\end{aligned}$$

Iterating the process with respect to all the variables we have immediately (6) for $0 \leq x_i \leq \varepsilon$. A similar argument applies if some of the x_i are negative.

The inequality (6) implies therefore that if $x \in \text{int}(\text{dom } f)$, then f is bounded in a neighbourhood of x and therefore, using Part 1, f is continuous at x.

Step 2: It therefore remains to show that f is locally Lipschitz in the interior of the domain of f. We follow here Ekeland-Témam [1]. Let $x \in \text{int}(\text{dom } f)$. By continuity of f at x, there exist $\delta > 0$, $\alpha, \beta \in \mathbb{R}$ such that

$$\|y - x\| = \max\{|y_i - x_i| : i = 1, \ldots, n\} \le 2\delta$$
$$\Rightarrow -\infty < \alpha \le f(y) \le \beta < +\infty . \tag{8}$$

Let z and z_1 be such that

$$\|z_1 - z\|, \ \|z_1 - x\| \le \delta . \tag{9}$$

Observe that (9) implies that $\|z - x\| \le 2\delta$ and (8) and (9) lead therefore to

$$\|z - z_1\| \le \delta \Rightarrow f(z) - f(z_1) \le \beta - \alpha . \tag{10}$$

Let $\varepsilon > 0$, combining (2) and (10) we have immediately

$$\|z - z_1\| \le \frac{\delta\varepsilon}{(\beta - \alpha)n} \Rightarrow |f(z) - f(z_1)| \le \varepsilon . \tag{11}$$

Choosing $\varepsilon\delta = \|z - z_1\|(\beta - \alpha)n$ we have from (9) and (11) that

$$\|z - z_1\| \le \delta \Rightarrow |f(z) - f(z_1)| \le \frac{(\beta - \alpha)n}{\delta}\|z - z_1\| . \tag{12}$$

Now let z_2 be such that $\|z_2 - x\| \le \delta$. Let $u_1, u_2, \ldots, u_M \in [z_1, z_2]$ (the segment in $\mathbb{R}^n$ with endpoints z_1 and z_2) be such that $u_1 = z_1, u_2, \ldots, u_M = z_2$ and

$$\|u_m - u_{m+1}\| \le \delta, \quad m = 1, \ldots, M - 1 .$$

Using (12) we immediately get

$$\|u_m - u_{m+1}\| \le \delta \Rightarrow |f(u_m) - f(u_{m+1})| \le \frac{(\beta - \alpha)n}{\delta}\|u_m - u_{m+1}\| . \tag{13}$$

Summing (13) we obtain (using also (9))

$$\|z_1 - x\|, \ \|z_2 - x\| \le \delta \Rightarrow |f(z_1) - f(z_2)| \le \frac{(\beta - \alpha)n}{\delta}\|z_1 - z_2\|$$

and hence the result. $\qquad\qquad\qquad\qquad\qquad\qquad\qquad\qquad\qquad\qquad\quad\square$

REMARK. If we let in the theorem $f : X \to \bar{\mathbf{R}} = \mathbf{R} \cup \{+\infty\}$ be convex and f be bounded in a neighbourhood of a point x, then f is also locally Lipschitz in $\mathrm{int}(\mathrm{dom}\, f)$, even though X might be an infinite dimensional Banach space.

PROOF of the remark. We decompose the proof into two steps.

Step 1: We first show (cf. Ekeland-Témam [1]) that if $y \in \mathrm{int}(\mathrm{dom}\, f)$ then f is bounded in a neighbourhood of y, i.e. there exist $\varepsilon > 0$ and $\beta \in \mathbf{R}$ so that

$$\|u - y\| \leq \varepsilon \Rightarrow f(u) \leq \beta \ . \tag{14}$$

From the hypothesis, we have that there exist $\delta > 0$ and $\alpha \in \mathbf{R}$ so that

$$\|z - x\| \leq \delta \to f(z) \leq \alpha \ . \tag{15}$$

We now choose $\eta > 0$ so small that if we let

$$w = y + \eta(y - x)$$

then the segment $[w, y] \subset \mathrm{int}(\mathrm{dom}\, f)$. We next use the convexity of f to deduce that

$$
\begin{aligned}
f(u) &= f\left(\frac{\eta}{1 + \eta} \left(\frac{(1 + \eta)u - w}{\eta} \right) + \frac{1}{1 + \eta} w \right) \\
&\leq \frac{\eta}{1 + \eta} f\left(\frac{(1 + \eta)u - w}{\eta} \right) + \frac{1}{1 + \eta} f(w) \ .
\end{aligned}
\tag{16}
$$

We finally choose $\varepsilon(1 + \eta) = \eta\delta$ so that if $\|u - y\| \leq \varepsilon$, then

$$
\begin{aligned}
\left\| \frac{(1 + \eta)u - w}{\eta} - x \right\| &= \frac{1}{\eta} \|(1 + \eta)u - y - \eta(y - x) - \eta x\| \\
&= \frac{1 + \eta}{\eta} \|u - y\| \\
&\leq \frac{1 + \eta}{\eta} \varepsilon = \delta \ .
\end{aligned}
$$

Using the above inequalities, (15) and (16), we immediately get

$$\|u - y\| \leq \varepsilon \Rightarrow f(u) \leq \frac{\eta}{1 + \eta}\alpha + \frac{1}{1 + \eta} f(w) \equiv \beta \ ,$$

which is the claimed result.

Step 2: To show that f is in fact locally Lipschitz in $\text{int}(\text{dom } f)$ is easily deduced from Step 1 in the same way as in the proof of Step 2 of the above theorem. $\square$

2.2.2 Duality and Hahn-Banach Theorem

In this section we present different versions of Hahn-Banach theorem and we show how this leads to the notion of duality as well as the connection between duality of convex functions and the Gâteaux differentiability of such functions.

2.2.2.1 Hahn-Banach Theorem

We give here only the geometrical versions of Hahn-Banach theorem. We first start with some definitions.

Definitions. Let X be a Banach space.

i) A *hyperplane H* is a set of the form

$$H = \{x \in X : f(x) = \alpha\}$$

where f is a non zero linear functional over X and $\alpha \in \mathbb{R}$.

ii) A hyperplane H, defined by $f(x) = \alpha$, is said to *separate* (respectively to *separate strictly*) the sets $A, B \subset X$ if $f(x) \leq \alpha$ for every $x \in A$ and $f(x) \geq \alpha$ for every $x \in B$ (respectively, if there exists $\varepsilon > 0$ such that $f(x) \leq \alpha - \varepsilon$ for every $x \in A$ and $f(x) \geq \alpha + \varepsilon$ for every $x \in B$).

REMARK. Note that the hyperplane H, defined by $f(x) = \alpha$, is closed if and only if f is continuous.

We are now in position to state the theorem.

Theorem 2.4 (Hahn-Banach). *Let X be a Banach space.*

i) *Let $A, B \subset X$ be non empty, disjoint and convex. Let A be open, then there exists a closed hyperplane which separates A and B.*

ii) *Let $A, B \subset X$ be non empty, disjoint and convex. Let A be closed and B compact then there exists a closed hyperplane which separates strictly A and B.*

iii) *Every closed convex set is the intersection of the closed half spaces which contain it.*

2.2.2.2 Duality

We now use the above theorem to introduce the notion of duality and of dual maps which were introduced by Fenchel [1] and Moreau [1], see also Rockafellar [1], Ekeland-Témam [1]. This notion plays a central role in convex analysis.

Definitions. Let X be a vector space, X^* its dual and $\langle \bullet ; \bullet \rangle$ denotes the bilinear canonical form over $X \times X^*$. Let $f : X \to \mathbb{R} \cup \{\pm\infty\}$.

i) The function $f^* : X^* \to \mathbb{R} \cup \{\pm\infty\}$ defined by

$$f^*(x^*) = \sup_{x \in X} \{\langle x; x^* \rangle - f(x)\}$$

is called the *conjugate*, or *polar*, function of f.

ii) The function $f^{**} : X \to \mathbb{R} \cup \{\pm\infty\}$ defined by

$$f^{**}(x) = \sup_{x^* \in X^*} \{\langle x; x^* \rangle - f^*(x^*)\}$$

is called the *biconjugate*, or *bipolar*, of f.

iii) The function $Cf : X \to \mathbb{R} \cup \{\pm\infty\}$ defined by

$$Cf = \sup\{g \leq f : g \text{ convex}\}$$

is called the *(lower) convex envelope of f*.

REMARK. In finite dimensional space, the notion of duality is also known as the Legendre transformation.

Before giving some examples, we state some important properties of these functions established, essentially, by Fenchel [1] and Moreau [1].

Theorem 2.5. *Let $f : X \to \bar{\mathbb{R}} = \mathbb{R} \cup \{+\infty\}$, then*

i) f^* is convex and lower semicontinuous,

ii) if f is convex and lower semicontinuous, then $f^* \not\equiv +\infty$,

iii) in general

$$f^{**} \leq Cf \leq f$$

and if f is convex and lower semicontinuous, then

$$f^{**} = Cf = f \ .$$

In particular if f takes only finite values then $f^{**} = Cf$.

iv) In general

$$f^{***} = f^* \ .$$

EXAMPLES.

i) *Characteristic function*: Let $E \subset X$ and

$$\chi_E = \begin{cases} 0 & \text{if } x \in E \\ +\infty & \text{if } x \notin E \ . \end{cases}$$

We then have

$$\chi_E^*(x^*) \,=\, \sup_{x \in E}\{\langle x; x^*\rangle\}$$

which is known as the *support function* of E. Applying again the duality
we obtain

$$\chi_E^{**}(x) \,=\, \chi_{\overline{co}E}(x)$$

where coE (resp. $\overline{co}E$) denotes the convex hull (resp. the closed convex
hull) of E. Finally we also have

$$C\chi_E(x) \,=\, \chi_{coE}(x) \ .$$

In particular if $X = \mathbb{R}$, $E = (0,1)$, we get

$$\chi_{(0,1)} \,=\, C\chi_{(0,1)} \,>\, \chi_{[0,1]} \,=\, \chi_{(0,1)}^{**} \ .$$

ii) Let $X = \mathbb{R}$, $\alpha > 1$ and

$$f(x) \,=\, \frac{1}{\alpha}|x|^\alpha$$

then

$$f^*(x^*) \,=\, \frac{1}{\alpha'}|x^*|^{\alpha'}$$

where $\frac{1}{\alpha} + \frac{1}{\alpha'} = 1$.

iii) Similarly if X is a Banach space and

$$f(x) \,=\, \frac{1}{\alpha}\|x\|_X^\alpha$$

then

$$f^*(x^*) \,=\, \frac{1}{\alpha'}\|x^*\|_{X^*}^{\alpha'} \ .$$

iv) Let $A \in \mathbb{R}^{n^2}$ (the set of $n \times n$ matrices) and $f(A) = \det A$, then $f^*(A^*) \equiv +\infty$ and $f^{**}(A) \equiv -\infty$.

We now turn to the proof of Theorem 2.5 and we follow here the proof of
Brézis [2].

PROOF.

i) Since $x^* \rightarrow \langle x; x^*\rangle - f(x)$ is convex and lower semicontinuous (in fact
   continuous), then f^{**} is convex and lower semicontinuous.

ii) Note first that if $f \equiv +\infty$, then $f^* \equiv -\infty$ and the result is proved. So
    we may assume that there exists $x_0 \in \mathrm{dom}(f)$. We next let $a_0 < f(x_0)$
    and we apply Hahn-Banach theorem to $A = \{(x, f(x)) : x \in X\}$ and
    $B = \{(x_0, a_0)\}$. We then obtain that there exists a closed hyperplane over

$X \times \mathbb{R}$ defined by $\langle (x,a); (x^*; a^*) \rangle = \langle x; x^* \rangle + aa^* = \alpha$ which separates strictly A and B, i.e.

$$\langle x; x^* \rangle + f(x)a^* > \alpha \text{ for every } x \in X$$
$$\langle x_0; x^* \rangle + a_0 a^* < \alpha \; . \tag{1}$$

Taking $x = x_0$ in (1) we immediately get

$$\langle x_0; x^* \rangle + f(x_0)a^* > \alpha > \langle x_0; x^* \rangle + a_0 a^*$$

and hence $a^* > 0$. We therefore deduce immediately from (1) that

$$\langle x; -\frac{1}{a^*} x^* \rangle - f(x) < -\frac{\alpha}{a^*} \tag{2}$$

and thus taking the supremum in (2) we obtain the result, i.e. $f^* \not\equiv +\infty$.

iii) We proceed in three steps.

Step 1: Observe first that f^{**} is convex and lower semicontinuous and that, by definition, $f(x) \geq \langle x; x^* \rangle - f^*(x^*)$, hence $f^{**} \leq f$. The first inequality, $f^{**} \leq Cf \leq f$, follows then immediately.

Step 2: We next reduce the problem to the case where $f \geq 0$. We may assume without loss of generality that $f \not\equiv +\infty$. Choosing $x^* \in \text{dom}(f^*)$, which is non empty as seen in Part 2 of this theorem, and defining

$$g(x) = f(x) - \langle x; x^* \rangle + f^*(x^*)$$

we obtain that $g \geq 0$, convex, lower semicontinuous and $g \not\equiv +\infty$. Observe also that

$$g^{**}(x) = f^{**}(x) - \langle x; x^* \rangle + f^*(x^*) \; .$$

Therefore the result, $f = f^{**}$, will follow from the corresponding result for g.

Step 3: We therefore may assume that $f \geq 0$, convex, lower semicontinuous and $f \not\equiv +\infty$. In view of Step 1 we only need to show that $f^{**} \geq f$. We proceed by contradiction and we assume that there exists $x_0 \in X$ such that

$$0 \leq f^{**}(x_0) < f(x_0) \; . \tag{3}$$

Applying Hahn-Banach theorem to $A = \text{epi} f$ and $B = \{(x_0, f^{**}(x_0))\}$ we have that there exists a hyperplane $\langle x; x^* \rangle + aa^* = \alpha$ which separates strictly A and B, i.e.,

$$\langle x; x^* \rangle + aa^* > \alpha \text{ for every } (x,a) \in \text{epi} f \tag{4}$$
$$\langle x_0; x^* \rangle + f^{**}(x_0)a^* < \alpha \; . \tag{5}$$

We then immediately deduce $a^* \geq 0$, if we choose in (4) $x \in \mathrm{dom}\, f$ and if we let $a \to +\infty$. We then let $\varepsilon > 0$ and use the positivity of f and (4) to get

$$\langle x; x^* \rangle + f(x)(a^* + \varepsilon) > \alpha \text{ for every } x \in \mathrm{dom}\, f$$

and hence

$$\langle x; -\frac{x^*}{a^* + \varepsilon} \rangle - f(x) < -\frac{\alpha}{a^* + \varepsilon} \text{ for every } x \in \mathrm{dom}\, f \ .$$

The last inequality implies that $f^*(-\frac{x^*}{a^*+\varepsilon}) \leq -\frac{\alpha}{a^*+\varepsilon}$. Using the definition of f^{**} we have therefore

$$f^{**}(x_0) \geq \langle x_0; -\frac{x^*}{a^* + \varepsilon} \rangle - f^*(-\frac{x^*}{a^* + \varepsilon}) \geq \langle x_0; -\frac{x^*}{a^* + \varepsilon} \rangle + \frac{\alpha}{a^* + \varepsilon} \ .$$

Thus

$$\langle x_0; x^* \rangle + f^{**}(x_0)(a^* + \varepsilon) \geq \alpha \ .$$

Using the arbitrariness of ε and (5) we have a contradiction and this concludes Step 3.

iv) We now want to show that $f^{***} = f^*$. Since we always have $f^{**} \leq f$, we deduce that $f^{***} \geq f^*$. Furthermore from the definition of duality we have for every $x \in X$, $x^* \in X^*$,

$$\langle x; x^* \rangle - f^{**}(x) \leq f^*(x^*)$$

and hence, taking the supremum in the left hand side, we obtain $f^{***} \leq f^*$. $\qquad\square$

2.2.2.3 Duality for Gâteaux Differentiable Functions

We now establish a relationship between convexity and monotonicity of Gâteaux derivatives and we follow here the presentation of Ekeland-Témam [1]. We recall first

Definition. Let X be a Banach space and $f : X \to R$. We define the derivative of f at a point x in the direction y as the limit, if it exists,

$$f'(x, y) \equiv \lim_{\substack{\lambda \to 0 \\ \lambda > 0}} \frac{f(x + \lambda y) - f(x)}{\lambda} \ .$$

A function f is said to be *Gâteaux differentiable* at x, and is denoted $f'(x) \in X^*$, if the above limit exists for every $y \in X$ and

$$f'(x, y) \equiv \langle y; f'(x) \rangle \ .$$

We now have the following theorem.

Theorem 2.6. *Let $f : X \to \mathbb{R}$ be Gâteaux differentiable.*
Part 1: The following conditions are equivalent
i) f is convex,
ii) for every $x, y \in X$

$$f(y) \geq f(x) + \langle y - x; f'(x) \rangle \ .$$

iii) For every $x, y \in X$

$$\langle y - x; f'(y) - f'(x) \rangle \geq 0 \ .$$

Part 2: Let f be convex, then

$$f(x) + f^*(f'(x)) = \langle x; f'(x) \rangle, \ \text{for every } x \in X \ .$$

PROOF. The proof is standard.
Part 1: i) $\Rightarrow$ ii) Let $\lambda > 0$, we have from the convexity of f that

$$\frac{1}{\lambda}[f(x + \lambda(y - x)) - f(x)] \leq f(y) - f(x) \ .$$

Letting $\lambda \to 0$, we have immediately ii).
ii) $\Rightarrow$ i) we have from the inequality ii) that, for $\lambda \in [0, 1]$,

$$f(x) \geq f(\lambda x + (1 - \lambda)y) + \langle x - (\lambda x + (1 - \lambda)y); f'(\lambda x + (1 - \lambda)y) \rangle$$

$$f(y) \geq f(\lambda x + (1 - \lambda)y) + \langle y - (\lambda x + (1 - \lambda)y); f'(\lambda x + (1 - \lambda)y) \rangle \ .$$

Multiplying the first equation by λ and the second by $(1 - \lambda)$ and adding them we obtain the convexity of f.
ii) $\Rightarrow$ iii) using the inequality ii) we have

$$f(y) \geq f(x) + \langle y - x, f'(x) \rangle$$

$$f(x) \geq f(y) + \langle x - y, f'(y) \rangle \ .$$

Combining these two inequalities we have

$$\langle y - x; f'(y)\rangle \geq f(y) - f(x) \geq \langle y - x; f'(x)\rangle$$

and thus the result.

iii) $\Rightarrow$ ii) let $\lambda \in [0,1]$ and consider

$$\phi(\lambda) = f(x + \lambda(y - x)) .$$

Observe that

$$\phi'(\lambda) - \phi'(0) = \langle y - x; f'(x + \lambda(y - x)) - f'(x)\rangle$$

$$= \frac{1}{\lambda}[\langle x + \lambda(y - x) - x; f'(x + \lambda(y - x)) - f'(x)\rangle] \geq 0$$

where we have used iii). Therefore integrating the inequality we have

$$\phi(\lambda) \geq \phi(0) + \lambda\phi'(0)$$

and thus letting $z = x + \lambda(y - x)$, we have

$$f(z) \geq f(x) + \langle z - x; f'(x)\rangle .$$

Part 2: We want to show that, for every $x \in X$,

$$f(x) + f^*(f'(x)) = \langle x; f'(x)\rangle .$$

From the definition of f^* we have immediately that

$$f(x) + f^*(f'(x)) \geq \langle x; f'(x)\rangle . \tag{1}$$

Using now Part 1 and the convexity of f we have

$$f(y) \geq f(x) + \langle y - x; f'(x)\rangle, \ \text{ for every } y \in X .$$

We therefore obtain

$$\langle x, f'(x)\rangle - f(x) \geq \langle y; f'(x)\rangle - f(y), \ \text{ for every } y \in X .$$

Taking the supremum over all $y \in X$, in the right hand side of the above inequality, leads to

$$\langle x; f'(x)\rangle - f(x) \geq f^*(f'(x)) . \tag{2}$$

Combining (1) and (2) we obtain the result. $\qquad\qquad\qquad\qquad\square$

Before concluding this section, we should observe that the results of Theorem 2.6 can be generalized to non differentiable convex functions.

Proposition 2.7. Let $f : X \to \mathbb{R}$ be continuous and convex. For every $x \in X$, there exists $x^* \in X^*$ such that

$$f(y) \geq f(x) + \langle y - x; x^* \rangle, \text{ for every } y \in X \ .$$

Moreover x and x^* are linked through the following

$$f(x) + f^*(x^*) = \langle x; x^* \rangle \ .$$

REMARK. It is this extension of Theorem 2.6 which has lead to the introduction of the notion of a subgradient. One calls x^* the subgradient of f at x. If f is Gâteaux differentiable then $x^* = f'(x)$.

PROOF.
Part 1: Let $x \in X$ be fixed. Since f is continuous, then int(epi f) (i.e., the interior of epi f) is non empty. The convexity of f implies then that int(epi f) is convex and non empty. Observe also that $(x, f(x)) \notin$ int(epi f). We may then use Hahn-Banach theorem to deduce that there exist a^*, $\alpha \in \mathbb{R}$ and $x_1^* \in X^*$ such that

$$\begin{cases} \langle y; x_1^* \rangle + a^* a \geq \alpha, \text{ for every } (y, a) \in \text{epi } f \\ \langle x; x_1^* \rangle + a^* f(x) = \alpha \ . \end{cases} \tag{1}$$

Note first that $a^* \geq 0$, since $(x, f(x) + 1) \in$ epi f. Moreover $a^* \neq 0$, otherwise $\langle y - x; x_1^* \rangle \geq 0$ for every $y \in X$ and this would imply that $x_1^* = 0$, as well as $a^* = 0$, which is absurd. Therefore $a^* > 0$ and we deduce from (1) and from the fact that $(y, f(y)) \in$ epi f,

$$\langle y; \frac{x_1^*}{a^*} \rangle + f(y) \geq \langle x; \frac{x_1^*}{a^*} \rangle + f(x) \ . \tag{2}$$

Letting $x^* = -x_1^*/a^*$ in (2) we have indeed proved Part 1 of the proposition.
Part 2: The proof is identical to that of Theorem 2.6. From the definition of f^* we immediately have

$$f(x) + f^*(x^*) \geq \langle x; x^* \rangle \ . \tag{3}$$

From Part 1 we have

$$\langle x; x^* \rangle - f(x) \geq \langle y; x^* \rangle - f(y), \text{ for every } y \in X \ . \tag{4}$$

Taking the supremum over all $y \in X$ in (4) we have immediately the reverse inequality of (3). This concludes the proof of the Proposition. $\qquad \square$

2.2.3 Carathéodory Theorem

We conclude this chapter with Carathéodory theorem, which gives a characterization of the convex hull of a subset of $\mathbb{R}^n$.

Theorem 2.8 (Carathéodory theorem). *Let $M \subset \mathbb{R}^n$. Let coM denote the convex hull of M; then*

$$coM = \{x \in \mathbb{R}^n : x = \sum_{i=1}^{n+1} \lambda_i x_i, \ x_i \in M, \ \lambda_i \geq 0 \text{ with } \sum_{i=1}^{n+1} \lambda_i = 1\} \ .$$

We give immediately an important corollary whose proof will be established in Theorem 1.1 of Chapter 5.

Corollary 2.9. *Let $f : \mathbb{R}^n \to \bar{\mathbb{R}} = \mathbb{R} \cup \{+\infty\}$ and*

$$Cf = \sup\{g \leq f : g \text{ convex}\}$$

then

$$Cf(x) = \inf \left\{ \sum_{i=1}^{n+1} \lambda_i f(x_i) : \sum_{i=1}^{n+1} \lambda_i x_i = x, \ \lambda_i \geq 0 \text{ with } \sum_{i=1}^{n+1} \lambda_i = 1 \right\} \ .$$

PROOF of the theorem. The proof is standard and we follow here Ioffe-Tihomirov [2]. We decompose the proof into two steps.

Step 1: Observe first that if I is an integer,

$$N_I = \left\{ x \in \mathbb{R}^n : \sum_{i=1}^{I} \lambda_i x_i = x, \ x_i \in M, \ \lambda_i \geq 0 \text{ with } \sum_{i=1}^{I} \lambda_i = 1 \right\}$$

and

$$N = \bigcup_{I \in \mathbb{N}} N_I$$

then obviously N is convex and $N \supset M$ and therefore $N \supset coM$. Conversely let $A \supset M$, A convex, then $N \subset A$ and therefore $N \subset coM$ and thus $N = coM$.

Step 2: We now show that in fact we have

$$N = \bigcup_{I=1}^{n+1} N_I \ .$$

Let $m \in N = coM$, then trivially $(1, m) \in \{1\} \times coM = co(\{1\} \times M)$. Applying Step 1 to $co(\{1\} \times M)$ we have that there exist I, an integer, $\lambda_i \geq 0$ with $\sum \lambda_i = 1$, $m_i \in M$ such that

$$\sum_{i=1}^{I} \lambda_i (1, m_i) = (1, m) . \tag{1}$$

We wish to show that we can take $I \leq n + 1$ in (1). Assume that $I > n + 1$, then there exist $\gamma_i \in \mathbb{R}$ not all zero such that

$$\sum_{i=1}^{I} \gamma_i (1, m_i) = 0 , \tag{2}$$

since $(1, m_i) \in \mathbb{R}^{n+1}$ and $I > n + 1$.

Let $T = \{i \in \{1, \ldots, I\} : \gamma_i > 0\}$. We may assume without loss of generality that $T \neq \emptyset$, otherwise replace γ_i by $-\gamma_i$. Let

$$\beta = \min_{i \in T} \left\{ \frac{\lambda_i}{\gamma_i} \right\} \tag{3}$$

$$\mu_i = \lambda_i - \beta \gamma_i . \tag{4}$$

We then deduce that

$$\mu_i \geq 0 \tag{5}$$

$$\sum_{i=1}^{I} \mu_i = 1 \tag{6}$$

$$\text{at least one of the } \mu_i = 0 ; \tag{7}$$

where (5) has been obtained trivially if $\gamma_i \leq 0$ and by (3) and (4) if $\gamma_i > 0$; similarly (6) follows from (1) and (2); and finally (7) holds if one takes the index $i \in T$ which corresponds to the minimum in (3). Furthermore

$$\sum_{i=1}^{I} \lambda_i (1, m_i) = \sum_{i=1}^{I} \mu_i (1, m_i) = (1, m) . \tag{8}$$

In view of (5), (6), (7) and (8), we have therefore reduced the number I to $(I - 1)$. Continuing this process up to $I = n + 1$ we have indeed established the theorem. $\qquad\square$

CHAPTER 3

General Setting and the Scalar Case

3.0 Introduction

In the first section of this chapter we start with abstract considerations. We let I be a functional defined over a Banach space X, i.e. $I : X \to \bar{\mathbb{R}} = \mathbb{R} \cup \{+\infty\}$ and we consider the minimization problem

$$(P) \qquad \qquad \inf\{I(u) : u \in X\} \ .$$

We first establish the existence of a minimum under the following conditions:

i) X is a reflexive Banach space

ii) I is (sequentially) weakly lower semicontinuous, i.e.

$$\liminf_{\nu \to \infty} I(u_\nu) \geq I(u) \text{ whenever } u_\nu \rightharpoonup u \text{ in } X$$

iii) I is coercive over X, i.e.

$$I(u) \geq \alpha \|u\| + \beta, \text{ for every } u \in X$$

and for some $\alpha > 0, \beta \in \mathbb{R}$.

However the second condition is in general hard to verify. We show nevertheless that the convex functionals are weakly lower semicontinuous.

Finally we give the classical first order condition that every minimum u should satisfy, namely that

$$I'(u) = 0$$

provided I has a Gâteaux derivative, denoted I'.

In the next sections we try to apply the above results to functionals of the type

$$I(u) = \int_\Omega f(x, u(x), \nabla u(x)) \, dx$$

where $\Omega \subset \mathbb{R}^n$ is a bounded open set, $u : \Omega \to \mathbb{R}^m$, $f \colon \Omega \times \mathbb{R}^m \times \mathbb{R}^{nm} \to \mathbb{R}$ is continuous and $X = W^{1,p}(\Omega; \mathbb{R}^m)$.

We shall therefore answer the following three questions:

1) when is I convex over $W_0^{1,p}(\Omega; \mathbb{R}^m)$?
2) when is I weakly lower semicontinuous over $W^{1,p}(\Omega; \mathbb{R}^m)$?
3) when is I Gâteaux differentiable ?

We shall answer the first question in Section 3.2 of this chapter, the second question in Section 3.3 as well as in Chapter 4 and the last question in Section 3.4.

Let us also recall the terminology introduced earlier. We speak of the *scalar case* whenever $m = 1$ or $n = 1$ and of the *vectorial case* whenever $m, n > 1$. In Sections 3.2 and 3.3 of this chapter we shall only deal with the scalar case, while Section 3.4 will be concerned with both cases.

In Section 3.2 we first establish that a necessary condition for I to be convex over $W_0^{1,p}(\Omega)$ is that $f(x, u, \bullet)$ be convex. We then study some particular cases, the most important being the one with no explicit dependence on u, i.e. $f(x, u, \xi) = f(x, \xi)$ where then the convexity of f is trivially sufficient for I to be convex. We then consider only the case $m = n = 1$, $\Omega = (0, 1)$ and study functions of the type:

i) $f(x, u, \xi) = a(u)\xi^2$
ii) $f(x, u, \xi) = g(u) + h(\xi)$.

We establish that the corresponding functional I is convex if and only if a is constant in the first case and in the second case if and only if

$$\pi^2 \inf\{h''(\xi) : \xi \in \mathbb{R}\} + \inf\{g''(u) : u \in \mathbb{R}\} \geq 0 \; .$$

In Section 3.3 we prove the main result of this chapter, namely that the functional I is weakly lower semicontinuous over $W^{1,p}(\Omega)$ if and only if $f(x, u, \bullet)$ is convex. So in particular the functional

$$I(u) = \int_0^1 ((u')^4 + (u^2 - 1)^2) \, dx$$

is weakly lower semicontinuous over $W_0^{1,4}(0, 1)$, but not convex.

We next consider *weakly continuous* functionals I, i.e. I and $-I$ are weakly lower semi-continuous. In view of the above result we have that a necessary and sufficient condition for I to be weakly continuous is that $f(x, u, \bullet)$ is affine in the last variable. Anticipating the result of Chapter 4, this should be contrasted with the vectorial case where functions such as $\det \nabla u$ (where $m = n > 1$) give rise to weakly continuous functionals I. We finally give a complete characterization of *invariant integrals*, i.e. integrals such that

$$I(u) \equiv \int_{\Omega} f(x, u(x), \nabla u(x))\, dx \equiv \text{constant}$$

for every $u \in W^{1,p}(\Omega)$ with $u = u_0$ on $\partial\Omega$. These integrals are a subclass of the weakly continuous integrals and are important in the field theories. We establish that I is invariant if and only if there exist $\Phi : \Omega \times \mathbf{R}^m \to \mathbf{R}^n$ (recall that either $m = 1$ or $n = 1$) and a function $\beta : \mathbf{R}^n \to \mathbf{R}$ such that

$$f(x, u(x), \nabla u(x)) = \text{div}\, \Phi(x, u(x)) + \beta(x) \ .$$

Finally in Section 3.4 we deduce immediately, from the above results, the existence of minima for the problem (P). We then study under what conditions on f the integral I is Gâteaux differentiable and we therefore establish the Euler equations. We conclude this section with an example illustrating the so called Lavrentiev phenomenon.

3.1 Abstract Results

3.1.1 Weak Lower Semicontinuous Functionals and Existence Theorems

We let X be a Banach space and we study the following problem:

$$(P) \qquad\qquad \inf\{I(u) : u \in X\}$$

where $I : X \to \bar{\mathbf{R}} = \mathbf{R} \cup \{+\infty\}$.

We first introduce the following definition.

Definition. $I : X \to \bar{\mathbf{R}} = \mathbf{R} \cup \{+\infty\}$ is said to be (sequentially) *weakly lower semicontinuous* over X if

$$\liminf_{\nu \to \infty} I(u_\nu) \geq I(u) \quad \text{whenever } u_\nu \rightharpoonup u \text{ in } X \ .$$

REMARK. From now on, we shall omit the word sequentially when we refer to weak lower semicontinuity.

We may now state the following.

Theorem 1.1. *Let X be a reflexive Banach space, $I : X \to \bar{\mathbb{R}}$ be weakly lower semicontinuous and coercive over X, i.e.*

$$I(u) \geq \alpha\|u\| + \beta$$

for every $u \in X$ and for some $\alpha > 0$, $\beta \in \mathbb{R}$. Assume also that there exists $\tilde{u} \in X$ with $I(\tilde{u}) < \infty$, then (P) has at least one solution $\bar{u} \in X$, i.e.

$$I(\bar{u}) = \inf(P) = \inf\{I(u) : u \in X\} \ .$$

PROOF. Let $\{u_\nu\}$ be a minimizing sequence for (P), i.e.

$$I(u_\nu) \to \inf(P) \ .$$

From the hypotheses we have that $\beta \leq \inf(P) \leq I(\tilde{u}) < \infty$. Using the coercivity of I, we may then deduce that there exists $K > 0$, independent of ν, such that

$$\|u_\nu\| \leq K \ .$$

Since X is reflexive we can extract a weakly convergent subsequence $\{u_{\nu'}\}$ such that

$$u_{\nu'} \rightharpoonup \bar{u} \ \text{in} \ X \ .$$

The weak lower semicontinuity of I implies then the result. $\square$

REMARK. The hypothesis on the reflexivity of X cannot be dropped in general. We give here a simple example (cf. Rudin [1], cf. also Section 4.1.1 of this chapter). Let

$$X = \left\{ u \in C([0,1]) : \int_0^{1/2} u(t)dt - \int_{1/2}^1 u(t)dt = 1 \right\}$$

and let

$$(P) \qquad\qquad \inf\{I(u) = \|u\|_{L^\infty} : u \in X\} \ .$$

It is obvious that

$$I(u) = \|u\|_{L^\infty} \geq \int_0^1 |u(t)|dt \geq \int_0^{1/2} u(t)dt - \int_{1/2}^1 u(t)dt = 1 \ .$$

Note that in fact

$$\inf(P) = \inf\{\|u\|_{L^\infty} : u \in X\} = 1 \ ,$$

since a minimizing sequence is for example

$$u_N(x) = \begin{cases} 1 & \text{if } x \in [0, \frac{1}{2} - \frac{1}{N}] \\ -Nx + \dfrac{N}{2} & \text{if } x \in [\frac{1}{2} - \frac{1}{N}, \frac{1}{2} + \frac{1}{N}] \\ -1 & \text{if } x \in [\frac{1}{2} + \frac{1}{N}, 1] \end{cases} .$$

However it is clear that no continuous function can satisfy

$$\|u\|_{L^\infty} = \int_0^1 |u(t)| dt = \int_0^{1/2} u(t) dt - \int_{1/2}^1 u(t) dt = 1 .$$

3.1.2 Convex Functionals

As mentioned in the introduction, it may be hard, in concrete examples, to determine whether a functional is weakly lower semicontinuous or not. We shall deal with this problem, in the next sections and chapters, in the context of the calculus of variations. There is however a large class of functionals which are weakly lower semicontinuous, namely those which are convex.

Theorem 1.2. *Let X be a Banach space and let $I : X \to \bar{\mathbb{R}} = \mathbb{R} \cup \{+\infty\}$ be convex and lower semicontinuous, then I is weakly lower semicontinuous.*

REMARK. The converse of the theorem is false and we shall see counterexamples in the next section.

PROOF. Let $u_\nu \rightharpoonup u$ in X, we wish to show that

$$L = \liminf_{\nu \to \infty} I(u_\nu) \geq I(u) . \tag{1}$$

Up to the extraction of a subsequence, still denoted by u_ν, we may assume that

$$L = \lim_{\nu \to \infty} I(u_\nu) . \tag{2}$$

If $L = +\infty$ the result is trivial. Note also that $L > -\infty$ in view of the following fact: I being convex and lower semicontinuous there exist (cf. Theorem 2.5 of Chapter 2) $u^* \in X^*$ and $\alpha \in \mathbb{R}$ such that

$$I(u) \geq \langle u; u^* \rangle + \alpha, \text{ for every } u \in X . \tag{3}$$

Since $u_\nu \rightharpoonup u$, we have that $\|u_\nu\|$ is uniformly bounded and therefore (3) implies that $L > -\infty$. We now let $\varepsilon > 0$ be fixed, then there exists $N = N(\varepsilon)$

such that

$$I(u_\nu) \le L + \varepsilon \text{ for every } \nu \ge N(\varepsilon) \ . \tag{4}$$

We now apply Theorem 1.2 of Chapter 2 (Mazur lemma) to the sequence $\{u_\nu\}_{\nu=N}^\infty$ to obtain that there exist $n = n(\varepsilon)$ an integer, $\alpha_i = \alpha_i(\varepsilon) \ge 0$, $N \le i \le n$, with $\sum \alpha_i = 1$ such that

$$\left\| u - \sum_{i=N}^{n} \alpha_i u_i \right\| \le \varepsilon \ . \tag{5}$$

We have therefore, from the convexity of I and (4), that

$$I\left(\sum_{i=N}^{n} \alpha_i u_i \right) \le \sum_{i=N}^{n} \alpha_i I(u_i) \le L + \varepsilon \ . \tag{6}$$

The lower semicontinuity of I combined with (5), (6) and the arbitrariness of ε lead immediately to (1) and the theorem. $\qquad\square$

3.1.3 First-Order Necessary Condition

As in the finite dimensional case, a necessary condition for $\bar{u}$ to be a minimum for I is that $I'(\bar{u}) = 0$.

Theorem 1.3. *Let X be a Banach space, $I : X \to \mathbb{R}$ be Gâteaux differentiable and let*

$$I(\bar{u}) = \inf\{I(u) : u \in X\} \ ,$$

then

$$\langle I'(\bar{u}); v \rangle = 0 \text{ for every } v \in X \ .$$

PROOF. Let for $v \in X$ and $t \in \mathbb{R}$

$$\varphi(t) = I(\bar{u} + tv) \ ;$$

then φ is differentiable. Since $\bar{u}$ is a minimum for I, we have that for every $v \in X$

$$\varphi(0) = \min\{\varphi(t) : t \in \mathbb{R}\} \ ,$$

thus

$$\varphi'(0) = \frac{d}{dt} I(\bar{u} + tv)\Big|_{t=0} = \langle I'(\bar{u}); v \rangle = 0 \ . \qquad\square$$

This necessary condition is usually not sufficient to ensure that $\bar{u}$ is a minimum. However it turns out to be sufficient for convex functionals, in fact we have

Theorem 1.4. *Let $I : X \to \mathbb{R}$ be convex, Gâteaux differentiable and assume that $\bar{u} \in X$ is such that*

$$\langle I'(\bar{u}); v \rangle = 0 \ \text{for every} \ v \in X \ ,$$

then

$$I(\bar{u}) = \inf\{I(u) : u \in X\} \ .$$

PROOF. Since I is convex and Gâteaux differentiable Theorem 2.6 of Chapter 2 implies that

$$I(v) \geq I(\bar{u}) + \langle I'(\bar{u}); v - \bar{u} \rangle \ \text{for every} \ v \in X$$

thus the result. $\qquad\qquad\square$

3.2 Convex Functionals

We now turn our attention to the functionals of the calculus of variations of the type

$$I(u) = \int_\Omega f(x, u(x), \nabla u(x)) \, dx$$

where

$(H1)$ $\Omega \subset \mathbb{R}^n$ is a bounded open set with Lipschitz boundary,

$(H2)$ $u : \Omega \to \mathbb{R}^m$ and hence $\nabla u \in \mathbb{R}^{nm}$,

$(H3)$ $f : \Omega \times \mathbb{R}^m \times \mathbb{R}^{nm} \to \mathbb{R}$ is continuous and satisfies

$$|f(x, u, \xi)| \leq a(x, |u|, |\xi|)$$

where a is increasing with respect to $|u|$ and $|\xi|$ and locally integrable in x.

We study in this section some necessary and sufficient conditions on f in order that I be convex over $W_0^{1,p}(\Omega; \mathbb{R}^m)$. We present here the results of Dacorogna [9].

Recall also that we restrict ourselves, in the whole section, to the scalar case where either $m = 1$ or $n = 1$.

3.2.1 Necessary Condition

We obtain immediately the following.

Theorem 2.1. *Let I and f be as above and assume that either $m = 1$ or $n = 1$. If I is convex over $W_0^{1,\infty}(\Omega; \mathbf{R}^m)$, then $f(x, u, \bullet)$ is convex for every $x \in \Omega$ and for every $u \in \mathbf{R}^m$.*

REMARK. Since the convexity of I over $W_0^{1,p}$ implies that over $W_0^{1,\infty}$, Theorem 2.1 holds therefore if I is convex over $W_0^{1,p}$, $p \geq 1$.

PROOF. Since f satisfies (H3) it is easy to see that I is lower semicontinuous over $W_0^{1,\infty}(\Omega; \mathbf{R}^m)$. Using Theorem 1.2 above we deduce that I is weak $*$ lower semi-continuous in $W_0^{1,\infty}$. Using Theorem 3.1 below we deduce immediately the theorem.   □

The above theorem is false in the vectorial case.

EXAMPLE. Let $m = n = 2$, $\Omega = \{x \in \mathbf{R}^2 : |x| < 1\}$ and

$$I(u) = \int_\Omega \det \nabla u(x)\, dx \ .$$

Then $I \equiv 0$ for every $u \in W_0^{1,\infty}(\Omega; \mathbf{R}^2)$ and hence I is convex over $W_0^{1,\infty}$, although $f(\xi) = \det \xi$ is not convex.

3.2.2 Sufficient Condition

We now turn to the converse of Theorem 2.1. We shall obtain sufficient conditions only in some special cases.

The first example we shall consider is the most important but the simplest one (recall that either $m = 1$ or $n = 1$).

Proposition 2.2. Let I and f be as above and assume furthermore that f is independent of u (i.e. $f(x, u, \xi) \equiv f(x, \xi)$) then I is convex over $W_0^{1,\infty}(\Omega; \mathbf{R}^m)$ if and only if $f(x, \bullet)$ is convex for every $x \in \Omega$.

REMARKS.

  i) The proof of the above proposition is trivial;
 ii) note also that many of the examples quoted in the introduction satisfy the hypotheses of the proposition.

We next consider the case where f depends on all its variables (x, u, ξ) but we restrict ourselves to the case $m = n = 1$ and $\Omega = (0, 1)$. We first give a trivial example showing that, in general, no convexity on the variable u can be infered from the convexity of I.

Proposition 2.3. Let $g : \mathbb{R} \to \mathbb{R}$ be continuous and let

$$f(x, u, \xi) = g(u)\xi$$

then

$$I(u) \equiv 0 \text{ for every } u \in W_0^{1,\infty}(0,1) \ .$$

REMARK. Note, however, that in the above example, there exists $\tilde{f} : (0,1) \times \mathbb{R} \times \mathbb{R} \to \mathbb{R}$, namely $\tilde{f} \equiv 0$, convex in the last two variables such that

$$I(u) \equiv \int_0^1 \tilde{f}(x, u(x), u'(x))\, dx, \text{ for every } u \in W_0^{1,\infty}(0,1) \ .$$

We now state two special cases where one can explicitly find a sufficient condition.

Proposition 2.4. Let $a \in C^\infty(\mathbb{R})$ be such that

$$a(u) \geq a_0 > 0 \text{ for every } u \in \mathbb{R}$$

and let for $n \geq 1$, n an integer,

$$f(x, u, \xi) = a(u)\xi^{2n}$$

then, I is convex over $W_0^{1,\infty}(0,1)$ if and only if a is constant.

Proposition 2.5. Let $g, h \in C^\infty(\mathbb{R})$ and

$$f(x, u, \xi) = g(u) + h(\xi) \ .$$

Let

$$g_0 = \inf\{g''(u) : u \in \mathbb{R}\}, \quad h_0 = \inf\{h''(\xi) : \xi \in \mathbb{R}\} \ ,$$

then,

i) there exist g non convex and h convex such that I is convex over $W_0^{1,\infty}(0,1)$; for example

$$g(u) = \frac{1}{2}(u^2 - 1)^2, \quad h(\xi) = \xi^2 \ ;$$

ii) I is convex over $W_0^{1,\infty}(0,1)$ if and only if $h_0 \geq 0$ and $\pi^2 h_0 + g_0 \geq 0$;

iii) *Case 1*: if $g_0 \geq 0$ and $h_0 \geq 0$, then $f(x, u, \xi) = g(u) + h(\xi)$ is convex in the variables (u, ξ);

Case 2: if $g_0 < 0$ and $\pi^2 h_0 + g_0 > 0$, then let

$$\varphi(x, u, \xi) = \sqrt{-g_0 h_0}\left(\tan\left[\sqrt{\frac{-g_0}{h_0}}\left(x - \frac{1}{2}\right)\right]\right) u\xi$$

$$- \frac{g_0}{2}\left(1 + \tan^2\left[\sqrt{\frac{-g_0}{h_0}}\left(x - \frac{1}{2}\right)\right]\right) u^2 \;.$$

Moreover, if

$$\Phi(x, u) = \frac{1}{2}\sqrt{-g_0 h_0}\left(\tan\left[\sqrt{\frac{-g_0}{h_0}}\left(x - \frac{1}{2}\right)\right]\right) u^2$$

then

$$\frac{d}{dx}\Phi(x, u(x)) = \varphi(x, u(x), u'(x)) \quad \text{a.e.}$$

for every $u \in W^{1,\infty}(0, 1)$. Furthermore

$$\tilde{f}(x, u, \xi) = g(u) + h(\xi) + \varphi(x, u, \xi)$$

is convex in the variables (u, ξ) for every $x \in [0, 1]$ and satisfies

$$I(u) \equiv \int_0^1 \tilde{f}(x, u(x), u'(x))\, dx$$

for every $u \in W_0^{1,\infty}(0, 1)$.

Case 3: If $g_0 < 0$ and $\pi^2 h_0 + g_0 = 0$, then $\tilde{f}$ defined as in Case 2 is convex in (u, ξ) for $x \in (0, 1)$ and the identity

$$I(u) \equiv \int_0^1 \tilde{f}(x, u(x), u'(x))\, dx$$

holds only if $u \in \mathcal{D}(0, 1) \equiv \{u \in C^\infty(0, 1)$, with compact support in $(0, 1)\}$.

REMARKS.

i) Note that the function $\varphi(x, u, \xi)$ is linear in ξ and satisfies

$$\int_0^1 \varphi(x, u(x), u'(x))\, dx \equiv 0 \text{ for every } u \in W_0^{1,\infty}(0, 1) \;;$$

such a functional is called an invariant integral (cf. Section 3.3.2 below);

ii) observe also that if $\pi^2 h_0 + g_0 = 0$ then the function φ is not defined at the boundary points $x = 0$ and 1.

Before proceeding with the proof of the two propositions we give a lemma whose proof is obvious.

Lemma 2.6. *Let $f : (0,1) \times \mathbf{R} \times \mathbf{R} \to \mathbf{R}$ be C^2. Let*

$$I(u) = \int_0^1 f(x, u(x), u'(x))\, dx \ .$$

For $\lambda \in [0,1]$, $u, v \in W_0^{1,\infty}(0,1)$, let

$$\psi(\lambda) = I(\lambda u + (1 - \lambda)v) - \lambda I(u) - (1 - \lambda)I(v) \ .$$

The three following conditions are then equivalent
i) I is convex over $W_0^{1,\infty}(0,1)$,
ii) ψ is convex in λ, for every $u, v \in W_0^{1,\infty}(0,1)$,
iii) for every $\lambda \in [0,1]$, $u, v \in W_0^{1,\infty}(0,1)$, one has

$$\psi''(\lambda) = \int_0^1 \Big[(u - v)^2 f_{uu}(x, \lambda(u - v) + v, \lambda(u' - v') + v')$$

$$+ 2(u - v)(u' - v')f_{u\xi} + (u' - v')^2 f_{\xi\xi} \Big]\, dx \geq 0 \ ,$$

where the following notations are adopted

$$f_{uu} = \frac{\partial^2 f}{\partial u^2}, \quad f_{u\xi} = \frac{\partial^2 f}{\partial u \partial \xi}, \quad f_{\xi\xi} = \frac{\partial^2 f}{\partial \xi^2} \ .$$

We now proceed with the proof of Proposition 2.4.

PROOF. The sufficiency part of the proposition is obvious, since clearly a constant implies I convex. We divide the proof of the necessity part into three steps.

Step 1: We let in the above lemma $w = u - v$ and $z = \lambda(u - v) + v$. We then define a new function

$$b(t) = (a(t))^{-\frac{1}{2n-1}} \ . \tag{1}$$

We shall prove below, using the notations of Lemma 2.6, that the convexity of I implies that

$$0 \le \psi''(\lambda) = \int_0^1 2n(2n-1)\frac{z'^{2n-2}}{(b(z))^{(2n-1)-1}}$$

$$\times \left\{ \left[b(z)\left(\frac{w}{b(z)}\right)' \right]^2 - \frac{w^2 z'^2 b''(z)}{2nb(z)} \right\} dx \ . \tag{2}$$

Step 2: In this step we show, by choosing appropriately the functions $w, z \in W_0^{1,\infty}(0,1)$, that (2) implies the concavity of b. i.e.

$$b''(t) \le 0, \ \text{ for every } t \in \mathbf{R} \ . \tag{3}$$

Step 3: The conclusion then follows immediately from (3), if one observes that the hypothesis $a(u) \ge a_0 > 0$ implies

$$0 < b(t) = (a(t))^{-\frac{1}{2n-1}} \le a_0^{-\frac{1}{2n-1}} \ .$$

The concavity of b and the fact that b is bounded above and below implies then that b, and therefore a, is constant.

It therefore remains to show only Steps 1 and 2.

Step 1: From Lemma 2.6 we have

$$\psi''(\lambda) = \int_0^1 \left\{ w^2 a''(z)z'^{2n} + 4nww'a'(z)z'^{2n-1} \right.$$

$$\left. + 2n(2n-1)w'^2 a(z)z'^{2n-2} \right\} dx$$

$$= \int_0^1 2n(2n-1)a(z)z'^{2n-2}\left\{ w'^2 + \frac{2a'(z)}{(2n-1)a(z)}z'ww' \right.$$

$$+ \left[\frac{a'(z)z'}{(2n-1)a(z)}w \right]^2 - \left[\frac{a'(z)z'}{(2n-1)a(z)}w \right]^2 + \frac{a''(z)z'^2}{2n(2n-1)a(z)}w^2 \left. \right\} dx$$

$$= \int_0^1 2n(2n-1)a(z)z'^{2n-2}\left\{ \left[w' + \frac{a'(z)z'}{(2n-1)a(z)}w \right]^2 \right. \tag{4}$$

$$\left. - \frac{w^2 z'^2}{2n(2n-1)^2(a(z))^2}(2n(a'(z))^2 - (2n-1)a''(z)a(z)) \right\} dx$$

$$= \int_0^1 2n(2n-1)a(z)z'^{2n-2}\left\{ \left[a^{\frac{-1}{2n-1}}\left(a^{\frac{1}{2n-1}}w \right)' \right]^2 \right.$$

$$\left. - \frac{w^2 z'^2 a^{\frac{1}{2n-1}}}{2n(2n-1)^2}\frac{2na'^2 - (2n-1)a''a}{a^{2+\frac{1}{2n-1}}} \right\} dx \ .$$

Using the notation introduced in (1) we have

$$b''(t) = -\frac{1}{2n-1}\left(a^{\frac{-1}{2n-1}-1}\ a'\right)'$$

$$= \frac{1}{(2n-1)^2}\frac{2na'^2 - (2n-1)a''a}{a^{2+\frac{1}{2n-1}}}\ .$$

(5)

Combining (4) and (5) we have indeed established (2) and thus Step 1.

Step 2: We now show that (2) implies that b is concave. Assume for contradiction that (3) does not hold, therefore there must exist $\alpha \in \mathbb{R}$ such that

$$b''(\alpha) > 0 \ .$$

(6)

In view of the continuity of b'', we may choose $\alpha \neq 0$. We now construct z and $w \in W_0^{1,\infty}(0,1)$ such that (2) is violated.

Construction of z. We define for N an integer

$$z(x) = \begin{cases} N\alpha x, & \text{if } x \in (0,\frac{1}{N}) \\[2mm] \alpha + N\alpha\left(x - \frac{k}{N} - \frac{m}{N^2}\right) & \text{if } x \in \bigcup_{m=0}^{N-1}\left(\frac{k}{N} + \frac{m}{N^2}, \frac{k}{N} + \frac{m}{N^2} + \frac{1}{2N^2}\right), \\ & \quad 1 \le k \le N-2 \\[2mm] \alpha - N\alpha\left(x - \frac{k}{N} - \frac{m+1}{N^2}\right) & \text{if } x \in \bigcup_{m=0}^{N-1}\left(\frac{k}{N} + \frac{m}{N^2} + \frac{1}{2N^2}, \frac{k}{N} + \frac{m+1}{N^2}\right), \\ & \quad 1 \le k \le N-2 \\[2mm] \alpha - N\alpha\left(x - \frac{N-1}{N}\right) & \text{if } x \in \left(\frac{N-1}{N}, 1\right) \end{cases}$$

We then have that $z \in W_0^{1,\infty}(0,1)$ and

$$\begin{cases} |z(x) - \alpha| \le \frac{|\alpha|}{2N} & \text{if } x \in \left(\frac{1}{N}, \frac{N-1}{N}\right) \\ |z'(x)| = N|\alpha| & \text{a.e. in } (0,1) \end{cases} \ .$$

(7)

Therefore if we fix $\varepsilon > 0$, there must exist N sufficiently large so that

$$|b''(z) - b''(\alpha)|, \ |b'(z) - b'(\alpha)|, \ |b(z) - b(\alpha)| \le \varepsilon \ ,$$

(8)

for every $x \in \left(\frac{1}{N}, \frac{N-1}{N}\right)$.

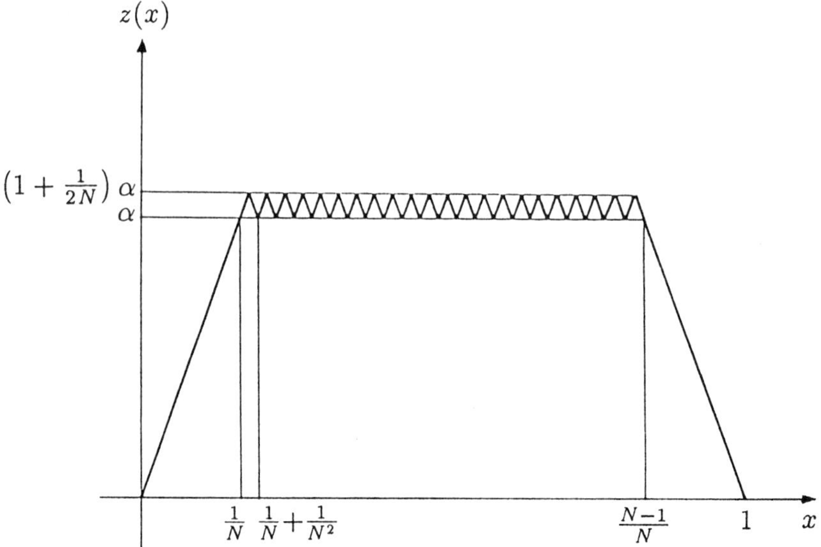

Figure 3.1

Construction of w. We choose w in such a way that

$$\frac{w(x)}{b(z(x))} = \begin{cases} 0 & \text{if } x \in \left(0, \frac{1}{N}\right) \\ \sin\left[\frac{\pi N}{N-2}\left(x - \frac{1}{N}\right)\right] & \text{if } x \in \left(\frac{1}{N}, \frac{N-1}{N}\right) \\ 0 & \text{if } x \in \left(\frac{N-1}{N}, 1\right) \end{cases} .$$

We now return to the inequality (2) applied to the above z and w, we have

$$0 \le \psi''(\lambda) = \int_{\frac{1}{N}}^{\frac{N-1}{N}} 2n(2n-1)\frac{N^{2n-2}|\alpha|^{2n-2}}{(b(z))^{\frac{1}{2n-1}}}$$

$$\left\{\left[b(z)\frac{\pi N}{N-2}\cos\left(\frac{\pi N}{N-2}\left(x - \frac{1}{N}\right)\right)\right]^2\right.$$

$$\left. - \frac{b(z)b''(z)N^2\alpha^2}{2n}\sin^2\left(\frac{\pi N}{N-2}\left(x - \frac{1}{N}\right)\right)\right\} dx .$$

Using (8) and denoting by K a constant depending on n, α and $b(\alpha)$, but not on N, we have, choosing N sufficiently large,

$$0 \le \psi''(\lambda) \le K[N^{2n-2} - b''(\alpha)N^{2n}] . \tag{9}$$

Letting N tend to infinity and using (6) we obtain that the right hand side of (9) is negative, which is a contradiction. This establishes Step 2 and thus the proposition. $\qquad\square$

We now turn to the proof of Proposition 2.5.

PROOF.

i) The first part is a direct consequence of the second part since if $g(u) = (u^2 - 1)^2/2$ and $h(\xi) = \xi^2$ then $g_0 = -2$ and $h_0 = 2$. Hence $\pi^2 h_0 + g_0 = 2(\pi^2 - 1) \geq 0$;

ii) as a consequence of Lemma 2.6 the convexity of I is equivalent to the positivity of the following expression, for every $\lambda \in [0,1]$ for every $u, v \in W_0^{1,\infty}(0,1)$,

$$\psi''(\lambda) = \int_0^1 [(u-v)^2 g''(\lambda(u-v)+v) + (u'-v')^2 h''(\lambda(u'-v')+v')]\, dx. \quad (1)$$

Sufficiency: The sufficiency of $h_0 \geq 0$ and $\pi^2 h_0 + g_0 \geq 0$ for the convexity of I is simple. Observe that

$$\psi''(\lambda) \geq \int_0^1 [(u-v)^2 g_0 + (u'-v')^2 h_0]\, dx \ .$$

Since h_0 is positive, we may use Poincaré-Wirtinger inequality (Theorem 1.7 of Chapter 2) to get that

$$\psi''(\lambda) \geq \int_0^1 (\pi^2 h_0 + g_0)(u-v)^2\, dx \ ,$$

thus the positivity of ψ'' and the result.

Necessity: We now assume that I is convex over $W_0^{1,\infty}(0,1)$. This implies that ψ'' defined by (1) is positive. The fact that $h_0 \geq 0$ results immediately from Theorem 2.1 above; we therefore only need to show that $\pi^2 h_0 + g_0 \geq 0$. We first introduce some notations we let in (1), $w = u - v$ and $z = \lambda(u - v) + v$, which both belong to $W_0^{1,\infty}(0,1)$. We may then rewrite (1) as

$$\psi''(\lambda) = \int_0^1 [w^2 g''(z) + w'^2 h''(z')]\, dx \geq 0 \ . \quad (2)$$

Note first that, if $g_0 \geq 0$, the result is trivial, so we may assume that $g_0 < 0$. We now construct, as in Proposition 2.4, w and z in an appropriate manner so that $\psi''(\lambda)$ is up to a multiplicative positive constant equal to $\pi^2 h_0 + g_0$, the positivity of $\pi^2 h_0 + g_0$ follows then immediately from (2).

We now fix an integer N, then there exist $\xi_0, u_0 \in \mathbf{R}$ such that

$$\begin{cases} 0 \leq h''(\xi_0) - h_0 \leq \dfrac{1}{N} \\ 0 \leq g''(u_0) - g_0 \leq \dfrac{1}{N} \end{cases} . \tag{3}$$

Construction of z. We let

$$z(x) = \begin{cases} Nu_0 x & \text{if } x \in \left(0, \frac{1}{N}\right) \\ u_0 + \xi_0 \left(x - \frac{k}{N}\right) & \text{if } x \in \left(\frac{k}{N}, \frac{k+1}{N} - \frac{1}{N^2}\right), \\ & \quad 1 \leq k \leq N - 3 \\ u_0 - \xi_0(N-1)\left(x - \frac{k+1}{N}\right) & \text{if } x \in \left(\frac{k+1}{N} - \frac{1}{N^2}, \frac{k+1}{N}\right), \\ & \quad 1 \leq k \leq N - 3 \\ u_0 & \text{if } x \in \left(\frac{N-2}{N}, \frac{N-1}{N}\right) \\ -Nu_0(x - 1) & \text{if } x \in \left(\frac{N-1}{N}, 1\right) \end{cases} .$$

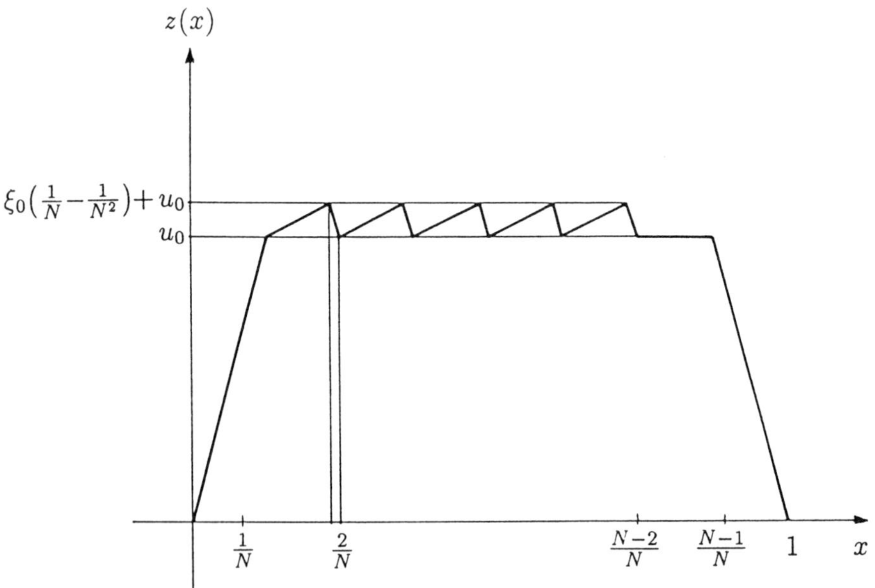

Figure 3.2

We then obviously have that $z \in W_0^{1,\infty}(0,1)$ and

$$\begin{cases} |z(x) - u_0| \leq |\xi_0| \left(\frac{1}{N} - \frac{1}{N^2}\right) & \text{if } x \in \left(\frac{1}{N}, \frac{N-1}{N}\right) \\ z'(x) \equiv \xi_0 & \text{if } x \in \bigcup_{k=1}^{N-3} \left(\frac{k}{N}, \frac{k+1}{N} - \frac{1}{N^2}\right) \end{cases} \qquad (4)$$

Fix $\varepsilon > 0$, we may then choose N sufficiently large so that

$$|g''(z(x)) - g''(u_0)| \leq \varepsilon \text{ for every } x \in \left(\frac{1}{N}, \frac{N-1}{N}\right) . \qquad (5)$$

Construction of w. We let

$$w(x) = \begin{cases} 0 & \text{if } x \in \left(0, \frac{1}{N}\right) \\ \sin\left[\frac{N\pi}{N-2}\left(x - \frac{1}{N}\right)\right] + a_{2k-1} & \text{if } x \in \left(\frac{k}{N}, \frac{k+1}{N} - \frac{1}{N^2}\right), \\ & \qquad 1 \leq k \leq N - 3 \\ a_{2k} & \text{if } x \in \left(\frac{k+1}{N} - \frac{1}{N^2}, \frac{k+1}{N}\right), \\ & \qquad 1 \leq k \leq N - 3 \\ -Na_{2(N-3)}\left(x - \frac{N-1}{N}\right) & \text{if } x \in \left(\frac{N-2}{N}, \frac{N-1}{N}\right) \\ 0 & \text{if } x \in \left(\frac{N-1}{N}, 1\right) \end{cases}$$

where for $1 \leq k \leq N - 3$, the a_k are chosen so that w is continuous, more precisely we choose

$$\begin{cases} a_1 = 0 \\ a_{2k} = a_{2k-1} + \sin\dfrac{Nk-1}{N(N-2)}\pi \\ a_{2k+1} = a_{2k} - \sin\dfrac{k\pi}{N-2} \end{cases} .$$

We have therefore that $w \in W_0^{1,\infty}(0,1)$. (See Figure 3.3).
We now observe that from the definitions of a_k we have

$$a_{2k} = \sum_{\nu=1}^{k} \sin\left[\frac{N\nu-1}{N(N-2)}\pi\right] - \sum_{\nu=1}^{k-1} \sin\left[\frac{\nu}{N-2}\pi\right]$$

$$= \sin\left[\frac{Nk-1}{N(N-2)}\pi\right] - 2\sin\left[\frac{\pi}{2N(N-2)}\right] \sum_{\nu=1}^{k-1} \cos\left[\frac{2N\nu-1}{2N(N-2)}\pi\right] .$$

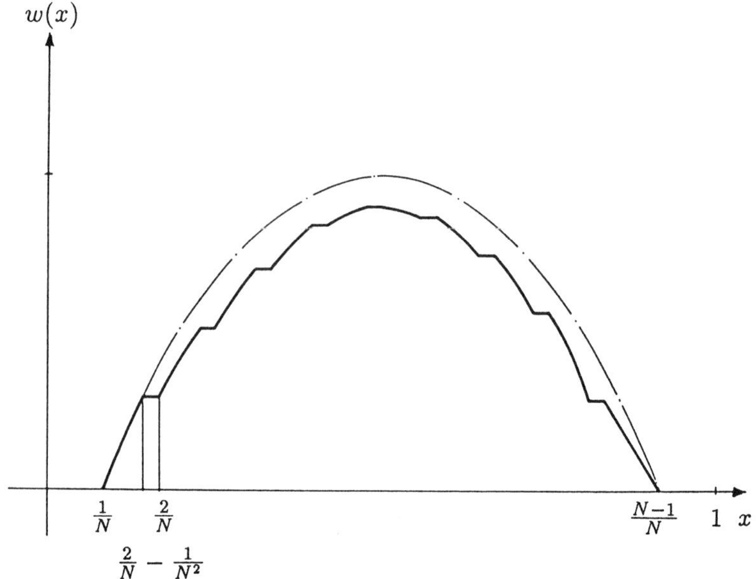

Figure 3.3. The *dotted* curve represents the function $f(x) = \sin\left[\frac{N\pi}{N-2}\left(x - \frac{1}{N}\right)\right]$, while the *plain* one represents the function w.

Similarly we obtain

$$a_{2k+1} = \sum_{\nu=1}^{k}\left\{\sin\left[\frac{N\nu - 1}{N(N-2)}\pi\right] - \sin\left[\frac{\nu}{N-2}\pi\right]\right\}$$

$$= -2\sin\left[\frac{\pi}{2N(N-2)}\right]\sum_{\nu=1}^{k}\cos\left[\frac{2N\nu - 1}{2N(N-2)}\pi\right] \ .$$

We then deduce that, K denoting a constant independent of N,

$$\begin{cases} |a_{2k+1}| \leq \dfrac{K}{N} \\ \left|a_{2k} - \sin\left[\dfrac{Nk - 1}{N(N-2)}\pi\right]\right| \leq \dfrac{K}{N} \end{cases} \cdot$$

In particular if $k = (N-3)$ we have that

$$|a_{2(N-3)}| \leq \left|\sin\left[\frac{N(N-3)-1}{N(N-2)}\pi\right]\right| + |a_{2N-5}| \leq \frac{2K}{N} \ . \tag{6}$$

We also have immediately that

$$\left| w(x) - \sin\left[\frac{N\pi}{N-2}\left(x - \frac{1}{N} \right) \right] \right| \to 0, \text{ as } N \to \infty, \; x \in \left(\frac{1}{N}, \frac{N-1}{N} \right).$$

Summarizing the properties of w, we have that for $\varepsilon > 0$ fixed, there exists N large enough such that, K denoting a constant independent of N,

$$\begin{cases} |w'| \le K \text{ a.e. in } (0,1) \\[2mm] \left| w(x) - \sin\left[\frac{N\pi}{N-2}\left(x - \frac{1}{N} \right) \right] \right| \le \varepsilon \quad x \in (0,1) \\[2mm] w'(x) = \frac{N\pi}{N-2}\cos\left[\frac{N\pi}{N-2}\left(x - \frac{1}{N} \right) \right] \text{ if } x \in \bigcup_{k=1}^{N-3}\left(\frac{k}{N}, \frac{k+1}{N} - \frac{1}{N^2} \right) \quad (7) \\[2mm] w'(x) = 0 \quad \text{if } x \in \bigcup_{k=1}^{N-3}\left(\frac{k+1}{N} - \frac{1}{N^2}, \frac{k+1}{N} \right). \end{cases}$$

We now return to (2) applied to the above, z and w, we have

$$0 \le \psi''(\lambda)$$

$$= \int_0^1 [w^2 g''(z) + w'^2 h''(z')]\, dx$$

$$= \int_{\frac{1}{N}}^{\frac{N-1}{N}} [w^2 g''(z) + w'^2 h''(z')]\, dx$$

$$= \int_{\frac{1}{N}}^{\frac{N-1}{N}} \sin^2\left[\frac{N\pi}{N-2}\left(x - \frac{1}{N} \right) \right] g''(z)\, dx$$

$$+ \int_{\frac{1}{N}}^{\frac{N-1}{N}} \left\{ w^2(x) - \sin^2\left[\frac{N\pi}{N-2}\left(x - \frac{1}{N} \right) \right] \right\} g''(z)\, dx$$

$$+ \sum_{k=1}^{N-3} \int_{\frac{k}{N}}^{\frac{k+1}{N} - \frac{1}{N^2}} \left(\frac{N\pi}{N-2} \right)^2 \cos^2\left[\frac{N\pi}{N-2}\left(x - \frac{1}{N} \right) \right] h''(\xi_0)\, dx$$

$$+ \int_{\frac{N-2}{N}}^{\frac{N-1}{N}} (Na_{2(N-3)})^2 h''(0)\, dx \;.$$

Using (5), (6), (7) and denoting by K a generic constant independent of N, we obtain for N sufficiently large

$$0 \leq \psi''(\lambda)$$

$$\leq -K\varepsilon + \int_{\frac{1}{N}}^{\frac{N-1}{N}} \sin^2\left[\frac{N\pi}{N-2}\left(x - \frac{1}{N}\right)\right] g''(z)\, dx$$

$$+ \left(\frac{N\pi}{N-2}\right)^2 h''(\xi_0) \sum_{k=1}^{N-3} \int_{\frac{k}{N}}^{\frac{k+1}{N} - \frac{1}{N^2}} \cos^2\left[\frac{N\pi}{N-2}\left(x - \frac{1}{N}\right)\right] dx \ .$$

Using (3) and (5) we have for N sufficiently large

$$0 \leq \psi''(\lambda)$$

$$\leq -K\varepsilon + g_0 \int_{\frac{1}{N}}^{\frac{N-1}{N}} \sin^2\left[\frac{N\pi}{N-2}\left(x - \frac{1}{N}\right)\right] dx$$

$$+ h_0 \left(\frac{N\pi}{N-2}\right)^2 \int_{\frac{1}{N}}^{\frac{N-1}{N}} \cos^2\left[\frac{N\pi}{N-2}\left(x - \frac{1}{N}\right)\right] dx$$

$$+ h_0 \left(\frac{N\pi}{N-2}\right)^2 \left\{ \sum_{k=1}^{N-3} \int_{\frac{k}{N}}^{\frac{k+1}{N} - \frac{1}{N^2}} \cos^2\left[\frac{N\pi}{N-2}\left(x - \frac{1}{N}\right)\right] dx \right.$$

$$\left. - \int_{\frac{1}{N}}^{\frac{N-1}{N}} \cos^2\left[\frac{N\pi}{N-2}\left(x - \frac{1}{N}\right)\right] dx \right\} \ .$$

Letting $N \to \infty$ and using the arbitrariness of ε we have indeed established that

$$0 \leq \pi^2 h_0 \int_0^1 \cos^2 \pi x\, dx + g_0 \int_0^1 \sin^2 \pi x\, dx$$

which implies immediately that $\pi^2 h_0 + g_0 \geq 0$. And this concludes the second part of the proposition.

iii) We now show that if I is convex over $W_0^{1,\infty}$ then there exists a convex integrand $\tilde{f}$ such that

$$I(u) = \int_0^1 \tilde{f}(x, u(x), u'(x))\, dx \tag{8}$$

for every $u \in W_0^{1,\infty}(0,1)$. First observe that Case 1 is trivial and we therefore consider only Case 2 and 3. We decompose the proof into two steps.

Step 1: We first show that

$$\tilde{f}(x, u, \xi) = g(u) + h(\xi) + \varphi(x, u, \xi)$$

is convex with respect to the last two variables. Observe that since $\pi^2 h_0 > -g_0 > 0$ (Case 2), then

$$-\frac{\pi}{2} < \sqrt{\frac{-g_0}{h_0}}\left(x - \frac{1}{2}\right) < \frac{\pi}{2} \quad \text{if} \quad x \in [0,1]$$

and if $\pi^2 h_0 = -g_0$ (Case 3), then the above inequality holds only if $x \in (0,1)$. Therefore $\tilde{f} : (0,1) \times \mathbb{R} \times \mathbb{R} \to \mathbb{R}$ is well defined. In order to show the convexity of $\tilde{f}$ we prove that the hessian of $\tilde{f}$ is positive definite, i.e. denoting by $\gamma = \sqrt{\frac{-g_0}{h_0}}\left(x - \frac{1}{2}\right)$

$$\nabla^2 \tilde{f} = \begin{pmatrix} \tilde{f}_{\xi\xi} & \tilde{f}_{u\xi} \\ \tilde{f}_{\xi u} & \tilde{f}_{uu} \end{pmatrix} = \begin{pmatrix} h''(\xi) & \sqrt{-g_0 h_0}\tan\gamma \\ \sqrt{-g_0 h_0}\tan\gamma & g''(u) - g_0(1 + \tan^2\gamma) \end{pmatrix}.$$

Since $h''(\xi) \geq h_0 \geq 0$, $g''(u) \geq g_0$ and $g_0 < 0$ it remains to show that

$$\det \nabla^2 \tilde{f} = h''(\xi)(g''(u) - g_0(1 + \tan^2\gamma)) + h_0 g_0 \tan^2\gamma \geq 0.$$

We have immediately that

$$\det \nabla^2 \tilde{f} \geq h_0(g_0 - g_0(1 + \tan^2\gamma)) + h_0 g_0 \tan^2\gamma = 0$$

and thus $\tilde{f}(x, \bullet, \bullet)$ is convex for every $x \in (0,1)$. This concludes Step 1.

Step 2: It remains therefore to show (8). Observe that if $u \in W^{1,\infty}(0,1)$, then

$$\tilde{f}(x, u, u') = g(u) + h(u')$$

$$+ \frac{d}{dx}\left[\frac{\sqrt{-g_0 h_0}}{2}\tan\left(\sqrt{\frac{-g_0}{h_0}}\left(x - \frac{1}{2}\right)\right)u^2\right], \quad \text{a.e. in } (0,1).$$

Hence (8) holds for every $u \in W_0^{1,\infty}(0,1)$ if $\pi^2 h_0 + g_0 > 0$ and only if $u \in \mathcal{D}(0,1)$ if $\pi^2 h_0 + g_0 = 0$. $\qquad\square$

3.3 Weak Lower Semicontinuity, Weak Continuity and Invariant Integrals

3.3.1 Weak Lower Semicontinuity

We show in this section that if

$$I(u) = \int_\Omega f(x, u(x), \nabla u(x)) \, dx$$

where $\Omega \subset \mathbb{R}^n$ is a bounded open set, $u : \Omega \to \mathbb{R}^m$ and $m = 1$ or $n = 1$ then I is weakly lower semicontinuous in $W^{1,p}$ if and only if $f(x, u, \bullet)$ is convex.

3.3.1.1 Necessity of Convexity

We first start with the necessary condition.

Theorem 3.1. *Let $\Omega \subset \mathbb{R}^n$ be a bounded open set, $f : \Omega \times \mathbb{R}^m \times \mathbb{R}^{nm} \to \mathbb{R}$ be continuous and satisfying*

$$(H3) \qquad\qquad\qquad |f(x, u, \xi)| \leq a(x, |u|, |\xi|)$$

for every $(x, u, \xi) \in \Omega \times \mathbb{R}^m \times \mathbb{R}^{nm}$, where a is increasing with respect to $|u|$ and $|\xi|$ and locally integrable in x. Assume that either $m = 1$ or $n = 1$. If I is weak $*$ lower semicontinuous in $W^{1,\infty}(\Omega; \mathbb{R}^m)$, then $f(x, u, \bullet)$ is convex.

REMARKS.

i) It is clear that if I is weakly lower semicontinuous in $W^{1,p}$, then it is weak $*$ lower semicontinuous in $W^{1,\infty}$ and thus the convexity of f is also necessary for $W^{1,p}$ weak lower semicontinuity.

ii) The theorem remains valid if the functional I is lower semicontinuous for every sequence

$$\begin{cases} u_\nu \overset{*}{\rightharpoonup} \bar{u} & \text{in } W^{1,\infty}(\Omega; \mathbb{R}^m) \\ u_\nu = \bar{u} & \text{on } \partial\Omega \end{cases}$$

since in the proof of Theorem 3.1 we use such sequences.

iii) The hypothesis of continuity of f can be weakened for example by considering Carathéodory functions (cf. below for a definition of such functions).

iv) A theorem of the above type has been proved under various kinds of hypotheses by Tonelli [1], Mac Shane [1], Cacciopoli-Scorza-Dragoni [1], Morrey [2], Ioffe [1-3], Cesari [1-4], Berkowitz [1,2], Olech [2,3], Marcellini-Sbordone [2], Buttazzo [1,2].

In the proof of the theorem we shall need the two following lemmas.

Lemma 3.2. *Let X be a metric space, $Y \subset X$ and $\alpha > 0$. Let $f : Y \to \mathbb{R}$ be Lipschitz with constant α, i.e.*

$$|f(y_1) - f(y_2)| \leq \alpha \|y_1 - y_2\|, \text{ for every } y_1, y_2 \in Y .$$

Then f can be extended to X as a Lipschitz function with constant α.

REMARK. The above lemma is sometimes known as *Mac Shane lemma*.

PROOF. Define for $x \in X$

$$\tilde{f}(x) = \sup\{f(y) - \alpha\|x - y\| : y \in Y\} . \tag{1}$$

Note that if $x \in Y$, then, since f is Lipschitz over Y, we have that the supremum in (1) is attained for $y = x$. Hence $\tilde{f}$ is an extension of f. We now wish to show that $\tilde{f}$ is Lipschitz over X with constant α. Let $x_1, x_2 \in X$ and assume without loss of generality that $\tilde{f}(x_1) - \tilde{f}(x_2) \geq 0$, then

$$0 \leq \tilde{f}(x_1) - \tilde{f}(x_2)$$

$$\leq \sup_{y_1 \in Y} \{f(y_1) - \alpha\|x_1 - y_1\|\} - \sup_{y_2 \in Y} \{f(y_2) - \alpha\|x_2 - y_2\|\}$$

$$\leq \sup_{y_1 \in Y_1} \{\alpha\|x_2 - y_1\| - \alpha\|x_1 - y_1\|\} \leq \alpha\|x_1 - x_2\| .$$

The lemma is indeed established.   □

We now give a lemma which constitutes the first step in the proof of the theorem.

Lemma 3.3. *Let Ω, f and I be as in the theorem. Assume that I is weak $*$ lower semicontinuous in $W^{1,\infty}(\Omega; \mathbb{R}^m)$, then*

$$\frac{1}{\operatorname{meas} D} \int_D f(x_0, u_0, \xi_0 + \nabla\varphi(y))\, dy \geq f(x_0, u_0, \xi_0)$$

for every cube $D \subset \Omega$, for every $(x_0, u_0, \xi_0) \in \Omega \times \mathbb{R}^m \times \mathbb{R}^{nm}$ and for every $\varphi \in W_0^{1,\infty}(D; \mathbb{R}^m)$.

REMARKS.

i) Note that, contrary to Theorem 3.1, we do not assume that either $m = 1$ or $n = 1$. The lemma is therefore valid also in the vectorial case, i.e. $m, n \geq 1$. In Chapter 4, we shall call a function f satisfying the above inequality, *quasiconvex*;

ii) the lemma remains valid if we assume the functional I to be lower semi-continuous for every sequence

$$\begin{cases} u_\nu \overset{*}{\rightharpoonup} \bar{u} & \text{in } W^{1,\infty}(\Omega; \mathbf{R}^m) \\ u_\nu = \bar{u} & \text{on } \partial\Omega; \end{cases}$$

since in the proof of Lemma 3.3, we use such sequences;
iii) the above lemma is essentially due to Morrey [1,2], and has been refined by Meyers [1], Acerbi-Fusco [1], Silvermann [1,2].

PROOF of Lemma 3.3. For the sake of illustration we first prove the lemma for the case $f(x, u, \xi) \equiv f(\xi)$ and then return to the general case.

Part 1: Let $f(x, u, \xi) \equiv f(\xi)$, D be a cube contained in Ω and $\xi_0 \in \mathbf{R}^{nm}$. Let $\varphi \in W_0^{1,\infty}(D; \mathbf{R}^m)$ be extended by periodicity from D to $\mathbf{R}^n$. Following Theorem 1.5 of Chapter 2 we have that, since $\varphi = 0$ on ∂D,

$$\frac{1}{\nu}\varphi(\nu x) \overset{*}{\rightharpoonup} 0 \text{ in } W_0^{1,\infty}(D; \mathbf{R}^m) \ .$$

Defining

$$\bar{u}(x) = \xi_0 x$$

$$u_\nu(x) = \begin{cases} \bar{u}(x) & \text{if } x \in \Omega - D \\ \bar{u}(x) + \dfrac{1}{\nu}\varphi(\nu x) & \text{if } x \in D \end{cases}$$

we have therefore

$$u_\nu \overset{*}{\rightharpoonup} \bar{u} \text{ in } W^{1,\infty}(\Omega, \mathbf{R}^m) \ .$$

Observe also that

$$I(u_\nu) = \int_\Omega f(\nabla u_\nu(x))\, dx = \int_{\Omega-D} f(\xi_0)\, dx + \int_D f(\xi_0 + \nabla\varphi(\nu x))\, dx$$

$$= f(\xi_0) \operatorname{meas}(\Omega - D) + \frac{1}{\nu^n}\int_{\nu D} f(\xi_0 + \nabla\varphi(y))\, dy$$

$$= f(\xi_0) \operatorname{meas}(\Omega - D) + \int_D f(\xi_0 + \nabla\varphi(y))\, dy \ ,$$

where we have used in the last equality the periodicity of φ. Taking the limit in the above identity and using the weak $*$ lower semicontinuity of I we have indeed obtained

$$\frac{1}{\operatorname{meas} D}\int_D f(\xi_0 + \nabla\varphi(y))\, dy \geq f(\xi_0) \ .$$

Part 2: We now proceed with the general case and we follow the proofs of Morrey [1] and Meyers [1]. In order to avoid a recurrent constant, we assume, without loss of generality, that $0 \in \Omega$. Let D be a cube of edge length $\alpha > 0$ sufficiently small so that

$$D = \{x = (x^1, \ldots, x^n) \in \mathbf{R}^n \quad 0 < x^i < \alpha, \ i = 1, \ldots, n\} \subset \Omega \ .$$

For h an integer we let

$$Q_h \equiv x_0 + \frac{1}{h}D = \left\{x \in \mathbf{R}^n : x_0^i < x^i < x_0^i + \frac{\alpha}{h}, \ i = 1, \ldots, n\right\} \quad (1)$$

where $x_0 \subset \Omega$. We choose h sufficiently large so that $Q_h \subset \Omega$. Let $\varphi \in W_0^{1,\infty}(D; \mathbf{R}^m)$. Extend φ by periodicity from D to $\mathbf{R}^n$ and define

$$\varphi_{\nu,h}(x) = \begin{cases} \dfrac{1}{\nu h}\varphi(\nu h(x - x_0)) & \text{if } x \in Q_h \\ 0 & \text{if } x \notin Q_h \end{cases} \ . \quad (2)$$

Fixing h we have, using Theorem 1.5 of Chapter 2,

$$\varphi_{\nu,h} \overset{*}{\rightharpoonup} 0 \text{ in } W_0^{1,\infty}(\Omega; \mathbf{R}^m) \cap W_0^{1,\infty}(Q_h; \mathbf{R}^m), \text{ as } \nu \to \infty \ .$$

Defining

$$\begin{cases} \bar{u}(x) = u_0 + \xi_0(x - x_0) \\ u_\nu(x) = \bar{u}(x) + \varphi_{\nu,h}(x) \end{cases} \quad (3)$$

we get

$$u_\nu \overset{*}{\rightharpoonup} \bar{u} \text{ in } W_0^{1,\infty}(\Omega, \mathbf{R}^m) \ . \quad (4)$$

We now split Q_h into cubes $Q_{h,j}^\nu$ of length $\alpha/\nu h$ and denote by x_j, $0 \leq j \leq \nu^n - 1$, the corner of $Q_{h,j}^\nu$ closest to x_0 (cf. diagram below). Therefore

$$Q_h = \bigcup_{j=0}^{\nu^n - 1} Q_{h,j}^\nu = \bigcup_{j=0}^{\nu^n - 1} \left(x_j + \frac{1}{\nu h}D\right) \ . \quad (5)$$

We now consider

$$I(u_\nu) = \int_\Omega f(x, u_\nu(x), \nabla u_\nu(x))\, dx$$

$$= \int_{\Omega - Q_h} f(x, \bar{u}(x), \nabla \bar{u}(x))\, dx + \int_{Q_h} f(x, u_\nu(x), \nabla u_\nu(x))\, dx$$

$$= \int_{\Omega - Q_h} f(x, \bar{u}(x), \nabla \bar{u}(x))\, dx + \sum_{j=0}^{\nu^n - 1} \int_{Q_{h,j}^\nu} f(x, u_\nu(x), \nabla u_\nu(x))\, dx \qquad (6)$$

$$= \int_{\Omega - Q_h} f(x, \bar{u}(x), \nabla \bar{u}(x))\, dx + \sum_{j=0}^{\nu^n - 1} \int_{Q_{h,j}^\nu} f(x_j, \bar{u}(x_j), \nabla u_\nu(x))\, dx$$

$$+ \sum_{j=0}^{\nu^n - 1} \int_{Q_{h,j}^\nu} [f(x, u_\nu(x), \nabla u_\nu(x)) - f(x_j, \bar{u}(x_j), \nabla u_\nu(x))]\, dx \ .$$

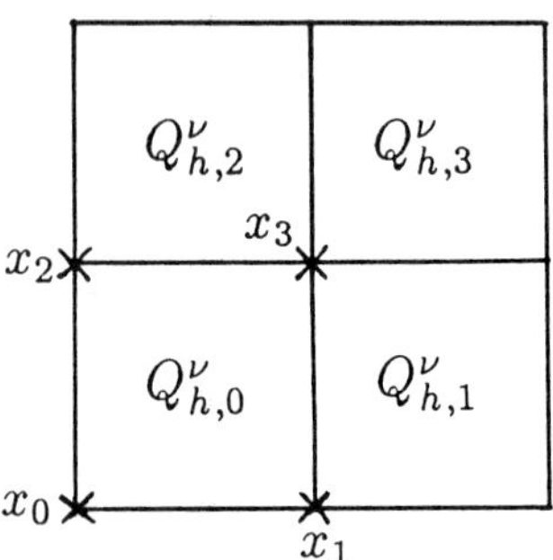

Figure 3.4. $(n = 2, \ \nu = 2)$

The continuity of f, the convergence $u_\nu \overset{*}{\rightharpoonup} \bar{u}$ in $W^{1,\infty}$, $(H3)$, and Lebesgue dominated convergence theorem imply that, for h fixed,

$$\lim_{\nu \to \infty} \sum_{j=0}^{\nu^n - 1} \int_{Q_{h,j}^\nu} [f(x, u_\nu(x), \nabla u_\nu(x)) - f(x_j, \bar{u}(x_j), \nabla u_\nu(x))]\, dx = 0 \ . \qquad (7)$$

It then remains to estimate the second term in (6), i.e.

$$\sum_{j=0}^{\nu^n - 1} \int_{Q_{h,j}^\nu} f(x_j, \bar{u}(x_j), \nabla u_\nu(x))\, dx$$

$$= \sum_{j=0}^{\nu^n - 1} \int_{x_j + \frac{1}{\nu h} D} f(x_j, u_0 + \xi_0(x_j - x_0), \xi_0 + \nabla \varphi(\nu h(x - x_0)))\, dx \qquad (8)$$

$$= \sum_{j=0}^{\nu^n - 1} \frac{1}{(\nu h)^n} \int_D f(x_j, u_0 + \xi_0(x_j - x_0), \xi_0 + \nabla \varphi(y + \nu h(x_j - x_0)))\, dy$$

where we have used (3), (5) and performed a change of variables $y = \nu h(x - x_j)$. Using finally the periodicity of φ we find that for h fixed

$$\sum_{j=0}^{\nu^n - 1} \int_{Q_{h,j}^\nu} f(x_j, \bar{u}(x_j), \nabla u_\nu(x)) \, dx$$

$$= \sum_{j=0}^{\nu^n - 1} \frac{1}{(\nu h)^n} \int_D f(x_j, u_0 + \xi_0(x_j - x_0), \xi_0 + \nabla\varphi(y)) \, dy \ .$$

We then immediately deduce that

$$\lim_{\nu \to \infty} \sum_{j=0}^{\nu^n - 1} \int_{Q_{h,j}^\nu} f(x_j, \bar{u}(x_j), \nabla u_\nu(x)) \, dx$$

$$= \frac{1}{\operatorname{meas} D} \int_{Q_h} \int_D f(x, u_0 + \xi_0(x - x_0), \xi_0 + \nabla\varphi(y)) \, dy dx \qquad (9)$$

$$= \frac{1}{\operatorname{meas} D} \int_{Q_h} \int_D f(x, \bar{u}(x), \xi_0 + \nabla\varphi(y)) \, dy dx \ .$$

Collecting (6), (7) and (9) and using the weak $*$ lower semicontinuity of I we have

$$\liminf_{\nu \to \infty} I(u_\nu) = \int_{\Omega - Q_h} f(x, \bar{u}(x), \nabla\bar{u}(x)) \, dx$$

$$+ \frac{1}{\operatorname{meas} D} \int_{Q_h} \int_D f(x, \bar{u}(x), \xi_0 + \nabla\varphi(y)) \, dy dx$$

$$\geq I(\bar{u}) = \int_\Omega f(x, \bar{u}(x), \nabla\bar{u}(x)) \, dx$$

and hence

$$\frac{1}{\operatorname{meas} Q_h} \int_{Q_h} \int_D f(x, u_0 + \xi_0(x - x_0), \xi_0 + \nabla\varphi(y)) \, dy dx$$

$$\geq \frac{\operatorname{meas} D}{\operatorname{meas} Q_h} \int_{Q_h} f(x, u_0 + \xi_0(x - x_0), \xi_0) \, dx \ .$$

Letting $h \to \infty$ we immediately obtain the claimed result. $\qquad\square$

PROOF of Theorem 3.1. We want to show that

$$f(x_0, u_0, \lambda\alpha + (1 - \lambda)\beta) \leq \lambda f(x_0, u_0, \alpha) + (1 - \lambda)f(x_0, u_0, \beta) \qquad (1)$$

for every $x_0 \in \Omega$, $u_0 \in \mathbb{R}^m$, $\alpha, \beta \in \mathbb{R}^{nm}$ and $\lambda \in [0,1]$. Recall also that we are now assuming that either $m = 1$ or $n = 1$. We first construct a function φ whose gradient takes essentially the values α and β, we shall then conclude.

Step 1: Let $a, b \in \mathbb{R}$ such that $D = (a,b)^n \subset \Omega$. Let N be a large fixed integer and divide (a,b) into intervals of length $(b-a)/2^N$ and subdivide each of these intervals into two intervals of length $\lambda(b-a)/2^N$ and $(1-\lambda)(b-a)/2^N$. We denote by I_N (respectively J_N) the union of the first (resp. the second) subintervals, then

$$\operatorname{meas} I_N = \lambda(b-a), \quad \operatorname{meas} J_N = (1-\lambda)(b-a) . \tag{2}$$

Assume first that $m = 1$ and $n > 1$ and let

$$D_1^N = I_N \times \left(a + \frac{1}{N}, b - \frac{1}{N}\right)^{n-1} , \tag{3}$$

$$D_2^N = J_N \times \left(a + \frac{1}{N}, b - \frac{1}{N}\right)^{n-1} .$$

Let $\varphi : D_1^N \cup D_2^N \to \mathbb{R}$ be defined as follows

$$\varphi(a, x_2, \ldots, x_n) = 0 \tag{4}$$

$$\nabla\varphi(x) = \begin{cases} (1-\lambda)(\alpha - \beta), & \text{if } x \in D_1^N \\ -\lambda(\alpha - \beta) & \text{if } x \in D_2^N \end{cases} \tag{5}$$

(if $n = 1$ and $m \geq 1$ choose directly $D_1^N = I_N$, $D_2^N = J_N$ and $\varphi : (a,b) \to \mathbb{R}^m$ with $\varphi(a) = 0$ and φ' as in (5)).

We then have that $\varphi(b, x_2, ..., x_n) = 0$ and φ is Lipschitz in $D_1^N \cup D_2^N$ with constant $K(K \leq |\alpha - \beta|)$ and $|\varphi(x)| \leq K/2^N$. We then define φ on ∂D by

$$\varphi(x) = 0 \text{ if } x \in \partial D . \tag{6}$$

We therefore have that $\varphi : D_1^N \cup D_2^N \cup \partial D \to \mathbb{R}$ is Lipschitz with constant K, to show this we must prove that

$$|\varphi(x) - \varphi(y)| \leq K|x - y|, \quad x, y \in D_1^N \cup D_2^N \cup \partial D . \tag{7}$$

Three cases may happen.

Case 1: $x, y \in D_1^N \cup D_2^N$ or $x, y \in \partial D$, then (7) is trivial.

Case 2: $x \in D_1^N \cup D_2^N$, $y = (y_1, \ldots, y_n) \in \partial D$ with $y_1 = a$ or b (suppose without loss of generality that $y_1 = b$). Then from Case 1 we have

$$|\varphi(x) - \varphi(y)| = |\varphi(x_1, x_2, \ldots, x_n) - \varphi(y_1, y_2, \ldots, y_n)|$$

$$= |\varphi(x_1, x_2, \ldots, x_n) - \varphi(b, x_2, \ldots, x_n)|$$

$$\leq K|x_1 - b| \leq K|x - y|$$

and thus (7).

Case 3: $x \in D_1^N \cup D_2^N$, $y \in \partial D$, $y_1 \neq a, b$. We then define $z \in [x, y] \cap \partial(D_1^N \cup D_2^N)$ (where $[x, y]$ denotes the segment joining x at y). We then have by definition

$$|z - y| \geq \frac{1}{N} \, . \tag{8}$$

Using (8) wc gct

$$|\varphi(x) - \varphi(y)| \leq |\varphi(x) - \varphi(z)| + |\varphi(z) - \varphi(y)| \leq K|x - z| + |\varphi(z)|$$

$$\leq K|x - z| + \frac{K}{2N} \leq K(|x - z| + |z - y|) = K|x - y| \, ,$$

which is precisely (7).

We may therefore use Mac Shane lemma to get $\varphi \in W_0^{1,\infty}(D; \mathbb{R}^m)$ such that $\|\nabla\varphi\|_{L^\infty} \leq K$, where K is independent of N and

$$\nabla\varphi(x) = \begin{cases} (1 - \lambda)(\alpha - \beta), & x \in D_1^N \text{ with} \\[4pt] & \text{meas } D_1^N = \lambda(b - a)\left(b - a - \frac{2}{N}\right)^{n-1} \\[8pt] -\lambda(\alpha - \beta), & x \in D_2^N \text{ with} \\[4pt] & \text{meas } D_2^N = (1 - \lambda)(b - a)\left(b - a - \frac{2}{N}\right)^{n-1} \end{cases} \, . \tag{9}$$

Step 2: We are now in a position to show (1), i.e. that f is convex. In view of Lemma 3.3 we have

$$\frac{1}{\text{meas } D} \int_D f(x_0, u_0, \lambda\alpha + (1 - \lambda)\beta + \nabla\varphi(x)) \, dx \tag{10}$$

$$\geq f(x_0, u_0, \lambda\alpha + (1 - \lambda)\beta)$$

where $\varphi \in W_0^{1,\infty}(D; \mathbb{R}^m)$ and satisfy (9). But

$$\int_D f(x_0, u_0, \lambda\alpha + (1 - \lambda)\beta + \nabla\varphi(x)) \, dx$$

$$= \int_{D_1^N} f(x_0, u_0, \alpha) \, dx + \int_{D_2^N} f(x_0, u_0, \beta) \, dx \tag{11}$$

$$+ \int_{D - D_1^N \cup D_2^N} f(x_0, u_0, \lambda\alpha + (1 - \lambda)\beta + \nabla\varphi(x)) \, dx \, .$$

Combining (9), (10), (11) and letting $N \to \infty$ we have indeed obtained (1) and thus the theorem. $\qquad\square$

3.3.1.2 Sufficiency of Convexity

Before giving the main result of this section, we need to introduce the following definition.

Definition. Let $\Omega \subset \mathbf{R}^n$ be an open set and let $f : \Omega \times \mathbf{R}^m \times \mathbf{R}^N \to \bar{\mathbf{R}} = \mathbf{R} \cup \{+\infty\}$. Then f is said to be a *Carathéodory function* if

i) $f(x, \bullet, \bullet)$ is continuous for almost every $x \in \Omega$,

ii) $f(\bullet, u, \xi)$ is measurable in x for every $(u, \xi) \in \mathbf{R}^m \times \mathbf{R}^N$.

Note the following useful property of Carathéodory functions (for a proof see for example Cesari [4], Ekeland-Témam [1]).

REMARK (*Scorza-Dragoni theorem*). Let $f : \Omega \times \mathbf{R}^m \times \mathbf{R}^N \to \bar{\mathbf{R}}$. The two following conditions are then equivalent:

i) f is a Carathéodory function;

ii) for every compact set $K \subset \Omega$ and every $\varepsilon > 0$ there exists a compact set $K_\varepsilon \subset K$ such that $\operatorname{meas}(K - K_\varepsilon) \leq \varepsilon$ and f restricted to $K_\varepsilon \times \mathbf{R}^m \times \mathbf{R}^N$ is continuous.

We now have the following.

Theorem 3.4. *Let Ω be a bounded open set of $\mathbf{R}^n$. Let $f : \Omega \times \mathbf{R}^m \times \mathbf{R}^N \to \bar{\mathbf{R}} = \mathbf{R} \cup \{+\infty\}$ be a Carathéodory function satisfying*

$$f(x, u, \xi) \geq \langle a(x); \xi \rangle + b(x)$$

for almost every $x \in \Omega$, for every $(u, \xi) \in \mathbf{R}^m \times \mathbf{R}^N$, for some $a \in (L^{q'}(\Omega))^N$, $1/q + 1/q' = 1$ (q defined below), $b \in L^1(\Omega)$ and where $\langle \bullet; \bullet \rangle$ denotes the scalar product in $\mathbf{R}^N$. Let

$$J(u, \xi) = \int_\Omega f(x, u(x), \xi(x))\, dx \ .$$

Assume that $f(x, u, \bullet)$ is convex and that

$$\begin{cases} u_\nu \to \bar{u} & \text{in } (L^p(\Omega))^m \ p \geq 1 \\ \xi_\nu \rightharpoonup \bar{\xi} & \text{in } (L^q(\Omega))^N \ q \geq 1 \end{cases} ,$$

then

$$\liminf_{\nu \to \infty} J(u_\nu, \xi_\nu) \geq J(\bar{u}, \bar{\xi}) \ .$$

REMARKS.

i) Observe first that the case of the calculus of variations, i.e. $\xi = \nabla u$, $u :$ $\mathbb{R}^n \to \mathbb{R}^m$, is contained in the above theorem. Choose $p = q$, $N = nm$. Thus the convexity of $f(x, u, \bullet)$ is sufficient in order to obtain the weak lower semicontinuity in $W^{1,p}(\Omega, \mathbb{R}^m)$ of

$$I(u) = J(u, \nabla u) = \int_\Omega f(x, u(x), \nabla u(x)) \, dx \ .$$

Therefore, summarizing the results of Theorem 3.1 and 3.4, we find that a necessary and sufficient condition for I to be weakly lower semicontinuous in $W^{1,p}$ is that $f(x, u, \bullet)$ be convex.

ii) There are some advantages in proving Theorem 3.4 as stated and not restricting the functional to the case of the calculus of variations; one of the reasons will be clearer in Chapter 4 (Theorem 2.10).

iii) The hypothesis of the theorem are nearly optimal. As mentioned above this theorem has a long history and we quote here only a few of the contributors Tonelli [1], Serrin [1,2], Morrey [2], Mac Shane [1], De Giorgi [3], Berkowitz [1,2], Cesari [1-4], Ioffe [2,3], Olech [2,3], Ekeland-Témam [1], Marcus-Mizel [1,2], Rockafellar [2], Marcellini-Sbordone [2], Eisen [2], Sbordone [1]. This theorem has also been generalized in many respects and we refer to the bibliography for more details.

iv) Comparing Theorem 3.4 involving weak lower semicontinuous functionals and the results of Section 2 of this chapter we find that for example

$$I(u) = \int_0^1 \left[(u'(x))^4 + ((u(x))^2 - 1)^2 \right] \, dx$$

is weakly lower semicontinuous over $W_0^{1,\infty}(0,1)$ without being convex over the same space.

We now proceed with the proof of the theorem and we follow here the one of De Giorgi [3].

PROOF. We decompose the proof into four steps.

Step 1: Replacing if necessary f by $\tilde{f}$

$$\tilde{f}(x, u, \xi) \equiv f(x, u, \zeta) - \langle a(x); \xi \rangle \quad b(x)$$

we may assume, without loss of generality (since $\langle a(x); \xi \rangle - b(x)$ is continuous with respect to the weak convergence of $\xi_\nu \rightharpoonup \bar{\xi}$), that

$$f(x, u, \xi) \geq 0, \quad (x, u, \xi) \in \Omega \times \mathbb{R}^m \times \mathbb{R}^N \ . \tag{1}$$

Step 2: Observe that if

$$L = \liminf_{\nu \to \infty} J(u_\nu, \xi_\nu)$$

then $L > -\infty$, since $f \geq 0$ by (1). We may also assume that $L < +\infty$, otherwise the theorem is trivial. Restricting if necessary to a subsequence, we may furthermore assume that

$$L = \lim_{\nu \to \infty} J(u_\nu, \xi_\nu) \ . \tag{2}$$

Step 3: We now fix $\varepsilon > 0$ and we wish to show that there exists a measurable set $\Omega_\varepsilon \subset \Omega$, ν_ε and a subsequence ν_j such that for every $\nu_j \geq \nu_\varepsilon$

$$\begin{cases} \mathrm{meas}\,(\Omega - \Omega_\varepsilon) < \varepsilon \\ \displaystyle\int_{\Omega_\varepsilon} \left| f(x, u_{\nu_j}(x), \xi_{\nu_j}(x)) - f(x, \bar{u}(x), \xi_{\nu_j}(x)) \right| \, dx \; < \; \varepsilon \,\mathrm{meas}\,\Omega \ . \end{cases} \tag{3}$$

This will be proved below.

Step 4: Assume that (3) has been established, we then show the theorem. Let

$$\chi_\varepsilon(x) = \begin{cases} 1 & \text{if } x \in \Omega_\varepsilon \\ 0 & \text{if } x \in \Omega - \Omega_\varepsilon \end{cases} .$$

Let

$$g(x, \xi) = \chi_\varepsilon(x) f(x, \bar{u}(x), \xi)$$

then $g : \Omega \times \mathbb{R}^N \to \bar{\mathbb{R}} = \mathbb{R} \cup \{+\infty\}$ is a Carathéodory function and $g(x, \bullet)$ is convex for almost all $x \in \Omega$. Furthermore

$$G(\xi) \equiv \int_\Omega g(x, \xi(x)) \, dx$$

is convex and lower semicontinuous over $(L^q(\Omega))^N$. Using Theorem 1.2 of Chapter 3, we deduce that if $\xi_{\nu_j} \rightharpoonup \bar{\xi}$ in $(L^q(\Omega))^N$ then

$$\liminf_{\nu_j \to \infty} G(\xi_{\nu_j}) = \liminf_{\nu_j \to \infty} \int_\Omega \chi_\varepsilon(x) \, f(x, \bar{u}(x), \xi_{\nu_j}(x)) \, dx$$

$$\geq G(\bar{\xi}) = \int_\Omega \chi_\varepsilon(x) \, f(x, \bar{u}(x), \bar{\xi}(x)) \, dx \ . \tag{4}$$

Therefore, using (3), we have, for ν_j sufficiently large, that

$$\int_{\Omega} \chi_{\varepsilon}(x)\, f(x, u_{\nu_j}(x), \xi_{\nu_j}(x))\, dx$$

$$\geq \int_{\Omega} \chi_{\varepsilon}(x)\, f(x, \bar{u}(x), \xi_{\nu_j}(x))\, dx$$

$$- \int_{\Omega} \chi_{\varepsilon}(x)\, \left| f(x, u_{\nu_j}(x), \xi_{\nu_j}(x)) - f(x, \bar{u}(x), \xi_{\nu_j}(x)) \right|\, dx$$

$$\geq \int_{\Omega} \chi_{\varepsilon}(x)\, f(x, \bar{u}(x), \xi_{\nu_j}(x))\, dx \; - \; \varepsilon \operatorname{meas} \Omega \; .$$

Using (1), i.e. the positivity of f, we have

$$\int_{\Omega} f(x, u_{\nu_j}(x), \xi_{\nu_j}(x))\, dx \;\geq\; \int_{\Omega} \chi_{\varepsilon}(x)\, f(x, \bar{u}(x), \xi_{\nu_j}(x))\, dx \; - \; \varepsilon \operatorname{meas} \Omega \; .$$

Letting $\nu_j \to \infty$ and using (4) we have

$$L \;=\; \liminf_{\nu_j \to \infty} \int_{\Omega} f(x, u_{\nu_j}(x), \xi_{\nu_j}(x))\, dx$$

$$\geq \int_{\Omega} \chi_{\varepsilon}(x)\, f(x, \bar{u}(x), \bar{\xi}(x))\, dx \; - \; \varepsilon \operatorname{meas} \Omega \; .$$

Letting $\varepsilon \to 0$, using the fact that $\operatorname{meas}(\Omega - \Omega_{\varepsilon}) \to 0$, Lebesgue monotone convergence theorem in the right hand side of the above inequality, we have indeed obtained the theorem.

Step 3: It therefore remains to construct Ω_{ε} with the property (3). Note first that since $u_{\nu} \to \bar{u}$ in $L^p(\Omega)$ and $\xi_{\nu} \rightharpoonup \bar{\xi}$ in $L^q(\Omega)$ we have that for every $\varepsilon > 0$, there exists $M_{\varepsilon} > 0$, which is independent of ν, such that if

$$K^1_{\varepsilon,\nu} = \{ x \in \Omega : |\bar{u}(x)|, |u_{\nu}(x)| \geq M_{\varepsilon} \}, \quad \text{then } \operatorname{meas} K^1_{\varepsilon,\nu} < \frac{\varepsilon}{6}$$

$$K^2_{\varepsilon,\nu} = \{ x \in \Omega : |\xi_{\nu}(x)| \geq M_{\varepsilon} \}, \quad \text{then } \operatorname{meas} K^2_{\varepsilon,\nu} < \frac{\varepsilon}{6}$$

for every ν. Hence if

$$\Omega^1_{\varepsilon,\nu} = \Omega - (K^1_{\varepsilon,\nu} \cup K^2_{\varepsilon,\nu})$$

then

$$\operatorname{meas}(\Omega - \Omega^1_{\varepsilon,\nu}) < \frac{\varepsilon}{3} \; . \tag{5}$$

Since f is a Carathéodory function, there exists (cf. Scorza-Dragoni theorem) $\Omega^2_{\varepsilon,\nu} \subset \Omega^1_{\varepsilon,\nu}$ a compact set such that $f(x, u, \xi)$ is continuous if $x \in \Omega^2_{\varepsilon,\nu}$, $|u| < M_{\varepsilon}$, $|\xi| < M_{\varepsilon}$ and

$$\operatorname{meas}(\Omega^1_{\varepsilon,\nu} - \Omega^2_{\varepsilon,\nu}) < \frac{\varepsilon}{3} \; . \tag{6}$$

We therefore have that there exists $\delta(\varepsilon) > 0$ such that

$$|u(x) - v(x)| < \delta(\varepsilon) \quad \Rightarrow \quad |f(x, u(x), \xi) - f(x, v(x), \xi)| < \varepsilon \,, \tag{7}$$

for every $x \in \Omega^2_{\varepsilon,\nu}$, every $|u|, |v| < M_\varepsilon$ and $|\xi| < M_\varepsilon$.

Having fixed $\delta(\varepsilon)$ in this way and using the fact that $u_\nu \to \bar{u}$ in L^p we can find $\nu_\varepsilon = \nu_{\varepsilon,\delta(\varepsilon)}$ such that if $\Omega^3_{\varepsilon,\nu} = \{x \in \Omega : |u_\nu(x) - \bar{u}(x)| < \delta(\varepsilon)\}$, then

$$\mathrm{meas}\,\{\Omega - \Omega^3_{\varepsilon,\nu}\} < \frac{\varepsilon}{3}, \ \text{ for every } \nu \geq \nu_\varepsilon \,. \tag{8}$$

Therefore letting

$$\Omega_{\varepsilon,\nu} = \Omega^2_{\varepsilon,\nu} \cap \Omega^3_{\varepsilon,\nu}$$

we have from (5), (6), (7) and (8)

$$\begin{cases} \mathrm{meas}\,(\Omega - \Omega_{\varepsilon,\nu}) < \varepsilon \\ \displaystyle\int_{\Omega_{\varepsilon,\nu}} |f(x, \bar{u}(x), \xi_\nu(x)) - f(x, u_\nu(x), \xi_\nu(x))|\, dx < \varepsilon\,\mathrm{meas}\,\Omega \end{cases} \tag{9}$$

for every $\nu \geq \nu_\varepsilon$. We now choose $\varepsilon_j = \varepsilon/2^j$, $j = 1, 2, \ldots$. We therefore have that (9) holds with ε and ν_ε replaced by $\varepsilon_j, \nu_{\varepsilon_j}$. We then choose any $\nu_j \geq \nu_{\varepsilon_j}$ with $\lim \nu_j = \infty$ and we let

$$\Omega_\varepsilon = \bigcap_{j=1}^{\infty} \Omega_{\varepsilon_j,\nu_j} \,.$$

We therefore have immediately (3) and this concludes Step 3 and the proof of the theorem. $\square$

3.3.2 Weak Continuity and Invariant Integrals

We show in this section that if

$$I(u) = \int_\Omega f(x, u(x), \nabla u(x))\, dx$$

then I is weakly continuous in $W^{1,p}$ if and only if $f(x, u, \bullet)$ is affine. In the second part of this section we show that invariant integrals, i.e. integrals which are constants whenever the boundary condition is fixed, can be fully characterized as those which are in divergence form.

3.3.2.1 Weak Continuity

Combining Theorem 3.1 and 3.4 we have immediately

Theorem 3.5. *Let $\Omega \subset \mathbf{R}^n$ be a bounded open set, $f : \Omega \times \mathbf{R}^m \times \mathbf{R}^{nm} \to \mathbf{R}$ be a continuous function satisfying*

$$|f(x, u, \xi)| \leq a(x, |u|, |\xi|) \text{ for every } (x, u, \xi) \in \Omega \times \mathbf{R}^m \times \mathbf{R}^{nm} \ ,$$

where a is increasing with respect to $|u|$ and $|\xi|$ and locally integrable in x. Assume that either $m = 1$ or $n = 1$ and let

$$I(u) = \int_\Omega f(x, u(x), \nabla u(x)) \, dx \ .$$

Then I is weak $*$ continuous in $W^{1,\infty}(\Omega; \mathbf{R}^m)$ if and only if $f(x, u, \bullet)$ is affine, i.e. there exist $g : \Omega \times \mathbf{R}^m \to \mathbf{R}^{nm}$ and $h : \Omega \times \mathbf{R}^m \to \mathbf{R}$ continuous such that

$$f(x, u, \xi) = \langle g(x, u); \xi \rangle + h(x, u) \ ,$$

where $\langle \bullet; \bullet \rangle$ denotes the scalar product in $\mathbf{R}^{nm}$.

REMARKS.

i) Similar results hold in $W^{1,p}$, provided one imposes some restrictions on g and h. In particular that $g(x, u) \in L^{p'}(\Omega)$ whenever $u \in W^{1,p}(\Omega)$.

ii) Note also that the result is strictly restricted to the scalar case. It is false in the vectorial case $(m, n > 1)$. We shall see in Chapter 4 that if $m = n$ and

$$I(u) = \int_\Omega \det \nabla u(x) \, dx$$

then I is weakly continuous although $f(\xi) = \det \xi$ is not affine in ξ.

iii) The necessity part of the theorem remains valid if the function I is continuous for every sequence

$$\begin{cases} u_\nu \overset{*}{\rightharpoonup} \bar{u} & \text{in } W^{1,\infty}(\Omega; \mathbf{R}^m) \\ u_\nu = \bar{u} & \text{on } \partial\Omega \end{cases}$$

since the proof is a direct consequence of Theorem 3.1.

PROOF. The necessity follows immediately from Theorem 3.1 applied to f, I and then to $-f$, $-I$. The sufficiency is also obvious since if $u_\nu \overset{*}{\rightharpoonup} \bar{u}$, in $W^{1,\infty}$, then $u_\nu \to \bar{u}$ in L^∞ and the conclusion follows from the continuity of g and h. $\qquad\square$

3.3.2.2 Invariant Integrals

We now turn our attention to invariant integrals, which are important in the field theories. Following Weyl [1] and Carathéodory [1], we give here a complete characterization of such integrals.

Theorem 3.6. *Let $\Omega \subset \mathbf{R}^n$ be a bounded open set with piecewise C^1 boundary. Let $f : \Omega \times \mathbf{R}^m \times \mathbf{R}^{nm} \to \mathbf{R}$ be a C^∞ function satisfying*

$$|f(x, u, \xi)| \leq a(x, |u|, |\xi|)$$

for every $(x, u, \xi) \in \Omega \times \mathbf{R}^m \times \mathbf{R}^{nm}$ and where a is as in Theorem 3.5. Let $m = 1$ or $n = 1$ and

$$I(u) = \int_\Omega f(x, u(x), \nabla u(x))\, dx \ .$$

The two following conditions are then equivalent:

i) I is invariant, i.e.

$$I(u) = \text{constant for every } u \in u_0 + W_0^{1,\infty}(\Omega; \mathbf{R}^m) \ ;$$

ii) there exist $\varphi : \Omega \times \mathbf{R}^m \to \mathbf{R}^n$, $\beta : \Omega \to \mathbf{R}$ C^∞ functions such that

$$f(x, u, \xi) = \left\langle \frac{\partial}{\partial u}\varphi(x, u); \xi \right\rangle + \text{div}_x\, \varphi(x, u) + \beta(x)$$

for every $(x, u, \xi) \in \Omega \times \mathbf{R}^m \times \mathbf{R}^{nm}$; where $\langle \bullet; \bullet \rangle$ denotes the scalar product in $\mathbf{R}^{nm}$ and for

$$\frac{\partial}{\partial u}\varphi(x, u) = \begin{cases} \left(\dfrac{\partial \varphi_1}{\partial u}, \dots, \dfrac{\partial \varphi_n}{\partial u}\right) & \text{if } m = 1 \\[2ex] \left(\dfrac{\partial \varphi}{\partial u_1}, \dots, \dfrac{\partial \varphi}{\partial u_m}\right) & \text{if } n = 1 \end{cases}$$

$$\text{div}_x\, \varphi(x, u) = \begin{cases} \displaystyle\sum_{i=1}^n \dfrac{\partial \varphi_i}{\partial x_i} & \text{if } m = 1 \\[2ex] \dfrac{\partial \varphi}{\partial x} & \text{if } n = 1 \end{cases} \ .$$

In particular if $\xi = \nabla u$, then

$$f(x, u, \nabla u) = \text{div}_x\, \varphi(x, u(x)) + \beta(x) \ .$$

REMARK. As mentioned above this result is striclty restricted to the scalar case; for the vectorial case, see Chapter 4, and Rund [1], Ericksen [1], Sivaloganathan [1].

PROOF. ii) $\Rightarrow$ i) Let f be as above then

$$I(u) = \int_\Omega \beta(x)\,dx + \int_\Omega \operatorname{div}_x\left(\varphi(x, u(x))\right)dx$$

and since $u = u_0$ on $\partial\Omega$, we have after an integration by part that I is constant.

i) $\Rightarrow$ ii) We assume here that $m = 1$ (the case $n = 1$ is treated similarly). Following Theorem 3.5 we have that if I is constant, then it is weak $*$ continuous and therefore there exist g and h such that

$$f(x, u, \xi) = \langle g(x, u); \xi \rangle + h(x, u) \ .$$

Note that since $f \in C^\infty$, then so are g and h. By hypothesis we have

$$I(u) = \int_\Omega \left(\sum_{i=1}^n g_i(x, u)\frac{\partial u}{\partial x_i} + h(x, u)\right)dx = \text{constant} \ .$$

Choosing $u \in u_0 + W_0^{1,\infty}(\Omega)$ and $v \in C_0^\infty(\Omega)$ we have that

$$\frac{d}{d\varepsilon}I(u + \varepsilon v)\Big|_{\varepsilon=0} = \int_\Omega \Bigg(\sum_{i=1}^n \frac{\partial}{\partial u}g_i(x, u)\frac{\partial u}{\partial x_i}v$$

$$+ \sum_{i=1}^n g_i(x, u)\frac{\partial v}{\partial x_i} + \frac{\partial}{\partial u}h(x, u)\,v\Bigg)dx$$

$$= \int_\Omega v\left[\sum_{i=1}^n \left(\frac{\partial}{\partial u}g_i(x, u(x))\frac{\partial u}{\partial x_i} - \frac{\partial}{\partial x_i}(g_i(x, u(x)))\right)\right.$$

$$\left. + \frac{\partial}{\partial u}h(x, u(x))\right]dx$$

$$= \int_\Omega v\left[-\sum_{i=1}^n \frac{\partial}{\partial x_i}g_i(x, u) + \frac{\partial}{\partial u}h(x, u)\right]dx \equiv 0 \ .$$

We then obtain that, for every $x \in \Omega$ and $u \in \mathbb{R}$,

$$\frac{\partial}{\partial u}h(x, u) = \operatorname{div}_x g(x, u) = \sum_{i=1}^n \frac{\partial}{\partial x_i}g_i(x, u) \ .$$

Let

$$\varphi_i(x_1,\ldots,x_n,u) = \int_0^u g_i(x_1,\ldots,x_n,s)\,ds, \quad i = 1,\ldots,n \ .$$

We have that if $\varphi = (\varphi_1,\ldots,\varphi_n)$, then

$$\frac{\partial}{\partial u}h(x,u) = \operatorname{div}_x\left(\frac{\partial\varphi}{\partial u}(x,u)\right) = \frac{\partial}{\partial u}(\operatorname{div}_x\varphi(x,u))$$

and thus

$$\begin{cases} h(x,u) = \beta(x) + \operatorname{div}_x\varphi(x,u) \\ g(x,u) = \dfrac{\partial}{\partial u}\varphi(x,u) \end{cases} \qquad . \qquad\qquad \Box$$

3.4 Existence Theorems and Euler Equations

In this section we first show how to apply the above results to the existence of minima. We also derive the Euler equations under various types of conditions. We then mention some regularity results, but we omit their proofs. We finally conclude with the Lavrentiev phenomenon.

3.4.1 Existence Theorems and Regularity Results

3.4.1.1 Existence Theorems

We are now in a position to show the existence of minima for our problem.

Theorem 4.1. *Let Ω be a bounded open set of $\mathbb{R}^n$ with Lipschitz boundary. Let $f : \Omega \times \mathbb{R}^m \times \mathbb{R}^{nm} \to \bar{\mathbb{R}} = \mathbb{R} \cup \{+\infty\}$ be a Carathéodory function satisfying the coercivity condition*

$$f(x,u,\xi) \geq a(x) + b|\xi|^p$$

for almost every $x \in \Omega$, for every $(u,\xi) \in \mathbb{R}^m \times \mathbb{R}^{nm}$ and for some $a \in L^1(\Omega)$, $b > 0$ and $p > 1$. Assume that $f(x,u,\bullet)$ is convex. Let

$$I(u) = \int_\Omega f(x,\acute{u}(x),\nabla u(x))\,dx \ .$$

Assume that there exists $\tilde{u} \in u_0 + W_0^{1,p}(\Omega;\mathbb{R}^m)$ such that $I(\tilde{u}) < \infty$, then

$$(P) \qquad\qquad \inf\{I(u) : u \in u_0 + W_0^{1,p}(\Omega;\mathbb{R}^m)\}$$

attains its minimum.

REMARKS.

i) The theorem is also valid in the vectorial case $m, n > 1$; but it can be
extended a great deal in this case, cf. Chapter 4;

ii) note that the *minimal surface* case where

$$f(x, u, \xi) = \sqrt{1 + |\xi|^2}$$

is not contained in the above theorem although f is convex, since the
coercivity condition holds only for $p = 1$ and then $W^{1,1}$ is not a reflexive
space. For the treatment of this problem we refer to Morrey [2], Ekeland-
Témam [1], Federer [1], Almgren [1,2], De Giorgi [1,2], Giusti [1] and the
references quoted there;

iii) the hypothesis $I(\tilde{u}) < \infty$ can be ensured if for example we impose a
growth condition on the function f, such as

$$f(x, u, \xi) \leq a_1(x) + b_1(|u|^q + |\xi|^p)$$

where $a_1 \in L^1(\Omega)$, $b_1 > 0$ and $q = \frac{np}{n-p}$ if $1 < p < n$ and no condition
on q if $p \geq n$;

iv) note that, in general, neither the convexity of f, nor the coercivity condition can be weakened. Examples of non existence of solutions in absence
of convexity are given in Chapter 5. We now give two examples (cf. Cesari
[4]) showing that in general the coercivity cannot be weakened either.

EXAMPLE 1. Let $m = n = 1$, $\Omega = (0, 1)$, $f(x, u, \xi) = x\xi^2$ and

$$(P) \quad \inf \left\{ I(u) = \int_0^1 x(u'(x))^2 \, dx \; : \; u(0) = 1, \, u(1) = 0, \, u \in W^{1,2}(0, 1) \right\}.$$

We now show that (P) has no minimum. Observe that since $I \geq 0$, then
$\inf(P) \geq 0$. Let N be an integer and

$$u_N(x) = \begin{cases} 1 & \text{if } 0 \leq x \leq \frac{1}{N} \\ -\dfrac{\log x}{\log N} & \text{if } \frac{1}{N} \leq x \leq 1 \end{cases}.$$

Observe that $u_N \in W^{1,2}(0, 1)$, $u_N(0) = 1$, $u_N(1) = 0$ and

$$I(u_N) = \frac{1}{\log N} \to 0, \quad \text{as } N \to \infty \, .$$

Therefore $\inf(P) = 0$. However if $u \in W^{1,2}(0,1)$ and is such that $I(u) = 0$, then $u' = 0$ a.e. and thus cannot satisfy the boundary conditions. Hence (P) has no global minimum.

EXAMPLE 2. Let $m = n = 1$, $\Omega = (0,1)$, $f(x,u,\xi) = (u^2 + \xi^2)^{1/2}$ and

$$(P) \qquad \inf\left\{ I(u) = \int_0^1 \left((u(x))^2 + (u'(x))^2 \right)^{1/2} dx : \right.$$
$$\left. u(0) = 0, \, u(1) = 1, \, u \in W^{1,1}(0,1) \right\} .$$

Since

$$I(u) = \int_0^1 (u^2 + u'^2)^{1/2} dx \geq \int_0^1 |u'| \, dx \geq \int_0^1 u' \, dx = 1 \qquad (1)$$

we have $\inf(P) \geq 1$. We now show that $\inf(P) = 1$. Let N be an integer and

$$u_N(x) = \begin{cases} 0 & \text{if } 0 \leq x \leq 1 - \frac{1}{N} \\ 1 + N(x - 1) & \text{if } 1 - \frac{1}{N} \leq x \leq 1 \end{cases} .$$

Observe that $u_N \in W^{1,1}(0,1)$, $u_N(0) = 0$, $u_N(1) = 1$ and satisfy

$$1 \leq I(u_N) \leq \frac{1}{N}\sqrt{1 + N^2} \to 1, \quad \text{as } N \to \infty .$$

Hence $\inf(P) = 1$. However it is clear, in view of (1), that no $u \in W^{1,1}(0,1)$ with $u(0) = 0$, $u(1) = 1$ can be such that $I(u) = 1$. Thus (P) has no solution.

PROOF of the theorem. In view of Theorem 3.4 and of Theorem 1.1 we only need to check that I is coercive over $W^{1,p}(\Omega; \mathbb{R}^m)$. From the coercivity condition on f we have for every $u \in u_0 + W_0^{1,p}$,

$$I(u) \geq \int_\Omega a(x) \, dx + b \int_\Omega |\nabla u(x)|^p \, dx . \qquad (1)$$

Using Poincaré inequality (Theorem 1.7, Chapter 2) we have

$$\|u - u_0\|_{L^p}^p \leq K(\|\nabla u\|_{L^p}^p + 1) .$$

We therefore deduce that, K denoting a generic constant independent of u,

$$\|u\|_{W^{1,p}}^p \leq K(1 + \|\nabla u\|_{L^p}^p) .$$

Returning to (1) we immediately get that

$$I(u) \geq \alpha \|u\|^p_{W^{1,p}} + \beta \ ,$$

where $\alpha > 0$ and $\beta \in \mathbf{R}$. Since $p > 1$, the space $W^{1,p}$ is reflexive and we may therefore apply Theorem 1.1 to get the result. $\square$

3.4.1.2 Some Regularity Results

We here just mention some results without proof. This is one of the famous problems of Hilbert and there is an extensive literature on this subject and we refer to Morrey [2], Ladyzhenskaya-Uraltseva [1], Giaquinta [1].

We just consider here the scalar case $m = 1$ (the case $n = 1$ is simpler), we shall mention similar results for the vectorial case in Chapter 4. The following result can be found in Ladyzhenskaya-Uraltseva [1], Morrey [2] and has been improved in some aspects by Giaquinta-Giusti [1]. The following theorem can be found in Giaquinta [1].

Theorem 4.2. *Let*

$$I(u) = \int_\Omega f(x, u(x), \nabla u(x)) \, dx$$

where $f : \Omega \times \mathbf{R} \times \mathbf{R}^n \to \mathbf{R}$ is a Carathéodory function satisfying

$$|\xi|^p - b|u|^\gamma - c(x) \leq f(x, u, \xi) \leq \mu|\xi|^p + b|u|^\gamma + c(x)$$

where $c \in L^s(\Omega)$ with $s > n/p$, $\mu \geq 1$, $b \geq 0$ and either $1 \leq p < n$ and $1 \leq \gamma < \frac{np}{n-p}$ or $p = n$ and $\gamma \geq 1$. Assume that $u \in W^{1,p}_{loc}(\Omega)$ is such that

$$I(u) \leq I(u + \varphi)$$

for every $\varphi \in W^{1,p}(\Omega)$ with $\operatorname{supp} \varphi \subset\subset \Omega$, then u is locally Hölder continuous (in particular u is locally bounded).

REMARKS.

i) In the case $p > n$, the above result is trivial by the Sobolev imbedding theorem;

ii) if one looks for minima of I subject to Dirichlet boundary condition $u = u_0$ on $\partial\Omega$, the same result is true up to the boundary provided u_0 is itself Hölder continuous, or bounded, and the boundary of Ω is sufficiently regular (for more details see the above references);

iii) the above result is still true if u, instead of being a local minimum, merely satisfies $I(u) \leq QI(u + \varphi)$ for some $Q > 0$, such a u is called a Q-*minimum* (cf. Giaquinta-Giusti [1]);

iv) note also that no convexity hypothesis is required on f.

We shall see below that the regularity of the solution can be further improved, but this will result from the regularity of the solutions of Euler equations and we therefore first need to derive them.

3.4.2 Euler Equations

3.4.2.1 Euler Equations

We now compute $I'(u)$, the Gâteaux derivative of

$$I(u) = \int_\Omega f(x, u(x), \nabla u(x)) \, dx \ .$$

We first introduce some notations. If ξ is a matrix in $\mathbf{R}^{nm}$ we write

$$\xi = \begin{pmatrix} \xi_1^1 & \cdots & \xi_n^1 \\ \vdots & & \vdots \\ \xi_1^m & \cdots & \xi_n^m \end{pmatrix} = (\xi_\alpha^i)_{1 \le i \le m,\, 1 \le \alpha \le n}$$

and therefore if $u : \mathbf{R}^n \to \mathbf{R}^m$, then

$$\nabla u = \begin{pmatrix} \dfrac{\partial u_1}{\partial x_1} & \cdots & \dfrac{\partial u_1}{\partial x_n} \\ \vdots & & \vdots \\ \dfrac{\partial u_m}{\partial x_1} & \cdots & \dfrac{\partial u_m}{\partial x_n} \end{pmatrix} .$$

We now proceed with a formal derivation of the *Euler equations*. Recall that in this section we consider the general case, $m, n \ge 1$.

Proposition 4.3. *Let Ω be a bounded open set with Lipschitz boundary. Let $f : \Omega \times \mathbf{R}^m \times \mathbf{R}^{nm} \to \mathbf{R}$ be a C^2 function. Assume that $\bar{u} \in C^2(\Omega; \mathbf{R}^m)$ is a solution of*

$$(P) \qquad \inf \left\{ I(u) = \int_\Omega f(x, u(x), \nabla u(x)) \, dx \ : \ u \in u_0 + W_0^{1,p}(\Omega; \mathbf{R}^m) \right\}$$

then $\bar{u}$ satisfies, for every $x \in \Omega$,

$$(E) \qquad \begin{aligned} I'(\bar{u}) &= \left[-\sum_{\alpha=1}^n \frac{\partial}{\partial x_\alpha}\left(\frac{\partial f}{\partial \xi_\alpha^i}(x, \bar{u}, \nabla \bar{u}) \right) + \frac{\partial}{\partial u_i} f(x, \bar{u}, \nabla \bar{u}) \right]_{1 \le i \le m} \\ &= \left[-\mathrm{div}\left(\frac{\partial f}{\partial \xi^i}(x, \bar{u}, \nabla \bar{u}) \right) + \frac{\partial}{\partial u_i} f(x, \bar{u}, \nabla \bar{u}) \right]_{1 \le i \le m} = 0 \ . \end{aligned}$$

REMARKS.

i) Note that if $n = 1$, then (E) is reduced to a system of ordinary differential equations. If $m = 1$ it is reduced to a single partial differential equation. While if $m, n > 1$ (E) is a system of partial differential equations;

ii) note also that if $m = 1$ and if $f(x, u, \bullet)$ is convex then we must have, provided f is C^2,

$$\sum_{i,j=1}^{n} \frac{\partial^2 f}{\partial \xi_i \partial \xi_j} \lambda_i \lambda_j \geq 0$$

for every $\lambda \in \mathbf{R}^n$, which in the context of a single partial differential equation is the usual *ellipticity condition*.

PROOF of the theorem. Since $\bar{u}$ is a minimum of (P) we must have

$$I(\bar{u}) \leq I(\bar{u} + \varepsilon \varphi)$$

for every $\varphi \in C_0^2(\Omega; \mathbf{R}^m)$ and for every $\varepsilon \in \mathbf{R}$. It is clear that the hypotheses on f, $\bar{u}$ and φ ensure that the functional $I(\bar{u} + \varepsilon \varphi)$ considered as a function of ε is differentiable and since $\bar{u}$ is a minimum we must have

$$\frac{d}{d\varepsilon} I(\bar{u} + \varepsilon \varphi) \Big|_{\varepsilon=0} = \langle I'(\bar{u}); \varphi \rangle$$

$$= \int_\Omega \left\{ \sum_{i=1}^{m} \sum_{\alpha=1}^{n} \frac{\partial f}{\partial \xi_\alpha^i}(x, \bar{u}, \nabla \bar{u}) \frac{\partial \varphi_i}{\partial x_\alpha} + \frac{\partial f}{\partial u_i}(x, \bar{u}, \nabla \bar{u}) \varphi_i \right\} dx$$

$$= 0 \ .$$

Since f and $\bar{u}$ are C^2, we may integrate by part once more and use the arbitrariness of $\varphi \in C_0^2(\Omega; \mathbf{R}^m)$ to get the result. $\qquad \square$

The result obtained above is purely formal in the sense that in general it is unnatural to prescribe that the minimum $\bar{u}$ be C^2. The existence theorems obtained above, indicate only that $\bar{u}$ is in a Sobolev space. Therefore we need to impose some restrictions on the behaviour of $\frac{\partial f}{\partial \xi}$ and $\frac{\partial f}{\partial u}$ so that $I'(u)$ be well defined for functions $u \in W^{1,p}$. There are several conditions which ensure the differentiability of I and which are reasonable. All of them are based on Hölder inequality and the Sobolev imbedding theorem.

We start by quoting three kinds of hypotheses.

Definitions. Let $u : \mathbf{R}^n \to \mathbf{R}^m$, $p \geq 1$, and denote by

$$\operatorname{grad}_u f = \left(\frac{\partial f}{\partial u_i} \right)_{1 \leq i \leq m} , \qquad \operatorname{grad}_\xi f = \left(\frac{\partial f}{\partial \xi_\alpha^i} \right)_{1 \leq i \leq m, 1 \leq \alpha \leq n} .$$

I) We say that f satisfies growth condition (I) if
  Case 1: $p > n$ then

$$|\mathrm{grad}_u \, f|, \ |\mathrm{grad}_\xi \, f| \le \alpha(x) + \beta|\xi|^p \qquad (I_1)$$

for every $|u| \le R$, R a constant, $\alpha = \alpha(R) \in L^1(\Omega)$, $\beta = \beta(R) \ge 0$.
  Case 2: $p = n$, then

$$|\mathrm{grad}_u \, f|, \ |\mathrm{grad}_\xi \, f| \le \alpha(x) + \beta(|u|^q + |\xi|^p) \qquad (I_2)$$

for $q \ge 1$.
  Case 3: $1 \le p < n$, then (I_2) is satisfied only for $1 \le q \le \frac{np}{n-p}$.

II) We say that f satisfies growth condition (II) if
  Case 1: $p > n$, then for every $|u| \le R$, R a constant,

$$\begin{cases} |\mathrm{grad}_\xi \, f| \le \alpha_1(x) + \beta(1 + |\xi|^{p-1}) \\ |\mathrm{grad}_u \, f| \le \alpha_2(x) + \beta(1 + |\xi|^p) \end{cases} \qquad (II_1)$$

with $\alpha_1 \in L^{p/(p-1)}(\Omega)$, $\alpha_2 \in L^1(\Omega)$, $\beta \ge 0$.
  Case 2: $p = n$, then

$$\begin{cases} |\mathrm{grad}_\xi \, f| \le \alpha_1(x) + \beta(|u|^{q_1} + |\xi|^{p-1}) \\ |\mathrm{grad}_u \, f| \le \alpha_2(x) + \beta(|u|^{r_1} + |\xi|^{r_2}) \end{cases} \qquad (II_2)$$

with $\beta \ge 0$, $\alpha_1 \in L^{p/(p-1)}(\Omega)$, $\alpha_2 \in L^s(\Omega)$, $s > 1$, $q_1, r_1 \ge 1$ and $1 \le r_2 < p$.
  Case 3: $1 \le p < n$, then (II_2) is satisfied with $\beta \ge 0$, $\alpha_1 \in L^{p/(p-1)}(\Omega)$, $\alpha_2 \in L^{np/(np-n+p)}$, $1 \le q_1 \le \frac{n(p-1)}{n-p}$, $1 \le r_1 \le \frac{np}{n-p} - 1$ and $1 \le r_2 \le p - 1 + \frac{p}{n}$.

III) We say that f satisfies growth condition (III) if (II_1) holds (with no restriction on p with respect to n).

REMARKS.

i) Growth condition (II) (resp. (III)) is sometimes called *controllable growth condition* (resp. *natural growth condition*), cf. Ladyzenskaya-Uraltseva [1], Morrey [2], Giaquinta [1];

ii) note that if $1 \le p \le n$ then (II) is a stronger hypothesis than (III), since we have only $1 \le r_2 < p$ for the growth condition on $\mathrm{grad}_u f$ for (II) while $r_2 = p$ is allowed in (III);

iii) (III) is more natural than (II) in the sense that if one considers

$$I(u) = \int_\Omega a(u)|\nabla u|^2 \, dx$$

with $0 < \alpha_1 \le a(u)$, $a'(u) \le \alpha_2 < \infty$, $p = 2$ and $n \ge 2$ then $\alpha_1 |\xi|^2 \le |\text{grad}_u f| \le \alpha_2 |\xi|^2$ and therefore (III) is satisfied while (II) is not.

We now prove the main theorem of this section which is only based on several applications of Hölder inequality and Sobolev imbedding theorem.

Theorem 4.4. *Let $f \in C^1(\Omega \times \mathbf{R}^m \times \mathbf{R}^{nm})$ and let for $\varphi : \Omega \to \mathbf{R}^m$*

$$(E_w) \qquad \langle I'(u); \varphi \rangle = \int_\Omega \sum_{i=1}^m \Bigg\{ \sum_{\alpha=1}^n \left[\frac{\partial}{\partial \xi_\alpha^i} f(x, u, \nabla u) \frac{\partial \varphi_i}{\partial x_\alpha} \right] + \frac{\partial}{\partial u_i} f(x, u, \nabla u) \varphi_i \Bigg\} \, dx \ .$$

Assume that $\bar{u}$ is a minimum for I in $u_0 + W_0^{1,p}(\Omega; \mathbf{R}^m)$, $p \ge 1$ and that $I(\bar{u}) < \infty$.

(I) Assuming growth condition (I), then

$$\langle I'(\bar{u}); \varphi \rangle = 0 \ \text{for every} \ \varphi \in C_0^\infty(\Omega; \mathbf{R}^m) \ .$$

(II) Assuming growth condition (II), then

$$\langle I'(\bar{u}); \varphi \rangle = 0 \ \text{for every} \ \varphi \in W_0^{1,p}(\Omega; \mathbf{R}^m) \ .$$

(III) Assuming growth condition (III) and that $\bar{u} \in W^{1,p} \cap L^\infty$ then

$$\langle I'(\bar{u}); \varphi \rangle = 0 \ \text{for every} \ \varphi \in W_0^{1,p} \cap L^\infty \ .$$

REMARK. It is clear that (I) is less interesting than (II). (III) is important in view of the regularity result (Theorem 4.2).

PROOF. In view of Theorem 1.3 of this chapter the only thing to be checked is that I is Gâteaux differentiable, i.e. that (E_w) is well defined.

We shall prove that (E_w) is well defined only in the case of growth condition (II), the others are proved similarly.

Observe first that

$$|\langle I'(u); \varphi \rangle| \le \sum_{i=1}^m \Bigg\{ \int_\Omega \sum_{\alpha=1}^n \left[\left| \frac{\partial}{\partial \xi_\alpha^i} f(x, u, \nabla u) \right| \left| \frac{\partial \varphi_i}{\partial x_\alpha} \right| \right] + \int_\Omega \left| \frac{\partial}{\partial u_i} f(x, u, \nabla u) \right| |\varphi_i| \Bigg\} \, dx \ . \tag{1}$$

In the sequel K will denote a generic constant.

Case 1: $p > n$, then $u, \varphi \in L^\infty$ (Sobolev theorem) and using (II_1) we have

$$|\langle I'(u); \varphi \rangle| \leq K \left\{ \int_\Omega (|\alpha_1(x)| + \beta + \beta|\nabla u|^{p-1}) |\nabla \varphi| \, dx \right.$$

$$\left. + \int_\Omega (|\alpha_2(x)| + \beta + \beta|\nabla u|^p) |\varphi| \, dx \right\} \ .$$

Using Hölder inequality we find

$$|\langle I'(u); \varphi \rangle| \leq K \left\{ 1 + (\|\alpha_1\|_{L^{p/p-1}} + \beta\|\nabla u\|_{L^p}^{p-1}) \|\nabla \varphi\|_{L^p} + \|\nabla u\|_{L^p}^p \|\varphi\|_{L^\infty} \right\}$$

and thus (E_w) is well defined.

Case 2: $p = n$, then $u, \varphi \in L^q$ with $q \geq 1$ using (II_2) we have

$$|\langle I'(u); \varphi \rangle| \leq K \left\{ \int_\Omega (|\alpha_1(x)| + \beta|u|^{q_1} + \beta|\nabla u|^{p-1}) |\nabla \varphi| \, dx \right.$$

$$\left. + \int_\Omega (|\alpha_2(x)| + \beta|u|^{r_1} + \beta|\nabla u|^{r_2}) |\varphi| \, dx \right\} \ . \tag{2}$$

Using Hölder inequality we have

$$|\langle I'(u); \varphi \rangle| \leq K \{ (\|\alpha_1\|_{L^{p/p-1}} + \beta\|u\|_{L^{q_1 p/p-1}}^{q_1} + \beta\|\nabla u\|_{L^p}^{p-1}) \|\nabla \varphi\|_{L^p}$$

$$+ (\|\alpha_2\|_{L^{q/q-1}} + \beta\|u\|_{L^{r_1 q/q-1}}^{r_1} + \beta\|\nabla u\|_{L^{r_2 q/q-1}}^{r_2}) \|\varphi\|_{L^q} \} \tag{3}$$

and thus (E_w) is well defined.

Case 3: $1 \leq p < n$, then $u, \varphi \in L^{np/n-p}$ and (3) still holds with $q = \frac{np}{n-p}$. Observe that:

$$\begin{cases} q_1 \dfrac{p}{p-1} \leq \dfrac{n(p-1)}{n-p} \cdot \dfrac{p}{p-1} = \dfrac{np}{n-p} = q \\[2ex] r_1 \dfrac{q}{q-1} = r_1 \dfrac{np}{np-n+p} \leq \dfrac{np-n+p}{n-p} \cdot \dfrac{np}{np-n+p} = \dfrac{np}{n-p} = q \\[2ex] r_2 \dfrac{q}{q-1} = r_2 \dfrac{np}{np-n+p} \leq \dfrac{np-n+p}{n} \cdot \dfrac{np}{np-n+p} = p \end{cases}$$

and therefore (E_w) is well defined. $\qquad\qquad\qquad\qquad\qquad\qquad\qquad\qquad\qquad\qquad$ $\square$

3.4.4.2 Some More Regularity

The result of Theorem 4.2 can now be improved if one assumes some more properties on the function f. We shall not prove the following theorem and we refer to the above mentioned books (cf. in particular Morrey [2]). The improvements with respect to Theorem 4.2 are obtained using the Euler equations. We again restrict ourselves to the scalar case $m = 1$.

Theorem 4.5. *Let $f \in C^\infty(\Omega \times \mathbf{R} \times \mathbf{R}^n)$. Denoting by*

$$f_x = \left(\frac{\partial f}{\partial x_i}\right)_{1 \leq i \leq n}, \quad f_u = \left(\frac{\partial f}{\partial u}\right), \quad f_\xi = \left(\frac{\partial f}{\partial \xi_i}\right)_{1 \leq i \leq n}$$

and similarly for the higher derivatives. Let

$$I(u) = \int_\Omega f(x, u(x), \nabla u(x))\, dx \ .$$

I) If f satisfies for $V^2 = (1 + |u|^2 + |\xi|^2)$

$$\begin{cases} \alpha_1 V^p - \alpha_2 \leq f(x, u, \xi) \leq \alpha_3 V^p \\ |f_\xi|, |f_{\xi x}|, |f_u|, |f_{ux}| \leq \alpha_4 V^{p-1} \\ |f_{u\xi}|, |f_{uu}| \leq \alpha_5 V^{p-2} \\ \alpha_6 V^{p-2}|\lambda|^2 \leq \displaystyle\sum_{i,j=1}^n \frac{\partial^2 f}{\partial \xi_i \partial \xi_j}\lambda_i\lambda_j \leq \alpha_7 V^{p-2}|\lambda|^2 \end{cases} \tag{A}$$

where $\alpha_1, \alpha_6 > 0$ (if $f(x, u, \xi) = f(x, \xi)$, then take $V^2 = (1 + |\xi|^2)$ in (A)), then if $p \geq 2$ and if $\bar{u} \in W^{1,p}(\Omega)$ is a minimum for I in a class $K \subset W^{1,p}(\Omega)$, then $\bar{u} \in C^\infty_{loc}(\Omega)$ and if f is analytic, then $\bar{u}$ is analytic.
II) If f satisfies for every $|u| < R$, R a constant,

$$\begin{cases} V^2 = (1 + |\xi|^2) \\ \alpha_1 V^p - \alpha_2 \leq f(x, u, \xi) \leq \alpha_3 V^p \\ |f_\xi|, |f_{\xi u}|, |f_{\xi x}| \leq \alpha_4 V^{p-1} \\ |f_u|\,|f_{ux}|, |f_{uu}| \leq \alpha_5 V^p \\ \alpha_6 V^{p-2}|\lambda|^2 \leq \displaystyle\sum_{i,j=1}^n \frac{\partial^2 f}{\partial \xi_i \partial \xi_j}\lambda_i\lambda_j \leq \alpha_7 V^{p-2}|\lambda|^2 \end{cases} \tag{B}$$

where $\alpha_i = \alpha_i(R)$ and $\alpha_1, \alpha_6 > 0$, then if $p > 1$ and if $\bar{u} \in W^{1,p}(\Omega) \cap L^\infty(\Omega)$ is a minimum for I in a class $K \subset W^{1,p}(\Omega)$, then $\bar{u} \in C^\infty_{loc}(\Omega)$ and if f is analytic so is $\bar{u}$.

REMARKS.

i) The above results are striclty restricted to the scalar case, i.e. $u : \mathbf{R}^n \to \mathbf{R}^m$ with $m = 1$, they are false if $m > 1$, cf. Chapter 4;

ii) in view of Theorem 4.2 the requirement that, with (B), $\bar{u} \in W^{1,p} \cap L^\infty$ is not unnatural;

iii) note that the last condition in (A) and (B) is a kind of uniform convexity of f with respect to the variable ξ and it ensures the ellipticity of the Euler equation.

3.4.3 Lavrentiev Phenomenon

We conclude this chapter by presenting an example of the so-called *Lavrentiev phenomenon*. It illustrates that some of the hypotheses used in order to get existence results, to derive Euler equations or to obtain regularity results are optimal. In particular a careful choice of the space of admissible functions is necessary.

Theorem 4.6. *Let* $f(x, u, \xi) = (x - u^3)^2 \, \xi^6$ *and*

$$I(u) = \int_0^1 f(x, u(x), u'(x)) \, dx \ .$$

Let

$$\mathcal{W}_1 = \{u \in W^{1,\infty}(0,1) : u(0) = 0, \, u(1) = 1\}$$

$$\mathcal{W}_2 = \{u \in W^{1,1}(0,1) : u(0) = 0, \, u(1) = 1\}$$

then

$$\inf\{I(u) : u \in \mathcal{W}_1\} \geq \frac{7^2 \, 3^5}{2^{18} \, 5^5} = c_0 > 0 = \inf\{I(u) : u \in \mathcal{W}_2\} \ .$$

Moreover the minimum of I *over* $\mathcal{W}_2$ *is attained by* $u(x) = x^{1/3}$.

Remarks.

i) The first observation of this phenomenon is due to Lavrentiev [1]. The example presented here is essentially due to Mania [1]; see also Cesari [4], Ball-Mizel [1];

ii) it is interesting to note (cf. Ball-Mizel [1]) that if one uses the usual finite element methods (by taking piecewise affine functions, which are in $W^{1,\infty}$) one will not be able to detect the minimum of some integrals such as the one in the theorem;

iii) note also that one can show, cf. Ball-Mizel [1], a similar result to that of the theorem with a function such as

$$f(x, u, \xi) = (x^4 - u^6)^2 \, |\xi|^s + \varepsilon \, |\xi|^2$$

$\varepsilon > 0$, $s \geq 27$. This last example has the advantage to lead to a coercive integral in $W^{1,2}$, while this is not the case in the above theorem.

Before proceeding with the proof, we establish a preliminary lemma.

Lemma 4.7. *Let* $0 < \alpha < \beta < 1$ *and*

$$
\mathcal{W} = \left\{ u \in W^{1,\infty}(\alpha, \beta) : \frac{1}{4} x^{1/3} \leq u(x) \leq \frac{1}{2} x^{1/3} \right.
$$

$$
\left. \text{for every } x \in [\alpha, \beta]; \quad u(\alpha) = \frac{1}{4}\alpha^{1/3}, \quad u(\beta) = \frac{1}{2}\beta^{1/3} \right\}.
$$

If $f(x, u, \xi) - (x - u^3)^2 \, \xi^6$ *and*

$$
I(u) = \int_\alpha^\beta f(x, u(x), u'(x)) \, dx
$$

then

$$
I(u) \geq \frac{c_0}{\beta}
$$

for every $u \in \mathcal{W}$ *and for* $c_0 = 7^2 \, 3^5 \, 2^{-18} \, 5^{-5}$.

PROOF. Since $u(x) \leq \frac{1}{2} x^{1/3}$ we have

$$
1 - \frac{u^3}{x} \geq 1 - \frac{1}{x} \left(\frac{x^{1/3}}{2} \right)^3 = \frac{7}{2^3}, \quad \text{for every } x \in [\alpha, \beta] \ .
$$

We thus obtain

$$
I(u) = \int_\alpha^\beta x^2 \left(1 - \frac{u^3}{x} \right)^2 u'^6 \, dx \geq \frac{7^2}{2^6} \int_\alpha^\beta x^2 \, u'^6 \, dx \ . \tag{1}
$$

We next let

$$
y = r^{3/5} \quad \text{and} \quad u(r) = \tilde{u}(y) = \tilde{u}(r^{3/5})
$$

We immediately deduce that

$$
u'(x) = \tilde{u}'(y)\frac{dy}{dx} = \frac{3}{5}\tilde{u}'(y)x^{-2/5} = \frac{3}{5}\tilde{u}'(y)y^{-2/3} \ .
$$

Returning to (1) we have

$$I(u) \geq \frac{7^2}{2^6} \int_{\alpha^{3/5}}^{\beta^{3/5}} y^{10/3} \left(\frac{3}{5}\tilde{u}'(y)y^{-2/3} \right)^6 \left(\frac{5}{3}y^{2/3} \right) dy$$

$$\geq \frac{7^2\, 3^5}{2^6\, 5^5} \int_{\alpha^{3/5}}^{\beta^{3/5}} (\tilde{u}'(y))^6 \, dy \ .$$

Applying Jensen inequality to the right hand side we obtain

$$I(u) \geq \frac{7^2\, 3^5}{2^6\, 5^5} \frac{(\tilde{u}(\beta^{3/5}) - \tilde{u}(\alpha^{3/5}))^6}{(\beta^{3/5} - \alpha^{3/5})^5} = \frac{7^2\, 3^5}{2^6\, 5^5} \frac{\left(\frac{1}{2}\beta^{1/3} - \frac{1}{4}\alpha^{1/3} \right)^6}{(\beta^{3/5} - \alpha^{3/5})^5}$$

$$\geq \frac{7^2\, 3^5}{2^{12}\, 5^5} \frac{\beta^2 \left(1 - \frac{1}{2}(\frac{\alpha}{\beta})^{1/3} \right)^6}{\beta^3 \left(1 - (\frac{\alpha}{\beta})^{3/5} \right)^5} \ . \tag{2}$$

Observe finally that since $0 < \alpha < \beta$, then

$$\left(1 - \frac{1}{2}\left(\frac{\alpha}{\beta}\right)^{1/3} \right)^6 \geq \left(\frac{1}{2} \right)^6 \text{ and } \left(1 - \left(\frac{\alpha}{\beta}\right)^{3/5} \right)^{-5} \geq 1 \ . \tag{3}$$

Combining (2) and (3) we have indeed obtained the lemma. $\square$

PROOF of Theorem 4.6.

Step 1: We first prove that if $u \in \mathcal{W}_1$ then there exist $0 < \alpha < \beta < 1$ such that $u \in \mathcal{W}$ ($\mathcal{W}$ as in the lemma) i.e.

$$\begin{cases} u(\alpha) = \frac{1}{4}\alpha^{1/3}, \ u(\beta) = \frac{1}{2}\beta^{1/3} \\ \frac{1}{4}x^{1/3} \leq u(x) \leq \frac{1}{2}x^{1/3}, \text{ for every } x \in [\alpha, \beta] \end{cases} \ . \tag{1}$$

The existence of such α and β is easily seen (cf. diagram below). Let

$$A = \left\{ a \in (0,1) : u(a) = \frac{1}{4}a^{1/3} \right\}$$

$$B = \left\{ b \in (0,1) : u(b) = \frac{1}{2}b^{1/3} \right\} \ .$$

Since u is locally Lipschitz, $u(0) = 0$ and $u(1) = 1$, it follows that $A \neq \emptyset$

and $B \neq \emptyset$. Choose for example

$$\alpha = \max\{a : a \in A\}$$
$$\beta = \min\{b : b \in B \text{ and } b > \alpha\} \ .$$

It is then clear that α and β satisfy (1).

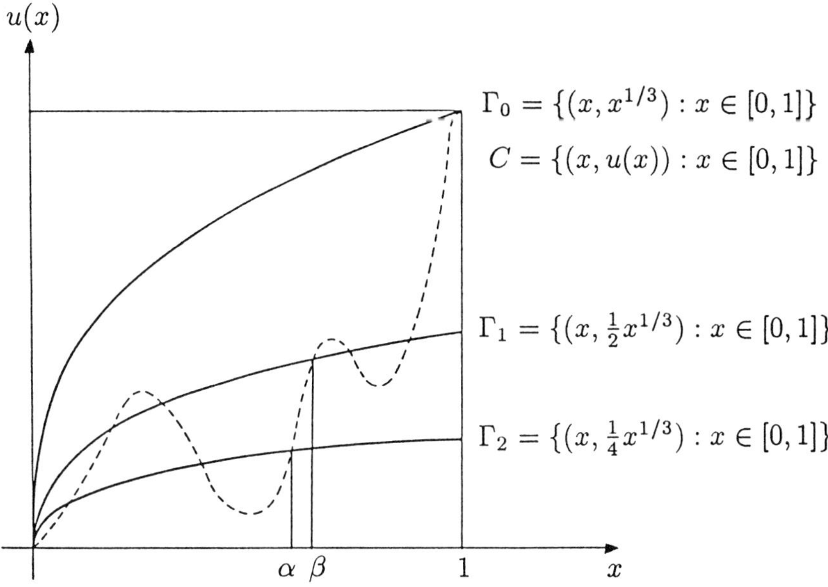

Figure 3.5

Step 2: We may therefore use the lemma to deduce that

$$I(u) = \int_0^1 (x - u^3)^2 \, u'^6 \, dx \geq \int_\alpha^\beta (x - u^3)^2 \, u'^6 \, dx \geq \frac{c_0}{\beta} > c_0 > 0 \ .$$

Step 3: The fact that $u(x) - x^{1/3}$ is the minimum of I over all $u \in \mathcal{W}_2$ is trivial.   □

CHAPTER 4

The Vectorial Case

4.0 Introduction

We now turn our attention to the vectorial case. Recall that

$$I(u) = \int_{\Omega} f(x, u(x), \nabla u(x)) \, dx$$

and $u : \Omega \subset \mathbb{R}^n \to \mathbb{R}^m$ (thus $\nabla u \in \mathbb{R}^{nm}$), with $n, m > 1$. While the convexity of f with respect to the last variable ∇u is playing the central role in the scalar case ($m = 1$ or $n = 1$), cf. Chapter 3., and is still sufficient, in the vectorial case, to ensure weak lower semicontinuity of I in $W^{1,p}(\Omega, \mathbb{R}^m)$, it is far from being a necessary condition. Such a condition is the so-called *quasiconvexity* introduced by Morrey. However it is hard to verify, in practice, if a given function f is quasiconvex, since it is not pointwise condition. Therefore one is lead to introduce a slightly weaker condition known as *rank one convexity* and a stronger condition, introduced by Ball, called *polyconvexity*. One can relate all these definitions through the following diagram (Fig. 4.1).

$$f \text{ convex} \quad \Rightarrow \quad f \text{ polyconvex} \quad \Rightarrow \quad f \text{ quasiconvex} \quad \Rightarrow \quad f \text{ rank one convex}$$

$$\Updownarrow \qquad\qquad\qquad\qquad \Updownarrow$$

$$I \text{ weakly lower} \qquad\quad \text{the Euler equations}$$

$$\text{semicontinuous} \qquad\quad I'(u) = 0 \text{ are elliptic}$$

Figure 4.1

We should again emphasize that in the scalar case all these notions are equivalent to the usual convexity condition.

In Section 4.1 we introduce the above notions and study their relationships. We give several examples and counter-examples.

In Section 4.2 we first establish that the quasiconvexity of f, with respect to the variable ∇u, is necessary and sufficient for I to be weakly lower semicontinuous in $W^{1,p}(\Omega; \mathbf{R}^m)$. We then establish a result which is typical of the vectorial case. While in the scalar case there is no nonlinear function f which is *weakly continuous*, i.e. I and $-I$ are weakly lower semicontinuous, there are such functions in the vectorial case. Following Morrey, Reshetnyak and Ball, we shall show that the *minors* of the matrix ∇u enjoy such a property. In particular if $m = n = 2$ we shall have that if

$$u_\nu \rightharpoonup u \text{ in } W^{1,p}(\Omega; \mathbf{R}^2), \ p > 2$$

then

$$\det \nabla u_\nu \rightharpoonup \det \nabla u \text{ in } L^{p/2}(\Omega) \ .$$

At the end of the second section we shall deduce, as in Chapter 3, existence theorems for the problem

$$(P) \qquad \inf\{I(u) : u \in u_0 + W_0^{1,p}(\Omega; \mathbf{R}^m)\} \ .$$

Finally we shall give some elementary properties of the determinants in an appendix.

4.1 Polyconvexity, Quasiconvexity and Rank One Convexity

4.1.1 Definitions and Properties

4.1.1.1 Definitions and Basic Relations

We first introduce some notations. If $A \in \mathbf{R}^{nm}$ we write

$$A = \begin{pmatrix} A_1^1 & \cdots & A_n^1 \\ \vdots & & \vdots \\ A_1^m & \cdots & A_n^m \end{pmatrix} = \begin{pmatrix} A^1 \\ \vdots \\ A^m \end{pmatrix} = (A_1, \ldots, A_n) = \left(A_\alpha^i\right)_{1 \leq i \leq m, \, 1 \leq \alpha \leq n}$$

In particular if $u : \mathbb{R}^n \to \mathbb{R}^m$ we write

$$
\nabla u = \begin{pmatrix} \dfrac{\partial u_1}{\partial x_1} & \cdots & \dfrac{\partial u_1}{\partial x_n} \\ \vdots & & \vdots \\ \dfrac{\partial u_m}{\partial x_1} & \cdots & \dfrac{\partial u_m}{\partial x_n} \end{pmatrix} .
$$

We may now define all the notions introduced above.

Definitions.

i) A function $f : \mathbb{R}^{nm} \to \bar{\mathbb{R}} = \mathbb{R} \cup \{+\infty\}$ is said to be *rank one convex* if

$$
f(\lambda A + (1 - \lambda)B) \le \lambda f(A) + (1 - \lambda)f(B) \tag{R}
$$

for every $\lambda \in [0, 1]$, $A, B \in \mathbb{R}^{nm}$ with rank $\{A - B\} \le 1$.

ii) A Borel measurable and locally integrable function $f : \mathbb{R}^{nm} \to \mathbb{R}$ is said to be *quasiconvex* if

$$
f(A) \le \frac{1}{\operatorname{meas} D} \int_D f(A + \nabla\varphi(x)) \, dx \tag{Q}
$$

for every bounded domain $D \subset \mathbb{R}^n$, for every $A \in \mathbb{R}^{nm}$ and for every $\varphi \in W_0^{1,\infty}(D; \mathbb{R}^m)$.

iii) A Borel measurable and locally integrable function $f : \mathbb{R}^{nm} \to \mathbb{R}$ is said to be *quasiaffine* if f and $-f$ are quasiconvex.

iv) A function $f : \mathbb{R}^{nm} \to \bar{\mathbb{R}} = \mathbb{R} \cup \{+\infty\}$ is said to be *polyconvex* if there exists $g : \mathbb{R}^{\tau(n,m)} \to \bar{\mathbb{R}}$ convex, such that

$$
f(A) = g(T(A)) \tag{P}
$$

where $T : \mathbb{R}^{nm} \to \mathbb{R}^{\tau(n,m)}$ is such that

$$
T(A) = (A, \operatorname{adj}_2 A, \ldots, \operatorname{adj}_{n \wedge m} A) .
$$

In the preceding definition, $\operatorname{adj}_s A$ stands for the matrix of all $s \times s$ minors of the matrix $A \in \mathbb{R}^{nm}$, $2 \le s \le n \wedge m = \min\{n, m\}$, and

$$
\tau(n, m) = \sum_{s=1}^{n \wedge m} \sigma(s)
$$

where

$$
\sigma(s) = \binom{m}{s}\binom{n}{s} = \frac{m! \, n!}{(s!)^2 \, (m - s)! \, (n - s)!} .
$$

REMARKS.

i) The terminology used in the definitions is of Morrey [1] for the second one and of Ball [2] for rank one convexity and polyconvexity. Note also that Ball calls quasiaffine functions, *null Lagrangians*;

ii) we have gathered in Section 4.3 some elementary facts about determinants and adj_s of matrices. Note that in the case $m = n = 2$, the notion of polyconvexity can be read as follows

$$\begin{cases} \tau(n,m) = \tau(2,2) = 5 \quad (\text{since } \sigma(1) = 4, \sigma(2) = 1) \\ T(A) = (A, \det A) \\ f(A) = g(A, \det A) \end{cases} \quad ;$$

iii) observe that, if we adopt the tensorial notation, the notion of rank one convexity can be read as follows

$$\varphi(t) = f(A + t\, a \otimes b) \qquad (R')$$

is convex in t for every $A \in \mathbb{R}^{nm}$ and for every $a \in \mathbb{R}^m$, $b \in \mathbb{R}^n$ where we have denoted by

$$a \otimes b = (a^i b_\alpha)_{1 \leq i \leq m, \, 1 \leq \alpha \leq n} \quad ;$$

iv) note that we did not give a definition of quasiconvex functions f which may take the value $+\infty$, contrary to polyconvexity and rank one convexity. There has been such definitions given by Dacorogna-Fusco [1] and Ball-Murat [1] in the case where f is allowed to take the value $+\infty$. However, although such definitions have been shown to be necessary for weak lower semicontinuity, it has not been proved that they were sufficient and this seems to be a difficult problem. The notion of quasiconvexity being useful only as an equivalent to weak lower semicontinuity we have disregarded the extension to the case $\bar{\mathbb{R}}$. While those of polyconvexity and rank one convexity will be shown to be useful;

v) Ball and Murat [1] have weakened the notion of quasiconvexity, introducing the notion of $W^{1,p}$-*quasiconvexity*, in their terminology the notion given above would be $W^{1,\infty}$-*quasiconvexity*;

vi) it is easily seen that in the definition of quasiconvexity, one can replace in (Q) the set of test functions $W_0^{1,\infty}$ by $C_0^\infty(D; \mathbb{R}^m)$;

vii) it is important to observe that in the definition of polyconvexity of a given function f, the associated function g (i.e. $f(A) = g(T(A))$) is in general not unique. For example, let $m = n = 2$,

$$A = \begin{pmatrix} A_1^1 & A_2^1 \\ A_1^2 & A_2^2 \end{pmatrix}$$

and

$$f(A) = |A|^2 = \left(A_1^1\right)^2 + \left(A_1^2\right)^2 + \left(A_2^1\right)^2 + \left(A_2^2\right)^2$$

$$= \left(A_1^1 - A_2^2\right)^2 + \left(A_2^1 + A_1^2\right)^2 + 2\det A \ .$$

Let $g_1, g_2 : \mathbb{R}^5 \to \mathbb{R}$ be defined by

$$g_1(A, a) = |A|^2$$

$$g_2(A, a) = \left(A_1^1 - A_2^2\right)^2 + \left(A_2^1 + A_1^2\right)^2 + 2a \ .$$

Then g_1 and g_2 are convex, $g_1 \neq g_2$ and

$$f(A) = g_1(T(A)) = g_1(A, \det A) = g_2(T(A)) = g_2(A, \det A) \ .$$

We shall see, after Theorem 1.3 (see also Kohn-Strang [1,2]), that using either Carathéodory theorem or Hahn-Banach theorem one can privilege one among the numerous functions g;

viii) one finally should note (cf. Meyers [1]) that in the definition of quasiconvexity if (Q) holds for a given domain $D \subset \mathbb{R}^n$, then (Q) holds for every such domain D.

PROOF of Remark viii). Assume that (Q) holds for one bounded domain D. We wish to show that for every bounded domain $E \subset \mathbb{R}^n$, one has

$$\int_E f(A + \nabla\varphi(x)) \, dx \geq f(A) \operatorname{meas} E \qquad (1)$$

$A \in \mathbb{R}^{nm}$ and $\varphi \in W_0^{1,\infty}(E; \mathbb{R}^m)$. Clearly there exist $x_0 \in \mathbb{R}^n$ and $\varepsilon > 0$ such that $x_0 + \varepsilon E \subset D$. Define $\psi \in W_0^{1,\infty}(D; \mathbb{R}^m)$ by

$$\psi(x) = \begin{cases} \varepsilon\varphi\left(\dfrac{x - x_0}{\varepsilon}\right) & \text{if } x \in x_0 + \varepsilon E \\ 0 & \text{if } x \in D - (x_0 + \varepsilon E) \end{cases}, $$

then

$$\int_E f(A + \nabla\varphi(y)) \, dy = \varepsilon^{-n} \int_{x_0 + \varepsilon E} f\left(A + \nabla\varphi\left(\frac{x - x_0}{\varepsilon}\right)\right) dx$$

$$= \varepsilon^{-n} \left\{ \int_D f(A + \nabla\psi(x)) \, dx - f(A) \operatorname{meas}(D - (x_0 + \varepsilon E)) \right\} . \qquad (2)$$

Using (Q) which is valid for D we get

$$\int_E f(A + \nabla\varphi(y)) \, dy \geq \varepsilon^{-n} f(A) \operatorname{meas}(x_0 + \varepsilon E) = f(A) \operatorname{meas} E \ . \qquad \square$$

In Section 4.1.2 we shall give examples (see Theorems 1.5, 1.7 and 1.10) of such functions, but before that we shall show the relationship between these notions. The following result is essentially due to Morrey [1,2] (see the diagram in the introduction).

Theorem 1.1.

i) Let $f : \mathbb{R}^{nm} \to \mathbb{R}$, then

$$f \text{ convex} \Rightarrow f \text{ polyconvex} \Rightarrow f \text{ quasiconvex} \Rightarrow f \text{ rank one convex}.$$

If $f : \mathbb{R}^{nm} \to \bar{\mathbb{R}} = \mathbb{R} \cup \{+\infty\}$, then

$$f \text{ convex} \Rightarrow f \text{ polyconvex} \Rightarrow f \text{ rank one convex} .$$

ii) If $m = 1$ or $n = 1$, then all these notions are equivalent.
iii) If $f \in C^2(\mathbb{R}^{nm})$ then rank one convexity is equivalent to Legendre-Hadamard (or ellipticity condition) condition

$$\sum_{i,j=1}^{m} \sum_{\alpha,\beta=1}^{n} \frac{\partial^2 f(A)}{\partial A_\alpha^i \partial A_\beta^j} \lambda^i \lambda^j \mu_\alpha \mu_\beta \geq 0 \qquad (L.H.)$$

for every $\lambda \in \mathbb{R}^m$, $\mu \in \mathbb{R}^n$, $A = \left(A_\alpha^i\right)_{1 \leq i \leq m, 1 \leq \alpha \leq n} \in \mathbb{R}^{nm}$.
iv) If $f : \mathbb{R}^{nm} \to \mathbb{R}$ is convex, polyconvex, quasiconvex or rank one convex, then f is locally Lipschitz.

REMARKS.

i) We shall show later that the implications, f polyconvex $\Rightarrow f$ convex and f quasiconvex $\Rightarrow f$ polyconvex are false; while it is an open problem to know whether f rank one convex implies f quasiconvex;
ii) the Legendre-Hadamard condition $(L.H.)$ is the usual inequality required for the Euler equations and is known in this case as ellipticity (cf. Nirenberg [1], Agmon-Douglas-Nirenberg [1]).

Before proceeding with the proof of the theorem we give a lemma involving some elementary properties of the determinants.

Lemma 1.2. *Let $A \in \mathbb{R}^{nm}$ and $T(A)$ be defined as above.*

i) For every $A, B \in \mathbb{R}^{nm}$ with rank $\{A - B\} \leq 1$ and for every $\lambda \in [0,1]$, then

$$T(\lambda A + (1 - \lambda)B) = \lambda T(A) + (1 - \lambda)T(B) .$$

ii) For every bounded domain $D \subset \mathbb{R}^n$, $A \in \mathbb{R}^{nm}$, $\varphi \in W_0^{1,\infty}(D; \mathbb{R}^m)$, then

$$T(A) = \frac{1}{\operatorname{meas} D} \int_D T(A + \nabla\varphi(x)) \, dx \ .$$

PROOF. The proof is elementary and can be found in the appendix (Proposition 3.1 and Theorem 3.2). We give here, for the sake of illustration, the proof in the case $m = n = 2$. We then have

$$A = \begin{pmatrix} A_1^1 & A_2^1 \\ A_1^2 & A_2^2 \end{pmatrix}$$

and

$$T(A) = (A; \det A) = \left(A_1^1; A_2^1; A_1^2; A_2^2; A_1^1 A_2^2 - A_2^1 A_1^2\right) \ .$$

i) We have, since rank $\{A - B\} \leq 1$, that there exist $a, b \in \mathbb{R}^2$ such that

$$B = A + a \otimes b = \begin{pmatrix} A_1^1 + a^1 b_1 & A_2^1 + a^1 b_2 \\ A_1^2 + a^2 b_1 & A_2^2 + a^2 b_2 \end{pmatrix} \ .$$

It is therefore trivial to see that

$$\det(\lambda A + (1 - \lambda)B) = \det(A + (1 - \lambda)a \otimes b)$$
$$= \lambda \det A + (1 - \lambda) \det B \ . \tag{1}$$

We then deduce that

$$T(\lambda A + (1 - \lambda)B) = (\lambda A + (1 - \lambda)B, \det(\lambda A + (1 - \lambda)B))$$
$$= \lambda T(A) + (1 - \lambda)T(B) \ ,$$

whenever rank $\{A - B\} \leq 1$.

ii) The proof is similar to the preceding one. Note first that if $\varphi \in C^2(D; \mathbb{R}^m)$, then

$$\det \nabla\varphi = \frac{\partial\varphi_1}{\partial x_1} \frac{\partial\varphi_2}{\partial x_2} - \frac{\partial\varphi_1}{\partial x_2} \frac{\partial\varphi_2}{\partial x_1}$$
$$= \frac{\partial}{\partial x_1}\left(\varphi_1 \frac{\partial\varphi_2}{\partial x_2}\right) - \frac{\partial}{\partial x_2}\left(\varphi_1 \frac{\partial\varphi_2}{\partial x_1}\right) \ . \tag{2}$$

We therefore obtain from (2) and after an integration by part that if $\varphi \in C_0^2(D; \mathbb{R}^m)$, then

$$\frac{1}{\operatorname{meas} D} \int_D \det(A + \nabla\varphi(x))\, dx$$

$$= \frac{1}{\operatorname{meas} D} \int_D \left[\det A + A_1^1 \frac{\partial \varphi_2}{\partial x_2} + A_2^2 \frac{\partial \varphi_1}{\partial x_1} \right.$$

$$\left. - A_2^1 \frac{\partial \varphi_2}{\partial x_1} - A_1^2 \frac{\partial \varphi_1}{\partial x_2} + \det \nabla\varphi \right]\, dx$$

$$= \det A \ . \tag{3}$$

By density, (3) holds also if $\varphi \in W_0^{1,\infty}(D; \mathbb{R}^m)$. We then deduce that for every $\varphi \in W_0^{1,\infty}(D; \mathbb{R}^m)$, we must have

$$\frac{1}{\operatorname{meas} D} \int_D T(A + \nabla\varphi(x))\, dx$$

$$= \frac{1}{\operatorname{meas} D} \left(\int_D (A + \nabla\varphi(x))\, dx, \int_D \det(A + \nabla\varphi(x))\, dx \right)$$

$$= T(A) \ . \qquad \square$$

We may now proceed with the proof of the theorem.

PROOF of Theorem 1.1.

i) *Part 1: f convex $\Rightarrow$ f polyconvex.* This implication is trivial.

Part 2: f polyconvex $\Rightarrow$ f quasiconvex. Since f is polyconvex, there exists $g : \mathbb{R}^{\tau(n,m)} \to \mathbb{R}$ convex, such that

$$f(A) = g(T(A)). \tag{P}$$

Using Lemma 1.2 and Jensen inequality we obtain

$$\frac{1}{\operatorname{meas} D} \int_D f(A + \nabla\varphi(x))\, dx = \frac{1}{\operatorname{meas} D} \int_D g(T(A + \nabla\varphi(x)))\, dx$$

$$\geq g \left(\frac{1}{\operatorname{meas} D} \int_D T(A + \nabla\varphi(x))\, dx \right) = g(T(A)) = f(A) \ ,$$

for every $D \subset \mathbb{R}^n$ bounded domain, for every $A \in \mathbb{R}^{nm}$ and for every $\varphi \in W_0^{1,\infty}(D; \mathbb{R}^m)$. The last inequality is precisely the definition of quasiconvexity.

Part 3: f quasiconvex $\Rightarrow$ f rank one convex. The proof is similar to that of Theorem 3.1 of Chapter 3. Recall that we want to show that

$$f(\lambda A + (1 - \lambda)B) \leq \lambda f(A) + (1 - \lambda)f(B) \tag{R}$$

for every $\lambda \in [0,1]$, $A, B \in \mathbb{R}^{nm}$ with rank $\{A - B\} \leq 1$. We divide the proof into two steps.

Step 1: Let $\lambda \in [0,1]$, $A, B \in \mathbf{R}^{nm}$ with rank $\{A - B\} \leq 1$, be fixed. Let $\varepsilon > 0$. Assume (cf. Step 2) that there exist $D \subset \mathbf{R}^n$ a bounded domain, $D_1, D_2 \subset D$ $(D_1 \cap D_2 = \emptyset)$ and $\varphi \in W_0^{1,\infty}(D; \mathbf{R}^m)$ such that

$$\begin{cases} |\operatorname{meas} D_1 - \lambda \operatorname{meas} D| \leq \varepsilon \\ |\operatorname{meas} D_2 - (1 - \lambda) \operatorname{meas} D| \leq \varepsilon \\ \nabla \varphi(x) = \begin{cases} (1 - \lambda)(A - B) & \text{if } x \in D_1 \\ -\lambda(A - B) & \text{if } x \in D_2 \end{cases} \\ \|\nabla \varphi\|_{L^\infty} \leq K = K(A, B) \end{cases} \quad . \tag{1}$$

We may then use the quasiconvexity of f to get

$$\int_D f(\lambda A + (1 - \lambda)B + \nabla \varphi(x))\, dx$$

$$= \int_{D_1} f(A)\, dx + \int_{D_2} f(B)\, dx$$

$$+ \int_{D-(D_1 \cup D_2)} f(\lambda A + (1 - \lambda)B + \nabla \varphi(x))\, dx$$

$$\geq f(\lambda A + (1 - \lambda)B) \operatorname{meas} D \ . \tag{2}$$

Using (1) and the arbitrariness of ε, we have indeed obtained that f is rank one convex.

Step 2: It therefore remains to show (1). Observe first that since rank $\{A - B\} \leq 1$, there exists an $m \times m$ invertible matrix R so that

$$A - B = R \begin{pmatrix} \alpha_1^1 & \cdots & \alpha_n^1 \\ & 0 & \end{pmatrix} \ . \tag{3}$$

We then choose $D = (0,1)^n$ and use the construction of Theorem 3.1 of Chapter 3 to deduce that for every $\varepsilon > 0$, there exist $D_1, D_2 \subset D$ $(D_1 \cap D_2 = \emptyset)$ and $\psi_1 \in W_0^{1,\infty}(D)$ such that

$$\begin{cases} |\operatorname{meas} D_1 - \lambda \operatorname{meas} D| \leq \varepsilon, \ |\operatorname{meas} D_2 - (1 - \lambda) \operatorname{meas} D| \leq \varepsilon \\ \nabla \psi_1(x) = \begin{cases} (1 - \lambda)\left(\alpha_1^1, \ldots, \alpha_n^1\right) & \text{if } x \in D_1 \\ \lambda\left(\alpha_1^1, \ldots, \alpha_n^1\right) & \text{if } x \in D_2 \end{cases} \\ \|\nabla \psi_1\|_{L^\infty} \leq K = K(\alpha) \end{cases} \quad . \tag{4}$$

Define then $\psi(x) = (\psi_1(x), 0, \ldots, 0) \in \mathbf{R}^m$ and

$$\varphi(x) = R\, \psi(x) \quad \text{for every } x \in D \ . \tag{5}$$

Combining (3), (4) and (5) we have indeed obtained (1).

Part 4: If we now consider the case where $f : \mathbb{R}^{nm} \to \bar{\mathbb{R}} = \mathbb{R} \cup \{+\infty\}$, the first implication f convex $\Rightarrow$ f polyconvex is still trivial. The implication f polyconvex $\Rightarrow$ f rank one convex is also elementary if we use Lemma 1.2. Since f is polyconvex, there exists $g : \mathbb{R}^{\tau(n,m)} \to \bar{\mathbb{R}}$ convex so that

$$f(A) = g(T(A)) \ .$$

Let $\lambda \in [0,1]$, $A, B \in \mathbb{R}^{nm}$ with $\operatorname{rank}\{A - B\} \leq 1$, then, using Lemma 1.2,

$$f(\lambda A + (1 - \lambda)B)$$

$$= g(T(\lambda A + (1 - \lambda)B)) = g(\lambda T(A) + (1 - \lambda)T(B))$$

$$\leq \lambda g(T(A)) + (1 - \lambda)g(T(B)) = \lambda f(A) + (1 - \lambda)f(B) \ ,$$

which is precisely the rank one convexity of f.

ii) The second statement of the theorem, asserting that if $m = 1$ or $n = 1$ then all the notions are equivalent, is trivial.

iii) We now assume that f is C^2 and rank one convex, that is

$$\varphi(t) = f(A + t\lambda \otimes \mu)$$

is convex in t for every $A \in \mathbb{R}^{nm}$, for every $\lambda \in \mathbb{R}^m$, $\mu \in \mathbb{R}^n$. Since φ is also C^2, we obtain immediately Legendre-Hadamard condition, by computing $\varphi''(t)$ and using the convexity of φ.

iv) The last part of Theorem 1.1 is an immediate consequence of Theorem 2.3 of Chapter 2, since a rank one convex function is evidently convex in each of its variables. $\qquad\qquad\square$

4.1.1.2 Further Properties of Polyconvex Functions

Before turning our attention to examples of quasiconvex functions, we give different characterizations of polyconvex functions, which are based on Carathéodory theorem and Hahn-Banach theorem. The next result is a slight extension of a theorem of Ball [2] (see Dacorogna [8]).

Theorem 1.3. *Let $f : \mathbb{R}^{nm} \to \bar{\mathbb{R}} = \mathbb{R} \cup \{+\infty\}$ and suppose that there exists a function $c : \mathbb{R}^{\tau(n,m)} \to \bar{\mathbb{R}}$ convex such that*

$$f(A) \geq c(T(A)) \tag{1}$$

for every $A \in \mathbb{R}^{nm}$.

Part 1: The following conditions are equivalent

i) f is polyconvex
ii) the following holds

$$\sum_{i=1}^{\tau+1} \lambda_i f(A_i) \geq f\left(\sum_{i=1}^{\tau+1} \lambda_i A_i\right) \tag{2}$$

whenever $A_i \in \mathbb{R}^{nm}$, $\lambda_i \geq 0$ with $\sum_{i=1}^{\tau+1} \lambda_i = 1$, satisfy

$$\sum_{i=1}^{\tau+1} \lambda_i T(A_i) = T\left(\sum_{i=1}^{\tau+1} \lambda_i A_i\right) \tag{3}$$

where $\tau = \tau(n,m)$.
In particular let $g : \mathbb{R}^\tau \to \bar{\mathbb{R}} = \mathbb{R} \cup \{+\infty\}$ be defined by

$$g(X) \equiv \inf\left\{\sum_{i=1}^{\tau+1} \lambda_i f(A_i) : \sum_{i=1}^{\tau+1} \lambda_i T(A_i) = X\right\} \tag{4}$$

then g is well defined and if f satisfies (2) and (3), then

$$f(A) = g(T(A)) \tag{5}$$

for every $A \in \mathbb{R}^{nm}$.
Part 2: Let $f : \mathbb{R}^{nm} \to \mathbb{R}$, i.e. f takes only finite values, then the following conditions are equivalent

i) f is polyconvex
iii) for every $A \in \mathbb{R}^{nm}$, there exists $\beta(A) \in \mathbb{R}^\tau$ such that

$$f(B) \geq f(A) + \langle \beta(A); T(B) - T(A) \rangle \tag{6}$$

for every $B \in \mathbb{R}^{nm}$ and where $\langle \bullet; \bullet \rangle$ denotes the scalar product in $\mathbb{R}^\tau$.
In particular, if $h : \mathbb{R}^\tau \to \bar{\mathbb{R}} = \mathbb{R} \cup \{+\infty\}$ is defined by

$$h(X) = \sup_{A \in \mathbb{R}^{nm}} \{\langle \beta(A); X - T(A) \rangle + f(A)\} \tag{7}$$

and f satisfies (6) then

$$f(A) = h(T(A)) \tag{8}$$

for every $A \in \mathbb{R}^{nm}$.

EXAMPLE. Let $m = n = 2$, then ii) is read

$$\begin{cases} \sum_{i=1}^{6} \lambda_i f(A_i) \geq f\left(\sum_{i=1}^{6} \lambda_i A_i\right) \\ \sum_{i=1}^{6} \lambda_i \det(A_i) = \det\left(\sum_{i=1}^{6} \lambda_i A_i\right) \end{cases}$$

and iii) is read

$$f(B) \geq f(A) + \langle \gamma(A); B - A \rangle + \delta(A)(\det B - \det A)$$

where $\gamma(A) \in \mathbb{R}^4$ and $\delta(A) \in \mathbb{R}$.

REMARKS.

i) The above theorem is a direct adaptation of Carathéodory theorem and Hahn-Banach theorem for polyconvex functions;

ii) the hypothesis (1) in the theorem will be used only to ensure that g defined in (4) does not take the value $-\infty$;

iii) the representation formula (4) is important for the following reasons;
- as already mentioned in the definition of the polyconvexity of a given function f, the associated convex function g is not unique. (4) allows to privilege one such function g. A similar remark can be done using (7), as observed by Kohn and Strang [1,2],
- if $f : \mathbb{R}^{nm} \to \mathbb{R}$, i.e. f takes only finite values, then g defined by (4) takes also finite values. This observation shows that the definition of polyconvexity, given at the beginning of Chapter 4, is equivalent to that given by Ball [2];

iv) in view of the above remark we can conclude that if f is polyconvex and takes only finite values then i), ii) and iii) of Theorem 1.3 are equivalent;

v) some other properties of polyconvex functions in the case $m = n = 2$ or $m = n = 3$ are given by Aubert [1].

PROOF of Theorem 1.3. The proof of Ball is based on a result of Buseman-Ewald-Shephard [1], Buseman-Shephard [1]. We give here a slight variation of Ball's proof (see Dacorogna [8]).

Part 1: **i)** $\Rightarrow$ **ii)** Since f is polyconvex, there exists $g : \mathbb{R}^\tau \to \bar{\mathbb{R}} = \mathbb{R} \cup \{+\infty\}$, $\tau = \tau(n, m)$, convex such that

$$f(A) = g(T(A)) \tag{9}$$

The convexity of g coupled with (3) gives immediately (2).

ii) $\Rightarrow$ **i)** Assume that (2) holds for every (λ_i, A_i), $1 \leq i \leq \tau + 1$, satisfying

(3). We wish to show that there exists $g : \mathbb{R}^{\tau(n,m)} \to \bar{\mathbb{R}}$ convex satisfying (9). Let $I \geq \tau + 1$ $(\tau = \tau(n,m))$ be an integer and for $X \in \mathbb{R}^{\tau}$ define

$$g_I(X) \equiv \inf \left\{ \sum_{i=1}^{I} \lambda_i f(A_i) : \lambda_i \geq 0, \sum_{i=1}^{I} \lambda_i = 1, \sum_{i=1}^{I} \lambda_i T(A_i) = X \right\} . \tag{10}$$

We shall prove that g_I satisfies (9) and that one can choose $I = \tau + 1$, without loss of generality, establishing hence (4). The proof is divided into four steps.

Step 1: We first show that g_I is well defined, to do this we must see that given $X \in \mathbb{R}^{\tau(n,m)}$ and $I \geq \tau + 1$, then there exist λ_i and A_i such that $\sum \lambda_i T(A_i) = X$. In view of Carathéodory theorem this is equivalent to show that

$$\mathrm{co}\, T(\mathbb{R}^{nm}) = \mathbb{R}^{\tau(n,m)} \tag{11}$$

where $\mathrm{co}\, M$ denotes the convex hull of M and

$$T(\mathbb{R}^{nm}) = \{ X \in \mathbb{R}^{\tau(n,m)} : \quad \text{there exists } A \in \mathbb{R}^{nm} \text{ s.t. } T(A) = X \} .$$

Step 2: We then prove that I can be taken to be $\tau + 1$ in (10) without loss of generality, and we therefore denote g_I by g (satisfying then (4)).

Step 3: We then show that g is convex.

Step 4: Finally, we establish that g satisfies (5).

Step 1: In order to establish (11), we proceed by contradiction. Assume that $\mathrm{co}\,(T(\mathbb{R}^{nm})) \neq \mathbb{R}^{\tau}$, then from a corollary of Hahn-Banach theorem (see Rockafellar [1], p. 99), there exist $0 \neq \alpha \in \mathbb{R}^{\tau}$, $\beta \in \mathbb{R}$ such that

$$\mathrm{co}\,(T(\mathbb{R}^{nm})) \subset V = \{ X \in \mathbb{R}^{\tau} : \langle \alpha; X \rangle \leq \beta \} \tag{12}$$

where $\langle \bullet; \bullet \rangle$ denotes the scalar product in $\mathbb{R}^{\tau}$, $\tau = \tau(n,m)$. Recall from the definition of polyconvexity, that

$$\tau(n,m) = \sum_{s=1}^{n \wedge m} \sigma(s)$$

where $\sigma(s) = \binom{m}{s}\binom{n}{s}$. We then let for $X \in \mathbb{R}^{\tau(n,m)}$

$$X = (X_1, X_2, \ldots, X_{n \wedge m}) \in \mathbb{R}^{\sigma(1)} \times \mathbb{R}^{\sigma(2)} \times \ldots \times \mathbb{R}^{\sigma(n \wedge m)} = \mathbb{R}^{\tau(n,m)}$$

and similarly for $\alpha \in \mathbb{R}^{\tau}$. We may then write

$$\langle \alpha; X \rangle = \sum_{s=1}^{n \wedge m} \langle \alpha_s; X_s \rangle . \tag{13}$$

Since $\alpha \neq 0$, there exist $t \in \{1, \ldots, n \wedge m\}$ such that $\alpha_t \neq 0$ while $\alpha_s = 0$ if $s < t$ (if $\alpha_1 \neq 0$, then take $t = 1$). We now show that (12) leads to a contradiction and therefore (11) holds. Let $A \in \mathbb{R}^{nm}$, and therefore $T(A) = (A, \mathrm{adj}_2 A, \ldots, \mathrm{adj}_{n \wedge m} A) \in T(\mathbb{R}^{nm}) \subset \mathrm{co}\, T(\mathbb{R}^{nm})$. We choose A such that t of the lines of A are arbitrary vectors of $\mathbb{R}^n$ and the other $(m - t)$ lines are zero, then

$$\langle \alpha; T(A) \rangle = \langle \alpha_t; \mathrm{adj}_t A \rangle \, , \tag{14}$$

the t lines being chosen such that the right hand side of (14) is non zero. Let $\lambda \in \mathbb{R}$ be arbitrary and multiply any of the t non zero lines of A by λ, and denote the obtained matrix by B. We then have $T(B) \in T(\mathbb{R}^{nm}) \subset \mathrm{co}\, T(\mathbb{R}^{nm})$ and

$$\langle \alpha; T(B) \rangle = \langle \alpha_t; \mathrm{adj}_t B \rangle = \lambda \langle \alpha_t; \mathrm{adj}_t A \rangle = \lambda \langle \alpha; T(A) \rangle \, .$$

Using (12) we deduce that $T(A), T(B) \in V$, i.e.

$$\begin{cases} \langle \alpha; T(A) \rangle \leq \beta \\ \langle \alpha; T(B) \rangle = \lambda \langle \alpha; T(A) \rangle \leq \beta \end{cases} .$$

The arbitrariness of λ and the fact that $\langle \alpha; T(A) \rangle \neq 0$ lead immediately to a contradiction. This completes Step 1.

Step 2: We now want to show that in (10) we can take $I = \tau + 1$. We first prove that there is no loss of generality if $I = \tau + 2$. Define

$$T(\mathrm{epi}\, f) = \{(T(A), a) \in \mathbb{R}^\tau \times \mathbb{R} : f(A) \leq a\} \subset \mathbb{R}^{\tau+1} \, .$$

We then trivially have that $(T(A_i), f(A_i)) \in T(\mathrm{epi} f)$ and therefore

$$\left(X, \sum_{i=1}^{I} \lambda_i f(A_i) \right) = \sum_{i=1}^{I} \lambda_i (T(A_i), f(A_i)) \in \mathrm{co}\, T(\mathrm{epi}\, f) \, .$$

Using Carathéodory theorem, we find that in (10) we can take $I = \tau + 2$. It now remains to reduce I from $\tau + 2$ to $\tau + 1$. This is also a standard procedure and we follow Ioffe-Tihomirov [1] (see also Rockafellar [1], Ekeland-Témam [1]). We show that given $X, T(A_i) \in \mathbb{R}^\tau, 1 \leq i \leq \tau + 2, f : \mathbb{R}^{nm} \to \bar{\mathbb{R}} = \mathbb{R} \cup \{+\infty\}$ and $\alpha_i \in \mathbb{R}$ with

$$\begin{cases} \alpha_i \geq 0, \ \displaystyle\sum_{i=1}^{\tau+2} \alpha_i = 1 \\ \displaystyle\sum_{i=1}^{\tau+2} \alpha_i T(A_i) = X \end{cases} , \tag{15}$$

then there exist β_i, $1 \leq i \leq \tau + 2$, such that

$$
\begin{cases}
\beta_i \geq 0, \sum_{i=1}^{\tau+2} \beta_i = 1, & \text{at least one of the } \beta_i = 0 \\
\sum_{i=1}^{\tau+2} \beta_i f(A_i) \leq \sum_{i=1}^{\tau+2} \alpha_i f(A_i)
\end{cases}
\qquad (16)
$$

It is clear that (16) will imply Step 2. Assume that all the $\alpha_i > 0$ in (15), otherwise (16) would be trivial. Since, by (15), $X \in \mathrm{co}\,\{T(A_1), \ldots, T(A_{\tau+2})\} \subset \mathbb{R}^\tau$, it results, from Carathéodory theorem that there exist $\tilde{\alpha}_i \geq 0$, $1 \leq i \leq \tau + 2$, with $\sum_{i=1}^{\tau+2} \tilde{\alpha}_i = 1$ and at least one of the $\tilde{\alpha}_i = 0$ such that

$$
\sum_{i=1}^{\tau+2} \tilde{\alpha}_i T(A_i) = X \ .
$$

We may assume without loss of generality that

$$
\sum_{i=1}^{\tau+2} \tilde{\alpha}_i f(A_i) > \sum_{i=1}^{\tau+2} \alpha_i f(A_i) \ , \qquad (17)
$$

otherwise choosing $\beta_i = \tilde{\alpha}_i$ we would have immediately (16). We then let

$$
J = \{i \in \{1, \ldots, \tau + 2\} : \alpha_i - \tilde{\alpha}_i < 0\} \ .
$$

Observe that $J \neq \emptyset$, since otherwise $\alpha_i \geq \tilde{\alpha}_i \geq 0$ for every $1 \leq i \leq \tau + 2$ and since at least one of the $\tilde{\alpha}_i = 0$, we would have a contradiction with $\sum_{i=1}^{\tau+2} \alpha_i = \sum_{i=1}^{\tau+2} \tilde{\alpha}_i = 1$ and the fact that $\alpha_i > 0$ for every i. We then define

$$
\lambda = \min_{i \in J} \left\{ \frac{\alpha_i}{\tilde{\alpha}_i - \alpha_i} \right\}
$$

and we have clearly $\lambda > 0$. Finally let

$$
\beta_i = \alpha_i + \lambda(\alpha_i - \tilde{\alpha}_i), \ 1 \leq i \leq \tau + 2 \ .
$$

We therefore have

$$
\beta_i \geq 0, \sum_{i=1}^{\tau+2} \beta_i = 1, \quad \text{at least one of the } \beta_i = 0 \ ,
$$

and from (17)

$$\sum_{i=1}^{\tau+2} \beta_i f(A_i) = \sum_{i=1}^{\tau+2} \alpha_i f(A_i) + \lambda \left(\sum_{i=1}^{\tau+2} (\alpha_i - \tilde{\alpha}_i) f(A_i) \right)$$

$$\leq \sum_{i=1}^{\tau+2} \alpha_i f(A_i)$$

We have therefore obtained (16) and this concludes Step 2. Since I can be taken to be $\tau + 1$, we will then denote g_I by g (i.e. (10) can be replaced by (4)).

Step 3: We now show that g is convex. Let $\lambda \in [0,1]$, $X, Y \in \mathbb{R}^\tau$. We want to prove that

$$\lambda g(X) + (1 - \lambda)g(Y) \geq g(\lambda X + (1 - \lambda)Y) \ .$$

Fix $\varepsilon > 0$. From (4) we deduce that there exist $\lambda_i \geq 0$ with $\sum_{i=1}^{\tau+1} \lambda_i = 1$, $\mu_i \geq 0$ with $\sum_{i=1}^{\tau+1} \mu_i = 1$, $A_i, B_i \in \mathbb{R}^{nm}$ such that

$$\lambda g(X) + (1 - \lambda)g(Y) + \varepsilon \geq \lambda \sum_{i=1}^{\tau+1} \lambda_i f(A_i) + (1 - \lambda) \sum_{i=1}^{\tau+1} \mu_i f(B_i) \ , \quad (18)$$

with

$$\sum_{i=1}^{\tau+1} \lambda_i T(A_i) = X, \ \sum_{i=1}^{\tau+1} \mu_i T(B_i) = Y \ . \tag{19}$$

Let for $1 \leq i \leq \tau + 1$

$$\begin{cases} \tilde{\lambda}_i = \lambda \lambda_i & C_i = A_i \\ \tilde{\lambda}_{i+\tau+1} = (1 - \lambda)\mu_i & C_{i+\tau+1} = B_i \end{cases} \ . \tag{20}$$

Then (18) and (19) can be rewritten as

$$\begin{cases} \lambda g(X) + (1 - \lambda)g(Y) + \varepsilon \geq \sum_{i=1}^{2\tau+2} \tilde{\lambda}_i f(C_i) & (21) \\ \\ \sum_{i=1}^{2\tau+2} \tilde{\lambda}_i T(C_i) = \lambda X + (1 - \lambda)Y \ . & (22) \end{cases}$$

Taking the infimum in the right hand side of (21) over all $\tilde{\lambda}_i$, C_i satisfying (22), using (10) and Step 2 we have

$$\lambda g(X) + (1 - \lambda)g(Y) + \varepsilon \geq g(\lambda X + (1 - \lambda)Y) \ ;$$

$\varepsilon > 0$ being arbitrary, we have indeed established the convexity of g.

Step 4: It now remains to show (5), i.e.

$$f(A) = g(T(A))$$

where g satisfies (4), i.e.

$$g(X) = \inf \left\{ \sum_{i=1}^{\tau+1} \lambda_i f(A_i) : \sum_{i=1}^{\tau+1} \lambda_i T(A_i) = X \right\} .$$

We have just shown that g is convex. Choosing $X = T(A)$ we have from (2) and (3) that the infimum in (4) is attained precisely by $f(A)$, hence (5) and Part 1.

Part 2: **i)** $\Rightarrow$ **iii)** Since f is polyconvex and finite we may use Part 1 to find $g : \mathbb{R}^\tau \to \mathbb{R}$ convex and finite satisfying (cf. (4))

$$\left\{ \begin{array}{l} f(A) = g(T(A)) \\ g(X) \equiv \inf \left\{ \displaystyle\sum_{i=1}^{\tau+1} \lambda_i f(A_i) : \sum_{i=1}^{\tau+1} \lambda_i T(A_i) = X \right\} \end{array} \right. .$$

Since g is convex and finite, it is continuous and therefore (see Proposition 2.7 of Chapter 2), for each $X \in \mathbb{R}^\tau$, there exists $\tilde{\beta}(X) \in \mathbb{R}^\tau$ such that

$$g(Y) \geq g(X) + \langle \tilde{\beta}(X); Y - X \rangle$$

for all $Y \in \mathbb{R}^\tau$. Choosing $Y = T(B)$, $X = T(A)$, $\beta(A) = \tilde{\beta}(T(A))$, we get (6), i.e.

$$f(B) \geq f(A) + \langle \beta(A); T(B) - T(A) \rangle .$$

iii) $\Rightarrow$ **i)** We define h as in (7), i.e.

$$h(X) = \sup_{A \in \mathbb{R}^{nm}} \left\{ \langle \beta(A); X - T(A) \rangle + f(A) \right\} .$$

h being a supremum of affine functions, it is convex. If $X = T(B)$ then (6) ensures that the supremum in (7) is attained by $f(B)$ and therefore $f(B) = h(T(B))$. $\qquad\qquad\square$

4.1.1.3 Further Properties of Rank One Convex Functions

There is no known equivalent to Theorem 1.3 for quasiconvex or rank one convex functions. We, nevertheless, give here a characterization of rank one convex functions which is in the same spirit as Part 1 of Theorem 1.3, but much weaker. It will turn out to be useful in Chapter 5.

To characterize rank one convex functions we give a property of matrices $A_i \in \mathbf{R}^{nm}$ which will play the same role as (3) of Theorem 1.3 for polyconvex functions. We follow here the presentation of Dacorogna [7].

Definition. *Let $\lambda_i > 0$ with $\sum_{i=1}^{N} \lambda_i = 1$ where N is an integer. Let $A_i \in \mathbf{R}^{nm}$, $1 \le i \le N$. (λ_i, A_i) are said to satisfy (H_N) if*

i) $N = 2$, *then* $\operatorname{rank}\{A_1 - A_2\} \le 1$

ii) $N > 2$, *then, up to a permutation,* $\operatorname{rank}\{A_1 - A_2\} \le 1$ *and if*

$$\begin{cases} \mu_1 = \lambda_1 + \lambda_2 & B_1 = \frac{\lambda_1 A_1 + \lambda_2 A_2}{\lambda_1 + \lambda_2} \\ \mu_i = \lambda_{i+1} & B_i = A_{i+1} \quad 2 \le i \le N - 1 \end{cases}$$

then (μ_i, B_i) satisfy (H_{N-1}).

EXAMPLES.

1) $N = 2$: $\lambda_1 + \lambda_2 = 1$, then (λ_1, A_1), (λ_2, A_2) satisfy (H_2) if and only if $\operatorname{rank}\{A_1 - A_2\} \le 1$.

2) $N = 3$: $\lambda_1 + \lambda_2 + \lambda_3 = 1$, then $(\lambda_i, A_i)_{1 \le i \le 3}$ satisfy (H_3) if, up to a permutation

$$\begin{cases} \operatorname{rank}\{A_1 - A_2\} \le 1 \\ \operatorname{rank}\left\{A_3 - \dfrac{\lambda_1 A_1 + \lambda_2 A_2}{\lambda_1 + \lambda_2}\right\} \le 1 \end{cases} \quad .$$

3) $N = 4$: $\sum_{i=1}^{4} \lambda_i = 1$, then $(\lambda_i, A_i)_{1 \le i \le 4}$ satisfy (H_4) if, up to a permutation, one of the following conditions holds, either,

$$\begin{cases} \operatorname{rank}\{A_1 - A_2\} \le 1 \\ \operatorname{rank}\left\{A_3 - \dfrac{\lambda_1 A_1 + \lambda_2 A_2}{\lambda_1 + \lambda_2}\right\} \le 1 \\ \operatorname{rank}\left\{A_4 - \dfrac{\lambda_1 A_1 + \lambda_2 A_2 + \lambda_3 A_3}{\lambda_1 + \lambda_2 + \lambda_3}\right\} \le 1 \end{cases}$$

or

$$\begin{cases} \operatorname{rank}\{A_1 - A_2\} \le 1, \quad \operatorname{rank}\{A_3 - A_4\} \le 1 \\ \operatorname{rank}\left\{\dfrac{\lambda_1 A_1 + \lambda_2 A_2}{\lambda_1 + \lambda_2} - \dfrac{\lambda_3 A_3 + \lambda_4 A_4}{\lambda_3 + \lambda_4}\right\} \le 1 \end{cases} \quad .$$

Proposition 1.4. *Let $f : \mathbf{R}^{nm} \to \bar{\mathbf{R}} = \mathbf{R} \cup \{+\infty\}$, then the following conditions are equivalent*

i) f rank one convex,

ii) the following holds

$$\sum_{i=1}^{N} \lambda_i f(A_i) \ge f\left(\sum_{i=1}^{N} \lambda_i A_i\right) \tag{1}$$

whenever $(\lambda_i, A_i)_{1 \le i \le N}$ satisfy (H_N).

PROOF.

ii) $\Rightarrow$ **i)**: This is trivial if one chooses $N = 2$ and rank $\{A_1 - A_2\} \leq 1$.

i) $\Rightarrow$ **ii)**: We establish (1) by induction.By definition of rank one convexity (1) holds for $N = 2$; assume therefore that the proposition is true for $N - 1$. Observe that

$$\sum_{i=1}^{N} \lambda_i f(A_i) = (\lambda_1 + \lambda_2) \left(\frac{\lambda_1}{\lambda_1 + \lambda_2} f(A_1) + \frac{\lambda_2}{\lambda_1 + \lambda_2} f(A_2) \right) + \sum_{i=3}^{N} \lambda_i f(A_i) .$$

If we now use the rank one convexity of f and the hypothesis (H_N) we get

$$\sum_{i=1}^{N} \lambda_i f(A_i) \geq (\lambda_1 + \lambda_2) f \left(\frac{\lambda_1 A_1 + \lambda_2 A_2}{\lambda_1 + \lambda_2} \right) + \sum_{i=3}^{N} \lambda_i f(A_i) .$$

Using again the rank one convexity of f, hypothesis (H_N) and the hypothesis of induction we have indeed established (1). $\qquad\square$

REMARK. The above result is much weaker than Theorem 1.3 in the sense that one cannot fix an upper bound on N. Two simple examples show that the situation is intrinsically more complicated for rank one convex functions.

EXAMPLE 1. Let $m = n = 2$,

$$A = \begin{pmatrix} 0 & 0 \\ 0 & 0 \end{pmatrix}, \quad B = \begin{pmatrix} 1 & 0 \\ 1 & 0 \end{pmatrix}, \quad C = \begin{pmatrix} 0 & -2 \\ 1/2 & 0 \end{pmatrix}, \quad D = \begin{pmatrix} -1/4 & 4 \\ 0 & 4 \end{pmatrix}$$

and

$$\begin{cases} \lambda_1 = \lambda_2 = \lambda_3 = \lambda_4 = \lambda_5 = \dfrac{1}{5} \\ A_1 = A, \ A_2 = B, \ A_3 = C, \ A_4 = D, \ A_5 = A \end{cases} .$$

It is then easy to see that $(\lambda_i, A_i)_{1 \leq i \leq 5}$ satisfy (H_5) since

$$\begin{cases} \det(A_1 - A_2) = 0 \\ \det \left\{ A_3 - \dfrac{\lambda_1 A_1 + \lambda_2 A_2}{\lambda_1 + \lambda_2} \right\} = 0 \\ \det \left\{ A_4 - \dfrac{\lambda_1 A_1 + \lambda_2 A_2 + \lambda_3 A_3}{\lambda_1 + \lambda_2 + \lambda_3} \right\} = 0 \\ \det \left\{ A_5 - \dfrac{\lambda_1 A_1 + \lambda_2 A_2 + \lambda_3 A_3 + \lambda_4 A_4}{\lambda_1 + \lambda_2 + \lambda_3 + \lambda_4} \right\} = 0 \end{cases} .$$

However if we put together A_1 and A_5 and if we consider

$$\begin{cases} \mu_1 = \lambda_1 + \lambda_5 = \dfrac{2}{5}, \ \mu_2 = \mu_3 = \mu_4 = \dfrac{1}{5} \\ B_1 = A, \ B_2 = B, \ B_3 = C, \ B_4 = D \end{cases}$$

then it is easy to see that $(\mu_i, B_i)_{1 \leq i \leq 4}$ do not satisfy (H_4). In other words if we use Proposition 1.4 we have this surprising result that if f is rank one convex then

$$f\left(\frac{2}{5}A + \frac{1}{5}B + \frac{1}{5}C + \frac{1}{5}D\right) \leq \frac{2}{5}f(A) + \frac{1}{5}f(B) + \frac{1}{5}f(C) + \frac{1}{5}f(D) \quad (1)$$

even though (μ_i, B_i) do not satisfy (H_4). In order to show (1) one has to use the commutativity and associativity of the addition, i.e.

$$f\left(\frac{1}{5}A + \frac{1}{5}B + \frac{1}{5}C + \frac{1}{5}D + \frac{1}{5}A\right)$$

$$\leq \frac{1}{5}f(A) + \frac{1}{5}f(B) + \frac{1}{5}f(C) + \frac{1}{5}f(D) + \frac{1}{5}f(A) \ .$$

EXAMPLE 2 (Casadio [1]). Let $m = n = 2$ and

$$\begin{cases} A_1 = \begin{pmatrix} -1 & 0 \\ 0 & 0 \end{pmatrix}, \ A_2 = \begin{pmatrix} 1 & 0 \\ 0 & -1 \end{pmatrix}, \ A_3 = \begin{pmatrix} 2 & 0 \\ 0 & 1 \end{pmatrix}, \ A_4 = \begin{pmatrix} 0 & 0 \\ 0 & 2 \end{pmatrix} \\ \alpha_1 = \dfrac{8}{15}, \ \alpha_2 = \dfrac{4}{15}, \ \alpha_3 = \dfrac{2}{15}, \ \alpha_4 = \dfrac{1}{15} \end{cases} .$$

Observe that

$$\begin{cases} \displaystyle\sum_{i=1}^{4} \alpha_i A_i = 0 \\ \operatorname{rank}\{A_i - A_j\} = 2, \quad \text{if } i \neq j \end{cases} \qquad (1)$$

Let

$$\begin{cases} B_1 = A_1, \ B_2 = A_2, \ B_3 = A_3, \ B_4 = A_4, \ B_5 = \begin{pmatrix} 0 & 0 \\ 0 & 0 \end{pmatrix} = \displaystyle\sum_{i=1}^{4} \alpha_i A_i \\ \lambda_1 = \dfrac{8}{16}, \ \lambda_2 = \dfrac{4}{16}, \ \lambda_3 = \dfrac{2}{16}, \ \lambda_4 = \dfrac{1}{16}, \ \lambda_5 = \dfrac{1}{16} \end{cases} .$$

Observe that $(\lambda_i, B_i)_{1 \leq i \leq 5}$ satisfy (H_5). Therefore, using Proposition 1.4, we obtain

$$f(0) = f\left(\sum_{i=1}^{5} \lambda_i B_i\right) \leq \sum_{i=1}^{5} \lambda_i f(B_i);$$

i.e.

$$16\, f(0) \leq 8\, f(A_1) + 4\, f(A_2) + 2\, f(A_3) + f(A_4) + f(B_5) \ .$$

Noting that $B_5 = 0$, we have, dividing the above inequality by 15, that

$$f(0) \ = \ f\left(\sum_{i=1}^{4} \alpha_i A_i\right) \ \leq \ \sum_{i=1}^{4} \alpha_i\, f(A_i) \ . \qquad (2)$$

We have therefore obtained (2), even though none of the $A_i - A_j$ is of rank one.

4.1.2 Examples

We have seen in Section 4.1.1 the definitions and the relations between the notions of convexity, polyconvexity, quasiconvexity and rank one convexity. In this section we give some examples of such functions. There will be four main results.

i) The first one will characterize completely the *quasi affine* functions (i.e. the functions f such that f and $-f$ are quasiconvex) by showing that they are linear combinations of *minors* of the matrix ∇u.

ii) The second one will study the particular case of *quadratic* functions f. The main result will be that in this case rank one convexity and quasiconvexity are equivalent. Note that the quadratic case is important in the sense that it leads to associated *linear* Euler equations, and therefore in the linear case the ellipticity of the Euler equations corresponds exactly to the weak lower semicontinuity of the associated variational problem.

iii) The third result will give some more examples. In particular we shall study functions of the type of the area (i.e. integrands of the minimal surfaces problems) and we shall characterize the quasiconvexity of such functions completely.

iv) The fourth result will concern counterexamples to the different implications (see the diagram in the introduction of Chapter 4).

4.1.2.1 Quasiaffine Functions

We start with a result established by Ball [2] which is an extension of results of Edelen [1], Ericksen [1], Rund [1]. It characterizes completely the quasiaffine functions (see also Vasilienko [1], Ball-Currie-Olver [1], Anderson-Duchamp [1], Sivaloganathan [1]).

Theorem 1.5. *Let* $f : \mathbb{R}^{nm} \to \mathbb{R}$. *The following conditions are then equivalent.*

i) f *is quasiaffine.*

ii) f is rank one affine, i.e. f and $-f$ are rank one convex, i.e.

$$f(\lambda A + (1 - \lambda)B) = \lambda f(A) + (1 - \lambda)f(B)$$

for every $\lambda \in [0,1]$, $A, B \in \mathbf{R}^{nm}$ with rank $\{A - B\} \le 1$.
ii') For every $A \in \mathbf{R}^{nm}$, $a \in \mathbf{R}^m$, $b \in \mathbf{R}^n$,

$$f(A + a \otimes b) = f(A) + \langle \nabla f(A); a \otimes b \rangle ,$$

where $\langle \bullet; \bullet \rangle$ denotes the scalar product in $\mathbf{R}^{nm}$.
iii) f is polyaffine, i.e. f and $-f$ are polyconvex.
iii') There exists $\beta \in \mathbf{R}^{\tau(n,m)}$ such that

$$f(A) = f(0) + \langle \beta; T(A) \rangle$$

for every $A \in \mathbf{R}^{nm}$ and where $\langle \bullet; \bullet \rangle$ denotes the scalar product in $\mathbf{R}^{\tau(n,m)}$.

EXAMPLES.

i) If $m = n = 2$, then the theorem asserts that the only quasiaffine functions are of the type

$$f(A) = f(0) + \langle \beta; A \rangle + \gamma \det A .$$

In particular the only fully non linear quasiaffine function is $\det A$.

ii) More generally if $n, m > 1$, then the only non linear quasiaffine functions are the $s \times s$ minors of the matrix $A \in \mathbf{R}^{nm}$, where $2 \le s \le n \wedge m = \min\{n, m\}$.

PROOF.
i) $\Rightarrow$ ii) results immediately from Theorem 1.1.
ii') $\Rightarrow$ ii) is trivial.
ii) $\Rightarrow$ ii') fix $A \in \mathbf{R}^{nm}$, $a \in \mathbf{R}^m$, $b \in \mathbf{R}^n$ and let for $t \in [0,1]$

$$\varphi(t) = f(A + ta \otimes b) .$$

Since f is rank one affine then φ is affine and thus $\varphi \in C^1$ and

$$\varphi(t) = \varphi(0) + t\varphi'(0) . \tag{1}$$

Since φ is differentiable, then, obviously, f is differentiable and the result follows from (1).
iii') $\Rightarrow$ iii) is trivial.
iii) $\Rightarrow$ i) follows from Theorem 1.1.

ii') $\Rightarrow$ **iii'**) this is the only non trivial implication. So let

$$A = \begin{pmatrix} A_1^1 & \cdots & A_n^1 \\ \vdots & & \vdots \\ A_1^m & \cdots & A_n^m \end{pmatrix} = \begin{pmatrix} A^1 \\ \vdots \\ A^m \end{pmatrix} = (A_1, \ldots, A_n) \; .$$

Assume also that f is such that

$$f(A + a \otimes b) - f(A) = \langle \nabla f(A); a \otimes b \rangle \; , \tag{2}$$

for every $A \in \mathbf{R}^{nm}$, $a \in \mathbf{R}^m$, $b \in \mathbf{R}^n$. We wish to show that there exists $\beta \in \mathbf{R}^{\tau(n,m)}$ such that

$$f(A) - f(0) = \langle \beta; T(A) \rangle, \text{ for every } A \in \mathbf{R}^{nm} \; . \tag{3}$$

We proceed by induction on m.

Step 1: $m = 1$. Since $m = 1$, (2) can be read as

$$f(A + B) - f(A) = \langle \nabla f(A); B \rangle$$

for every $A, B \in \mathbf{R}^n$. It is then trivial to see that the above identity implies that f is affine and therefore if we choose $\beta = \nabla f(0)$, we have immediately (3).

Step 1': $m = 2$. This step is unnecessary but we prove it for the sake of illustration. Let

$$A = \begin{pmatrix} A_1^1 & \cdots & A_n^1 \\ A_1^2 & \cdots & A_n^2 \end{pmatrix} = \begin{pmatrix} A^1 \\ A^2 \end{pmatrix} = (A_1, \ldots, A_n)$$

and for $a \in \mathbf{R}^2$, $b \in \mathbf{R}^n$

$$a \otimes b = \begin{pmatrix} a^1 b \\ a^2 b \end{pmatrix} = \begin{pmatrix} a^1 b_1 & \cdots & a^1 b_n \\ a^2 b_1 & \cdots & a^2 b_n \end{pmatrix} \; .$$

We want to show that if f is rank one affine, i.e.

$$f(A + a \otimes b) - f(A) = \langle \nabla f(A); a \otimes b \rangle \tag{4}$$

then there exists $\beta \in \mathbf{R}^{\tau(n,2)}$ such that

$$f(A) = f(0) + \langle \beta; T(A) \rangle \tag{5}$$

where

$$T(A) = (A, \mathrm{adj}_2 A) \in \mathbf{R}^{n \cdot 2} \times \mathbf{R}^{\binom{n}{2}} = \mathbf{R}^{\tau(n,2)} \; . \tag{6}$$

For the notations concerning $\mathrm{adj}_2 A$, see the appendix. But note that, up to a sign and the ordering, an element of the matrix $\mathrm{adj}_2 A$ is essentially $\det(A_k, A_l)$, $1 \leq k < l \leq n$. We then fix A^2 and choose $a^1 = 1$, $a^2 = 0$ in (4) and define

$$g(A^1) = f\begin{pmatrix} A^1 \\ A^2 \end{pmatrix} .$$

Then $g(A^1 + tb) = f\begin{pmatrix} A^1 + tb \\ A^2 \end{pmatrix}$ is affine in t and we may then use Step 1 to find $\gamma = \gamma(A^2) \in \mathbb{R}^n$ such that

$$g(A^1) = g(0) + \langle \gamma(A^2); A^1 \rangle = f\begin{pmatrix} 0 \\ A^2 \end{pmatrix} + \langle \gamma(A^2); A^1 \rangle . \tag{7}$$

Repeating the argument when $A^1 = 0$, for $f\begin{pmatrix} 0 \\ A^2 \end{pmatrix}$ we have

$$f\begin{pmatrix} 0 \\ A^2 \end{pmatrix} = f(0) + \langle \beta^2; A^2 \rangle . \tag{8}$$

Combining (7) and (8) we have

$$f\begin{pmatrix} A^1 \\ A^2 \end{pmatrix} = f(0) + \langle \beta^2; A^2 \rangle + \langle \gamma(A^2); A^1 \rangle . \tag{9}$$

Since f is rank one affine, it is affine (when A^1 is fixed) with respect to A^2 and therefore $\gamma(A^2) = (\gamma_1(A^2), \ldots, \gamma_n(A^2))$ is affine and hence there exist $\beta^1 = (\beta_1^1, \ldots, \beta_n^1) \in \mathbb{R}^n$, $\delta_1, \ldots, \delta_n \in \mathbb{R}^n$ such that

$$\gamma_l(A^2) = \beta_l^1 + \langle \delta_l; A^2 \rangle, l = 1, \ldots, n .$$

Returning to (9) we have therefore

$$f\begin{pmatrix} A^1 \\ A^2 \end{pmatrix} = f(0) + \langle \beta^1; A^1 \rangle + \langle \beta^2; A^2 \rangle + \sum_{l=1}^{n} A_l^1 \langle \delta_l; A^2 \rangle$$

$$= f(0) + \langle \beta^1; A^1 \rangle + \langle \beta^2; A^2 \rangle + \sum_{l=1}^{n} \sum_{\alpha=1}^{n} \delta_{l\alpha} A_l^1 A_\alpha^2 . \tag{10}$$

Since f is rank one affine we have from (10) that if

$$h(A) = \sum_{l=1}^{n} \sum_{\alpha=1}^{n} \delta_{l\alpha} A_l^1 A_\alpha^2$$

then h is rank one affine and therefore, using Lemma 1.6, we must have $\delta_{l\alpha} = -\delta_{\alpha l}$ and thus there exists $\varepsilon \in \mathbf{R}^{\binom{n}{2}}$ such that

$$h(A) = \sum_{1 \le l < \alpha \le n} \delta_{l\alpha} \left(A_l^1 A_\alpha^2 - A_\alpha^1 A_l^2 \right) = \langle \varepsilon; \mathrm{adj}_2 A \rangle . \tag{11}$$

Combining (10) and (11) we deduce (5) and this concludes Step $1'$.

Step 2: We now proceed with the general case. Assume that we have proved the theorem for every $l < m$. Fixing $A^2, \dots, A^m$ and using the fact that f is rank one affine, then f is affine in A^1, for $A^2, \dots, A^m$ fixed. Therefore there exist $\psi \begin{pmatrix} A^2 \\ \vdots \\ A^m \end{pmatrix} = (\psi_1, \dots, \psi_n)$ and $\chi \begin{pmatrix} A^2 \\ \vdots \\ A^m \end{pmatrix}$, such that

$$f(A) = \langle \psi \begin{pmatrix} A^2 \\ \vdots \\ A^m \end{pmatrix} ; A^1 \rangle + \chi \begin{pmatrix} A^2 \\ \vdots \\ A^m \end{pmatrix} . \tag{12}$$

Using the hypothesis of induction and proceeding as in Step $1'$ we find that

$$\begin{cases} \chi \begin{pmatrix} A^2 \\ \vdots \\ A^m \end{pmatrix} = f(0) + \langle \beta^0; T \begin{pmatrix} A^2 \\ \vdots \\ A^m \end{pmatrix} \rangle \\ \psi_l \begin{pmatrix} A^2 \\ \vdots \\ A^m \end{pmatrix} = \beta_l + \langle \gamma_l; T \begin{pmatrix} A^2 \\ \vdots \\ A^m \end{pmatrix} \rangle, \quad l = 1, \dots, n \end{cases} \tag{13}$$

for some $\beta^0, \gamma_1, \dots, \gamma_n \in \mathbf{R}^{\tau(n, m-1)}$ and $\beta^1 = (\beta_1, \dots, \beta_n) \in \mathbf{R}^n$. Combining (12) and (13) we have that

$$f(A) = f(0) + \langle \beta^0; T \begin{pmatrix} A^2 \\ \vdots \\ A^m \end{pmatrix} \rangle + \langle \beta^1; A^1 \rangle + \sum_{l=1}^{n} A_l^1 \langle \gamma_l; T \begin{pmatrix} A^2 \\ \vdots \\ A^m \end{pmatrix} \rangle$$

$$= f(0) + \langle \beta^0; T \begin{pmatrix} A^2 \\ \vdots \\ A^m \end{pmatrix} \rangle + \langle \beta^1; A^1 \rangle$$

$$+ \sum_{s=1}^{n \wedge (m-1)} \sum_{l=1}^{n} \sum_{\alpha=1}^{\binom{n}{s}} \sum_{i=1}^{\binom{m-1}{s}} \gamma_{l\alpha}^{is} A_l^1 \left(\mathrm{adj}_s \begin{pmatrix} A^2 \\ \vdots \\ A^m \end{pmatrix} \right)_\alpha^i . \tag{14}$$

Letting

$$h(A) = \sum_{l=1}^{n} A_l^1 \langle \gamma_l; T \begin{pmatrix} A^2 \\ \vdots \\ A^m \end{pmatrix} \rangle$$

$$= \sum_{s=1}^{n \wedge (m-1)} \sum_{l=1}^{n} \sum_{\alpha=1}^{\binom{n}{s}} \sum_{i=1}^{\binom{m-1}{s}} \gamma_{l\alpha}^{is} A_l^1 \left(\mathrm{adj}_s \begin{pmatrix} A^2 \\ \vdots \\ A^m \end{pmatrix} \right)_\alpha^i,$$

we deduce from the fact that f is rank one affine and from (14) that h is rank one affine. We may then use Lemma 1.6 to deduce that there exist $\delta_\beta^{js} \in \mathbf{R}$, $1 \le s \le n \wedge (m-1)$, $1 \le \beta \le \binom{n}{s+1}$, $1 \le j \le \binom{m}{s+1}$ such that

$$h(A) = \sum_{s=1}^{n \wedge (m-1)} \sum_{\beta=1}^{\binom{n}{s+1}} \sum_{j=1}^{\binom{m}{s+1}} \delta_\beta^{js} (\mathrm{adj}_{s+1} A)_\beta^j . \tag{15}$$

Combining (14) and (15) we have indeed found $\beta \in \mathbf{R}^{\tau(n,m)}$ such that

$$f(A) = f(0) + \langle \beta; T(A) \rangle ,$$

which is the claimed result. $\qquad\qquad\qquad\qquad\qquad\qquad\qquad\qquad\qquad\qquad\square$

In the above proof we have used the following lemma.

Lemma 1.6. *Let $A \in \mathbf{R}^{nm}$,*

$$A = \begin{pmatrix} A_1^1 & \cdots & A_n^1 \\ \vdots & & \vdots \\ A_1^m & \cdots & A_n^m \end{pmatrix} = \begin{pmatrix} A^1 \\ \vdots \\ A^m \end{pmatrix} = (A_1, \ldots, A_n).$$

Let for $1 \le s \le n \wedge (m-1)$,

$$h(A) = \sum_{l=1}^{n} \sum_{\alpha=1}^{\binom{n}{s}} \sum_{i=1}^{\binom{m-1}{s}} \gamma_{l\alpha}^{i} A_l^1 \left(\mathrm{adj}_s \begin{pmatrix} A^2 \\ \vdots \\ A^m \end{pmatrix} \right)_\alpha^i .$$

If h is rank one affine, i.e. $h(A + a \otimes b) = h(A) + \langle \nabla h(A); a \otimes b \rangle$, then there exist $\delta_\beta^j \in \mathbf{R}$, $1 \leq \beta \leq \binom{n}{s+1}$, $1 \leq j \leq \binom{m}{s+1}$ such that

$$h(A) = \sum_{\beta=1}^{\binom{n}{s+1}} \sum_{j=1}^{\binom{m}{s+1}} \delta_\beta^j \left(\mathrm{adj}_{s+1} \begin{pmatrix} A^1 \\ \vdots \\ A^m \end{pmatrix} \right)_\beta^j = \langle \delta; \mathrm{adj}_{s+1} A \rangle \ .$$

PROOF.

Part 1: As usual we start with the case $m = 2$, therefore $s = 1$ and

$$h(A) = \sum_{l=1}^{n} \sum_{\alpha=1}^{n} \gamma_{l\alpha} A_l^1 A_\alpha^2 \ .$$

Since h is rank one affine and quadratic then

$$\frac{d^2}{dt^2} h(A + t\, a \otimes b) = h(a \otimes b) = \sum_{l,\alpha=1}^{n} \gamma_{l\alpha}\, a^1\, a^2\, b_l\, b_\alpha = 0 \ ,$$

for every $a \in \mathbf{R}^2$, $b \in \mathbf{R}^n$. We therefore immediately deduce that $\gamma_{l\alpha} = -\gamma_{\alpha l}$ and hence

$$h(A) = \sum_{1 \leq l < \alpha \leq n} \gamma_{l\alpha} \left(A_l^1 A_\alpha^2 - A_\alpha^1 A_l^2 \right) = \sum_{1 \leq l < n \leq \alpha} \gamma_{l\alpha} \det \begin{pmatrix} A_l^1 & A_\alpha^1 \\ A_l^2 & A_\alpha^2 \end{pmatrix}$$

$$= \sum_{\beta=1}^{\binom{n}{2}} \delta_\beta (\mathrm{adj}_2 A)_\beta = \langle \delta; \mathrm{adj}_2 A \rangle \ ,$$

since $\mathrm{adj}_2 A$ is a vector of $\mathbf{R}^{\binom{n}{2}}$ composed of elements of the form $\det(A_l, A_\alpha)$, $1 \leq l < \alpha \leq n$ and therefore δ_β is essentially $\gamma_{l\alpha}$ with the appropriate sign.

Part 2: We now proceed with the general case. In fact we shall show a stronger version, we shall show that for every i, $1 \leq i \leq \binom{m-1}{s}$ there exists j, $1 \leq j \leq \binom{m}{s+1}$, and $\delta_\beta^j \in \mathbf{R}$, such that if

$$\tilde{h}(A) = \sum_{l=1}^{n} \sum_{\alpha=1}^{\binom{n}{s}} \gamma_{l\alpha}^i A_l^1 \left(\mathrm{adj}_s \begin{pmatrix} A^2 \\ \vdots \\ A^m \end{pmatrix} \right)_\alpha^i \tag{1}$$

is rank one affine, then

$$\tilde{h}(A) = \sum_{\beta=1}^{\binom{n}{s+1}} \delta_\beta^j \left(\mathrm{adj}_{s+1} \left(\begin{pmatrix} A^1 \\ \vdots \\ A^m \end{pmatrix} \right) \right)_\beta^j . \tag{2}$$

It is clear that (2) implies the lemma. Without loss of generality and only for notational convenience, we show the above result when $i = \binom{m-1}{s}$, then

$$\left(\mathrm{adj}_s \left(\begin{pmatrix} A^2 \\ \vdots \\ A^m \end{pmatrix} \right) \right)_\alpha^i = (-1)^{i+1} \left(\mathrm{adj}_s \left(\begin{pmatrix} A^2 \\ \vdots \\ A^{s+1} \end{pmatrix} \right) \right)_\alpha \quad 1 \le \alpha \le \binom{n}{s} .$$

We also write, for reason of simplicity, $\gamma_{l\alpha}^i = (-1)^{i+1}\gamma_{l\alpha}$ in this case. We have therefore to show that if

$$h_1(A) = \sum_{l=1}^{n} \sum_{\alpha=1}^{\binom{n}{s}} \gamma_{l\alpha} A_l^1 \left(\mathrm{adj}_s \left(\begin{pmatrix} A^2 \\ \vdots \\ A^{s+1} \end{pmatrix} \right) \right)_\alpha \tag{3}$$

is rank one affine then there exist $\delta_\beta \in \mathbb{R}$, $1 \le \beta \le \binom{n}{s+1}$, such that

$$h_1(A) = \sum_{\beta=1}^{\binom{n}{s+1}} \delta_\beta \left(\mathrm{adj}_{s+1} \left(\begin{pmatrix} A^1 \\ A^2 \\ \vdots \\ A^{s+1} \end{pmatrix} \right) \right)_\beta . \tag{4}$$

Recall that for given α, $1 \le \alpha \le \binom{n}{s}$, there exists a unique s-tuple $(\lambda_1, \lambda_2, \ldots, \lambda_s)$ with $1 \le \lambda_1 < \lambda_2 < \ldots < \lambda_s \le n$, such that

$$\left(\mathrm{adj}_s \left(\begin{pmatrix} A^2 \\ \vdots \\ A^{s+1} \end{pmatrix} \right) \right)_\alpha = (-1)^{\alpha+1} \det \begin{pmatrix} A_{\lambda_1}^2 & \cdots & A_{\lambda_s}^2 \\ \vdots & & \vdots \\ A_{\lambda_1}^{s+1} & \cdots & A_{\lambda_s}^{s+1} \end{pmatrix} . \tag{5}$$

We now fix an arbitrary $(s+1)$-tuple $(\lambda_1, \ldots, \lambda_{s+1})$, where $1 \le \lambda_1 < \lambda_2 < \ldots < \lambda_{s+1} \le n$ and we denote by β the associate integer (as in (5)); note that there are $\binom{n}{s+1}$ such $(s+1)$-tuple. Denote by α_1 the integer corresponding (as in (5)) to the s-tuple $(\lambda_1, \ldots, \lambda_s)$, by α_k the integer corresponding to the s-tuple $(\lambda_1, \ldots, \lambda_{k-1}, \lambda_{k+1}, \ldots, \lambda_{s+1})$, $2 \le k \le s$ and by α_{s+1} the integer corresponding to the s-tuple $(\lambda_2, \ldots, \lambda_{s+1})$. Finally let

$$
\begin{aligned}
X_\beta(A) =& \sum_{l_1=1}^{n} (-1)^{\alpha_1+1} \gamma_{l_1\alpha_1} A_{l_1}^1 \det
\begin{pmatrix}
A_{\lambda_1}^2 & \cdots & A_{\lambda_s}^2 \\
\vdots & & \vdots \\
A_{\lambda_1}^{s+1} & \cdots & A_{\lambda_s}^{s+1}
\end{pmatrix} \\[2mm]
&+ \sum_{k=2}^{s} \sum_{l_k=1}^{n} (-1)^{\alpha_k+1} \gamma_{l_k\alpha_k} A_{l_k}^1 \det
\begin{pmatrix}
A_{\lambda_1}^2 \cdots A_{\lambda_{k-1}}^2 & A_{\lambda_{k+1}}^2 \cdots A_{\lambda_{s+1}}^2 \\
\vdots & \vdots & \vdots & \vdots \\
A_{\lambda_1}^{s+1} \cdots A_{\lambda_{k-1}}^{s+1} & A_{\lambda_{k+1}}^{s+1} \cdots A_{\lambda_{s+1}}^{s+1}
\end{pmatrix} \\[2mm]
&+ \sum_{l_{s+1}=1}^{n} (-1)^{\alpha_{s+1}+1} \gamma_{l_{s+1}\alpha_{s+1}} A_{l_{s+1}}^1 \det
\begin{pmatrix}
A_{\lambda_2}^2 & \cdots & A_{\lambda_{s+1}}^2 \\
\vdots & & \vdots \\
A_{\lambda_2}^{s+1} & \cdots & A_{\lambda_{s+1}}^{s+1}
\end{pmatrix} . \quad (6)
\end{aligned}
$$

We then obviously have that

$$
h_1(A) = \sum_{\beta=1}^{\binom{n}{s+1}} X_\beta(A) . \tag{7}
$$

Since h_1 is rank one affine, then so is X_β. Therefore in order to show (4) it is then sufficient to find $\delta_\beta \in \mathbb{R}$, $1 \le \beta \le \binom{n}{s+1}$ such that

$$
X_\beta(A) = \delta_\beta \det
\begin{pmatrix}
A_{\lambda_1}^1 & \cdots & A_{\lambda_{s+1}}^1 \\
\vdots & & \vdots \\
A_{\lambda_1}^{s+1} & \cdots & A_{\lambda_{s+1}}^{s+1}
\end{pmatrix} . \tag{8}
$$

Since $X_\beta(A + t\, a \otimes b)$ is affine in t for every $A \in \mathbb{R}^{nm}$, $a \in \mathbb{R}^m$, $b \in \mathbb{R}^n$, we must have, choosing $b_l = 0$ if $l \ne \lambda_1, \ldots, \lambda_{s+1}$,

$$
\left\{
\begin{aligned}
& a^1 a^2 \det
\begin{pmatrix}
A_{\lambda_1}^3 & \cdots & A_{\lambda_{k-1}}^3 & A_{\lambda_{k+1}}^3 & \cdots & A_{\lambda_s}^3 \\
\vdots & & \vdots & \vdots & & \vdots \\
A_{\lambda_1}^{s+1} & \cdots & A_{\lambda_{k-1}}^{s+1} & A_{\lambda_{k+1}}^{s+1} & \cdots & A_{\lambda_s}^{s+1}
\end{pmatrix} \\
& \quad \times \Big[(-1)^{s+1} (-1)^{\alpha_k+1} \gamma_{l_k\alpha_k}\, b_{l_k}\, b_{\lambda_{s+1}} \\
& \qquad\qquad + (-1)^{k+1} (-1)^{\alpha_1+1} \gamma_{l_1\alpha_1}\, b_{l_1}\, b_{\lambda_k} \Big] = 0 \quad 2 \le k \le s \quad (9) \\[2mm]
& a^1 a^2 \det
\begin{pmatrix}
A_{\lambda_2}^3 & \cdots & A_{\lambda_s}^3 \\
\vdots & & \vdots \\
A_{\lambda_2}^{s+1} & \cdots & A_{\lambda_s}^{s+1}
\end{pmatrix}
\Big[(-1)^{\alpha_1+1} \gamma_{l_1\alpha_1}\, b_{l_1}\, b_{\lambda_1} \\
& \qquad\qquad + (-1)^{s+1} (-1)^{\alpha_{s+1}+1} \gamma_{l_{s+1}\alpha_{s+1}}\, b_{l_{s+1}}\, b_{\lambda_{s+1}} \Big] = 0 .
\end{aligned}
\right.
$$

Since the above identities must hold for every A, a, b we deduce that

$$\begin{cases} l_1 = \lambda_{s+1}, \; l_2 = \lambda_2, \ldots, l_s = \lambda_s, \; l_{s+1} = \lambda_1 \\ (-1)^s (-1)^{\alpha_k+1} \gamma_{\lambda_k \alpha_k} = (-1)^{k+1} (-1)^{\alpha_1+1} \gamma_{\lambda_{s+1}\alpha_1}, \; 2 \le k \le s \\ (-1)^s (-1)^{\alpha_{s+1}+1} \gamma_{\lambda_1 \alpha_{s+1}} = (-1)^{\alpha_1+1} \gamma_{\lambda_{s+1}\alpha_1} \end{cases} . \quad (10)$$

Combining (6) and (10) we have

$$X_\beta(A) = (-1)^{\alpha_1+1} \gamma_{\lambda_{s+1}\alpha_1} A^1_{\lambda_{s+1}} \det \begin{pmatrix} A^2_{\lambda_1} & \cdots & A^2_{\lambda_s} \\ \vdots & & \vdots \\ A^{s+1}_{\lambda_1} & \cdots & A^{s+1}_{\lambda_s} \end{pmatrix}$$

$$+ \sum_{k=2}^{s} (-1)^{\alpha_1+1} (-1)^{s+k+1} \gamma_{\lambda_{s+1}\alpha_1} A^1_{\lambda_k}$$

$$\times \det \begin{pmatrix} A^2_{\lambda_1} & \cdots & A^2_{\lambda_{k-1}} & A^2_{\lambda_{k+1}} & \cdots & A^2_{\lambda_{s+1}} \\ \vdots & & \vdots & \vdots & & \vdots \\ A^{s+1}_{\lambda_1} & \cdots & A^{s+1}_{\lambda_{k-1}} & A^{s+1}_{\lambda_{k+1}} & \cdots & A^{s+1}_{\lambda_{s+1}} \end{pmatrix}$$

$$+ (-1)^s (-1)^{\alpha_1+1} \gamma_{\lambda_{s+1}\alpha_1} A^1_{\lambda_1} \det \begin{pmatrix} A^2_{\lambda_2} & \cdots & A^2_{\lambda_{s+1}} \\ \vdots & & \vdots \\ A^{s+1}_{\lambda_2} & \cdots & A^{s+1}_{\lambda_{s+1}} \end{pmatrix} . (11)$$

Letting, in (11),

$$\delta_\beta = (-1)^{s+\alpha_1+1} \gamma_{\lambda_{s+1}\alpha_1}$$

we have indeed obtained (8). This completes the proof of the lemma.   □

4.1.2.2 Quadratic forms

We now turn our attention to the case where f is quadratic. This case is of particular interest since the associated Euler equations are linear. It has therefore received much attention. Let us first mention the theorem.

Theorem 1.7. *Let M be a symmetric matrix in $\mathbf{R}^{nm \cdot nm}$. Let*

$$f(A) = \langle MA; A \rangle$$

where $A \in \mathbf{R}^{nm}$ and $\langle \bullet; \bullet \rangle$ denotes the scalar product in $\mathbf{R}^{nm}$. Then

i) f is rank one convex if and only if f is quasiconvex,
ii) if $m = 2$ or $n = 2$, then

$$f \text{ polyconvex} \Leftrightarrow f \text{ quasiconvex} \Leftrightarrow f \text{ rank one convex}$$

iii) if $m, n \geq 3$, then in general

$$f \text{ rank one convex } \not\Rightarrow f \text{ polyconvex .}$$

REMARKS.

i) The first proof of i) of Theorem 1.7 was given by Van Hove [1,2], although it was implicitly known earlier. In a different setting i) was also proved by Murat [1,2], Murat and Tartar [1], Tartar [2];

ii) the second part of the theorem has received considerable attention. The question was raised in 1937 by Bliss and it received a progressive answer through the works of Albert [1], Reid [1], Mac Shane [2], Mac Shane-Hestenes [1], Terpstra [1], Serre [1], Marcellini [4]. The proof of ii) of Theorem 1.7 relies on an algebraic lemma whose importance is summarized in Uhlig [1];

iii) a counterexample to the third part of the theorem was produced by Terpstra [1] and later by Serre [1] (cf. also Ball [4]);

iv) we shall see, in Propositions 1.13 and 1.14, below, that the following holds

$$f \text{ rank one convex } \not\Rightarrow f \text{ polyconvex}$$

even if $m = n = 2$. Obviously, from Theorem 1.7, one sees that such an f cannot be quadratic;

v) note also that even if $m = n = 2$ and f is quadratic, then in general f polyconvex $\not\Rightarrow f$ convex as the trivial example $f(\xi) = \det \xi$ shows it.

Before proceeding with the proof of the theorem we mention two simple facts, the first one is

Lemma 1.8. *Let M be a symmetric matrix in $\mathbb{R}^{nm \cdot nm}$ and let*

$$f(A) = \langle MA; A \rangle$$

then

i) f is convex if and only if

$$f(A) \geq 0$$

for every $A \in \mathbb{R}^{nm}$,

ii) f is polyconvex if and only if there exists $\alpha \in \mathbb{R}^{\sigma(2)}$ such that

$$f(A) \geq \langle \alpha; \mathrm{adj}_2 A \rangle$$

for every $A \in \mathbb{R}^{nm}$ and where $\langle \bullet; \bullet \rangle$ denotes the scalar product in $\mathbb{R}^{\sigma(2)}$ and $\sigma(2) = \binom{m}{2}\binom{n}{2}$, *iii)* *f is quasiconvex if and only if*

$$\int_D f(\nabla\varphi(x))\,dx \geq 0$$

for every $D \subset \mathbf{R}^n$ bounded domain and for every $\varphi \in W_0^{1,\infty}(D;\mathbf{R}^m)$,
iv) *f is rank one convex if and only if*

$$f(a \otimes b) \geq 0$$

for every $a \in \mathbf{R}^m$, $b \in \mathbf{R}^n$.

PROOF of Lemma 1.8. i), iii) and iv) are trivial. The fact that

$$f(A) \geq \langle \alpha; \mathrm{adj}_2 A\rangle \tag{1}$$

implies that f is polyconvex follows immediately from the following observation.

Let

$$g(A) = f(A) - \langle \alpha; \mathrm{adj}_2 A\rangle$$

then by (1) and Part i) of the lemma, we deduce that g is convex. Thus $f(A) = g(A) + \langle \alpha; \mathrm{adj}_2 A\rangle$ is polyconvex.

Assume now that f is polyconvex. We wish to show that (1) holds for some $\alpha \in \mathbf{R}^{\sigma(2)}$. Using Theorem 1.3, bearing in mind that $f(0) = 0$, we find that there exists $\beta = (\beta_{\sigma(1)}, \beta_{\sigma(2)}, \ldots, \beta_{\sigma(n\wedge m)}) \in \mathbf{R}^{\tau(n,m)}$ such that

$$f(A) \geq \langle \beta; T(A)\rangle = \sum_{s=1}^{n\wedge m} \langle \beta_{\sigma(s)}; \mathrm{adj}_s A\rangle \ . \tag{2}$$

Multiplying A by $\varepsilon > 0$, we get

$$f(\varepsilon A) = \varepsilon^2 f(A) \geq \varepsilon \langle \beta_{\sigma(1)}; A\rangle + \varepsilon^2 \langle \beta_{\sigma(2)}; \mathrm{adj}_2 A\rangle + O(\varepsilon^3) \ . \tag{3}$$

Dividing by ε and letting $\varepsilon \to 0$, we obtain

$$\langle \beta_{\sigma(1)}; A\rangle \leq 0$$

for every $A \in \mathbf{R}^{nm}$, thus $\beta_{\sigma(1)} = 0$. Returning to (3), dividing by ε^2 and letting $\varepsilon \to 0$ we have indeed obtained (1) with $\alpha = \beta_{\sigma(2)}$. $\qquad\square$

The second important point that we wish to mention is the following

Lemma 1.9. *Let $\Omega \subset \mathbb{R}^n$ be a bounded open set. Let $\varphi \in C_0^\infty(\Omega; \mathbb{R}^m)$. Define for $\xi \in \mathbb{R}^n$*

$$\hat{\varphi}_\alpha(\xi) = \frac{1}{2\pi i} \int_{\mathbb{R}^n} \varphi_\alpha(x) e^{-2\pi i \xi \cdot x}\, dx, \quad 1 \le \alpha \le m \ .$$

Then

$$\widehat{\nabla\varphi} = 2\pi i (\hat{\varphi}_\alpha \xi_j)_{1 \le j \le n,\, 1 \le \alpha \le m} = 2\pi i\, \hat{\varphi} \otimes \xi \ ,$$

i.e. $\mathrm{rank}\{\mathrm{Re}(\widehat{\nabla\varphi})\}$, $\mathrm{rank}\{\mathrm{Im}(\widehat{\nabla\varphi})\} \le 1$.

PROOF. The proof follows immediately if an integration by part is performed. $\qquad\square$

REMARK. Lemma 1.9 explains in another way than that of Theorem 1.1 why matrices of rank one play such an important role in the theory developed in this chapter.

We now proceed with the proof of the theorem.

PROOF of Theorem 1.7.

i) Recall that

$$f(A) = \langle MA; A \rangle \ .$$

Theorem 1.1 implies that if f is quasiconvex then f is rank one convex. We now prove the converse. By Lemma 1.8 we have to show that

$$\int_\Omega \langle M\nabla\varphi(x); \nabla\varphi(x) \rangle\, dx \ge 0 \tag{1}$$

for every $\varphi \in C_0^\infty(\Omega; \mathbb{R}^m)$, provided

$$f(a \otimes b) = \langle M\, a \otimes b;\, a \otimes b \rangle \ge 0 \ . \tag{2}$$

We now extend f from $\mathbb{R}^{nm}$ to $\mathbb{C}^{nm}$ in the following manner

$$\tilde{f}(A) = \langle MA; \bar{A} \rangle \ , \tag{3}$$

where $\bar{A}$ denotes the complex conjugate of A. We then use *Plancherel formula* to get

$$\int_\Omega \langle M\nabla\varphi(x); \nabla\varphi(x) \rangle\, dx = \int_{\mathbb{R}^n} \langle M\nabla\varphi(x); \nabla\varphi(x) \rangle\, dx$$

$$= \int_{\mathbb{R}^n} \langle M\widehat{\nabla\varphi}(\xi); \overline{\widehat{\nabla\varphi}(\xi)} \rangle d\xi$$

$$= \int_{\mathbb{R}^n} \mathrm{Re}\langle M\widehat{\nabla\varphi}(\xi); \overline{\widehat{\nabla\varphi}(\xi)} \rangle d\xi \ . \tag{4}$$

Using Lemma 1.9 and (2) in (4), we obtain (1).

ii) We do not prove this result and we refer to the above mentioned proofs (cf. the remark above).

iii) We now want to show that if $m = n = 3$, then there exists f rank one convex which is not polyconvex. We give here an example due to Serre [1], (cf. also Ball [4]). Let

$$A = \begin{pmatrix} A_1^1 & A_2^1 & A_3^1 \\ A_1^2 & A_2^2 & A_3^2 \\ A_1^3 & A_2^3 & A_3^3 \end{pmatrix}$$

and let

$$f(A) = \left(A_1^1 - A_2^3 - A_3^2\right)^2 + \left(A_2^1 - A_1^3 + A_3^1\right)^2$$
$$+ \left(A_1^2 - A_1^3 - A_3^1\right)^2 + \left(A_2^2\right)^2 + \left(A_3^3\right)^2 . \tag{5}$$

We divide the proof into two steps.

Step 1: We first show that there exists $\varepsilon > 0$ such that

$$f(a \otimes b) - \varepsilon |a \otimes b|^2 \geq 0 \tag{6}$$

for every $a, b \in \mathbf{R}^3$ and where $|A|^2 \equiv \langle A; A \rangle$ denotes the usual norm. Lemma 1.8 will then ensure that

$$g(A) = f(A) - \varepsilon |A|^2 \tag{7}$$

is rank one convex. In Step 2 we then prove that this g is not polyconvex and this will end the proof of the theorem. We now let

$$\varepsilon_0 = \inf\{f(a \otimes b) : a, b \in \mathbf{R}^3, |a \otimes b| = 1\} . \tag{8}$$

Then, since $f \geq 0$ (cf. (5)), we have $\varepsilon_0 \geq 0$. In order to prove (6) it is sufficient to prove that $\varepsilon_0 > 0$. We proceed by contradiction and assume that $\varepsilon_0 = 0$. Observe that in (8) the minimum is trivially attained and therefore there exist $a, b \in \mathbf{R}^3$ such that

$$\begin{cases} f(a \otimes b) = \varepsilon_0 = 0 \\ |a \otimes b| = 1 \end{cases} . \tag{9}$$

Recall that

$$a \otimes b = \begin{pmatrix} a^1 b_1 & a^1 b_2 & a^1 b_3 \\ a^2 b_1 & a^2 b_2 & a^2 b_3 \\ a^3 b_1 & a^3 b_2 & a^3 b_3 \end{pmatrix} ,$$

therefore the first equation of (9) becomes

$$\begin{cases} a^1\, b_1 \;=\; a^2\, b_3 + a^3\, b_2 \\ a^1\, b_2 \;=\; a^3\, b_1 - a^1\, b_3 \\ a^2\, b_1 \;=\; a^1\, b_3 + a^3\, b_1 \\ a^2\, b_2 \;=\; 0 \\ a^3\, b_3 \;=\; 0 \end{cases} \tag{10}$$

We then show that (10) is in contradiction with the fact that $|a \otimes b| = 1$. To do so, we examine carefully (10) and separate the discussion in several cases.

Case 1: $a^2 = a^3 = 0$ (cf. the two last equations of (10)), then (10) becomes

$$\begin{cases} a^2 \;=\; a^3 \;=\; 0 \\ a^1\, b_1 \;=\; 0 \\ a^1\, b_2 \;=\; -a^1\, b_3 \\ a^2\, b_1 \;=\; a^1\, b_3 \end{cases} \tag{11}$$

Case 1a: $a^1 = 0$, therefore $a^1 = a^2 = a^3 = 0$ and hence $|a \otimes b| = 0$, contradiction.

Case 1b: $b_1 = 0$, hence from (11), $a^1\, b_3 = 0$ and thus $a^1\, b_2 = 0$. We then also conclude that $|a \otimes b| = 0$ and this is a contradiction.

Case 2: $a^2 = b_3 = 0$ (cf. the two last equations of (10)), then (10) becomes

$$\begin{cases} a^2 \;=\; b_3 \;=\; 0 \\ a^1\, b_1 \;=\; a^3\, b_2 \\ a^1\, b_2 \;=\; a^3\, b_1 \\ a^3\, b_1 \;=\; 0 \end{cases} \tag{12}$$

Case 2a: $a^3 = 0$, then $a^1\, b_1 = a^1\, b_2 = 0$ and therefore $|a \otimes b| = 0$, contradiction.

Case 2b: $b_1 = 0$, then $a^3\, b_2 = a^1\, b_2 = 0$ and therefore $|a \otimes b| = 0$, contradiction.

Similarly for the case $a^3 = b_2 = 0$ and $b_2 = b_3 = 0$. Thus $\varepsilon_0 > 0$ and hence Step 1, i.e. g defined by (7), is rank one convex for every $\varepsilon \leq \varepsilon_0$.

Step 2: We now show that g is not polyconvex. In view of Lemma 1.8 it is sufficient to show that for every $\alpha \in \mathbf{R}^9$, there exists $A \in \mathbf{R}^9$ such that

$$g(A) + \langle \alpha; \operatorname{adj}_2 A \rangle < 0 \ . \tag{13}$$

We shall show that (13) holds for matrices A of the following form

$$A = \begin{pmatrix} b+d & c-a & a \\ c+a & 0 & b \\ c & d & 0 \end{pmatrix} . \tag{14}$$

For such matrices we have $f(A) = 0$ and therefore

$$\begin{aligned} g(A) = -\varepsilon|A|^2 = -\varepsilon[(b+d)^2 + (c-a)^2 + a^2 \\ + (c+a)^2 + b^2 + c^2 + d^2] , \end{aligned} \tag{15}$$

and

$$\operatorname{adj}_2 A = \begin{pmatrix} -bd & bc & cd+ad \\ ad & -ac & -(bd+d^2-c^2+ac) \\ bc-ab & ac+a^2-b^2-bd & a^2-c^2 \end{pmatrix} .$$

Therefore

$$\begin{aligned} \langle \alpha; \operatorname{adj}_2 A \rangle = {} & -\alpha_1 bd + \alpha_2 bc + \alpha_3(cd+ad) \\ & + \alpha_4 ad - \alpha_5 ac - \alpha_6(bd+d^2-c^2+ac) \\ & + \alpha_7(bc-ab) + \alpha_8(ac+a^2-b^2-bd) + \alpha_9(a^2-c^2) . \end{aligned}$$

As in Step 1 we consider several cases:
Case 1: $\alpha_8 > 0$, then take $a = c = d = 0$ and $b \neq 0$, to get

$$\begin{aligned} g(A) + \langle \alpha; \operatorname{adj}_2 A \rangle = {} & -\varepsilon|A|^2 + \langle \alpha; \operatorname{adj}_2 A \rangle \\ = {} & -\varepsilon(2b^2) - \alpha_8 b^2 < 0 . \end{aligned}$$

Case 2: $\alpha_6 > 0$, then take $a = b = c = 0$ and $d \neq 0$, to get

$$g(A) + \langle \alpha; \operatorname{adj}_2 A \rangle = -\varepsilon(2d^2) - \alpha_6 d^2 < 0 .$$

We therefore can assume that $\alpha_8 \leq 0$ and $\alpha_6 \leq 0$.
Case 3: $\alpha_9 - \alpha_6 > 0$ ($\alpha_8 \leq 0$, $\alpha_6 \leq 0$), then take $a = b = d = 0$ and $c \neq 0$ to get

$$g(A) + \langle \alpha; \operatorname{adj}_2 A \rangle = -\varepsilon(3c^2) + (\alpha_6 - \alpha_9)c^2 < 0 .$$

We therefore assume $\alpha_8 \leq 0$, $\alpha_6 \leq 0$ and $\alpha_9 - \alpha_6 \leq 0$. From these three inequalities we deduce that $\alpha_8 + \alpha_9 \leq 0$, and then taking

$b = c = d = 0$ and $a \neq 0$, we get

$$g(A) + \langle \alpha; \mathrm{adj}_2 A \rangle = -\varepsilon(3a^2) + (\alpha_8 + \alpha_9)a^2 < 0 \; .$$

And this concludes the proof of the theorem. $\qquad\square$

4.1.2.3 Some More Examples

We now give some more examples (for the history of this theorem see the proof below).

Theorem 1.10. *Let* $f : \mathbb{R}^{nm} \to \mathbb{R}$.

i) Let $\Phi : \mathbb{R}^{nm} \to \mathbb{R}$ *be quasiaffine and* $g : \mathbb{R} \to \mathbb{R}$ *be such that*

$$f(A) = g(\Phi(A))$$

(in particular if $m = n$, *one can take* $\Phi(A) = \det A$), *then*

$$f \text{ polyconvex} \Leftrightarrow f \text{ quasiconvex} \Leftrightarrow f \text{ rank one convex} \Leftrightarrow g \text{ convex} \; .$$

ii) Minimal surfaces: let $m = n + 1$. *Recall that if* $A \in \mathbb{R}^{n(n+1)}$, *then*

$$\mathrm{adj}_n A = \left(\det \hat{A}^1, -\det \hat{A}^2, \ldots, (-1)^{k+1} \det \hat{A}^k, \ldots, (-1)^{n+2} \det \hat{A}^{n+1} \right)$$

where $\hat{A}^k$ *is the* $n \times n$ *matrix obtained from* A *by suppressing the* k-*th line. Let* $g : \mathbb{R}^{n+1} \to \mathbb{R}$ *be such that*

$$f(A) = g(\mathrm{adj}_n A)$$

then

$$f \text{ polyconvex} \Leftrightarrow f \text{ quasiconvex} \Leftrightarrow f \text{ rank one convex} \Leftrightarrow g \text{ convex} \; .$$

iii) For $A \in \mathbb{R}^{nm}$, *let* $|A|$ *denote the Euclidean norm, i.e.*

$$|A| = \left(\sum_{\alpha=1}^{n} \sum_{i=1}^{m} \left(A_\alpha^i \right)^2 \right)^{1/2} \; .$$

Let $g : \mathbb{R}_+ \to \mathbb{R}$ *be such that*

$$f(A) = g(|A|)$$

then

$$f \text{ convex} \Leftrightarrow f \text{ polyconvex} \Leftrightarrow f \text{ quasiconvex} \Leftrightarrow f \text{ rank one convex}$$

$$\Leftrightarrow g \text{ convex } and \; g(0) = \inf\{g(x) : x \geq 0\} \; .$$

iv) Let $m = n$, $1 \leq \alpha < 2n$, $h : \mathbf{R} \to \mathbf{R}$ be such that

$$f(A) = |A|^{\alpha} + h(\det A)$$

where $|A|$ denotes the Euclidean norm. Then

$$f \text{ polyconvex} \Leftrightarrow f \text{ quasiconvex} \Leftrightarrow f \text{ rank one convex} \Leftrightarrow h \text{ convex} \; .$$

v) Let $m = n = 2$, $A = \begin{pmatrix} A_1^1 & A_2^1 \\ A_1^2 & A_2^2 \end{pmatrix}$, $g_1 : \mathbf{R}^2 \to \mathbf{R}$, $g_2, h : \mathbf{R} \to \mathbf{R}$ be such that

$$f(A) = g_1\left(A_1^1, A_2^1\right) + g_2\left(A_1^2\right) + h(\det A) \; .$$

Then

$$f \text{ polyconvex} \Leftrightarrow f \text{ quasiconvex}$$

$$\Leftrightarrow f \text{ rank one convex} \Leftrightarrow g_1, g_2 \text{ and } h \text{ convex} \; .$$

vi) Let $m = n$, $p > 0$, $1 \leq s \leq n - 1$ and

$$f(A) = \begin{cases} \left(\dfrac{|\mathrm{adj}_s A|^{n/s}}{\det A}\right)^p & if \det A > 0 \\ +\infty & otherwise \end{cases}$$

where $|\bullet|$ denotes the Euclidean norm. Then

$$f \text{ polyconvex} \Leftrightarrow f \text{ rank one convex} \Leftrightarrow p \geq \frac{s}{n - s} \; .$$

Remarks.

i) In the minimal surfaces case, i.e. $m = n + 1$, it is clear that if $u : \mathbf{R}^n \to \mathbf{R}^{n+1}$, then $\mathrm{adj}_n \nabla u$ represents the normal to the surface. In the case $n = 2$, $u(x_1, x_2) = (u_1, u_2, u_3)$ we have

$$\mathrm{adj}_2 \nabla u = \left(\frac{\partial u_2}{\partial x_1}\frac{\partial u_3}{\partial x_2} - \frac{\partial u_2}{\partial x_2}\frac{\partial u_3}{\partial x_1}; \; \frac{\partial u_3}{\partial x_1}\frac{\partial u_1}{\partial x_2} - \frac{\partial u_1}{\partial x_1}\frac{\partial u_3}{\partial x_2}; \right.$$

$$\left. \frac{\partial u_1}{\partial x_1}\frac{\partial u_2}{\partial x_2} - \frac{\partial u_1}{\partial x_2}\frac{\partial u_2}{\partial x_1} \right) \; ;$$

ii) in the Case iv) of the theorem, the hypothesis $\alpha < 2n$ cannot be dropped. Indeed if $n = 2$ and $\alpha = 4$, then $f(A) = |A|^4 - 2(\det A)^2$ is even convex;

iii) the Case vi) is interesting in *elasticity* for *slightly compressible materials* (see the appendix);

iv) the are also results in the case $m = n$, interesting in elasticity, involving the eigenvalues $v_1, \ldots, v_n$ of the matrix $(A^t A)^{1/2}$ where A^t denotes the transpose of the matrix A. Let

$$V = \begin{pmatrix} v_1 & & 0 \\ & \ddots & \\ 0 & & v_n \end{pmatrix}$$

and f be *isotropic*, i.e.

$$f(A) = \psi(V, \mathrm{adj}_2 V, \ldots, \mathrm{adj}_{n-1} V, \det V) \ ,$$

then these results relate the polyconvexity (or rank one convexity) of ψ with that of f. For more details see Ball [2,3], Hill [1], Aubert-Tahraoui [3], Aubert [1], Knowles-Sternberg [1] and the appendix.

Before proceeding with the proof of the theorem, we mention an algebraic lemma which will be used for the minimal surfaces case. This lemma is stronger than needed but it will be fully used in Chapter 5. We shall prove it after the proof of the Theorem 1.10.

Lemma 1.11. *Let $0 < \lambda < 1$, $b, c \in \mathbb{R}^{n+1}$ and $A \in \mathbb{R}^{n(n+1)}$ be such that*

$$\mathrm{adj}_n A = \lambda b + (1 - \lambda)c \neq 0 \ .$$

Then there exist $B, C \in \mathbb{R}^{n(n+1)}$ such that

$$\begin{cases} A = \lambda B + (1 - \lambda)C \\ \mathrm{adj}_n B = b, \ \mathrm{adj}_n C = c \\ \mathrm{rank}\{B - C\} \leq 1 \end{cases} \ .$$

PROOF of Theorem 1.10.

i) This part of the theorem has been established in Dacorogna [4]. Let $\Phi : \mathbb{R}^{nm} \to \mathbb{R}$ be quasiaffine and

$$f(A) = g(\Phi(A)) \ .$$

The implications g convex $\Rightarrow f$ polyconvex $\Rightarrow f$ quasiconvex $\Rightarrow f$ rank one convex follow immediately from Theorem 1.1, therefore it remains to show that if f is rank one convex then g is convex. We may assume without loss of generality that Φ is not constant, otherwise the result is trivial. We want to show that for $\lambda \in (0,1)$, $\alpha, \beta \in \mathbb{R}$

$$g(\lambda \alpha + (1 - \lambda)\beta) \leq \lambda g(\alpha) + (1 - \lambda)g(\beta) \tag{1}$$

provided f is rank one convex. Following Theorem 1.5 we have that

$$\Phi(A) = a_0 + \langle a; T(A) \rangle$$

$$= a_0 + \langle a_1; A \rangle + \sum_{j=2}^{n \wedge m} \langle a_j; \mathrm{adj}_j A \rangle \tag{2}$$

where $a_0 \in \mathbb{R}$, $a_1 \in \mathbb{R}^{nm}$ and $a_j \in \mathbb{R}^{\sigma(j)}$ where $\sigma(j) = \binom{m}{j}\binom{n}{j}$. Since $\Phi(A) \not\equiv a_0$, then at least one of the a_j, $1 \leq j \leq n \wedge m$ is not zero. Let s be such that $a_s \neq 0$ but $a_{s-1} = a_{s-2} = \ldots = a_1 = 0$ (if $a_1 \neq 0$, we then take $s = 1$). Since $a_s \neq 0$ $(\in \mathbb{R}^{\sigma(s)})$ we have that at least one of the components of $a_s = (a_s^1, \ldots, a_s^{\sigma(s)})$ is non zero. Without loss of generality we take $a_s^{\sigma(s)} \neq 0$. Then choose $B, C \in \mathbb{R}^{nm}$ in the following way

$$B = \left(\begin{array}{ccc|ccc}
B_1^1 & \cdots & B_s^1 & B_{s+1}^1 & \cdots & B_n^1 \\
\vdots & & \vdots & \vdots & & \vdots \\
B_1^s & \cdots & B_s^s & B_{s+1}^s & \cdots & B_n^s \\
\hline
B_1^{s+1} & \cdots & B_s^{s+1} & B_{s+1}^{s+1} & \cdots & B_n^{s+1} \\
\vdots & & \vdots & \vdots & & \vdots \\
B_1^m & \cdots & B_s^m & B_{s+1}^m & \cdots & B_n^m
\end{array} \right)$$

$$= \left(\begin{array}{cc|c}
\dfrac{\alpha - a_0}{a_s^{\sigma(s)}} & 0 & \\
& & 0 \\
0 & 1 & \\
\hline
& 0 & 0
\end{array} \right)$$

and similarly for C except that we replace the first component by $\frac{\beta - a_0}{a_s^{\sigma(s)}}$. We then immediately have

$$\begin{cases} \Phi(B) = \alpha, \ \Phi(C) = \beta \\ \mathrm{rank}\{B - C\} \leq 1 \end{cases}, \tag{3}$$

since $a_j = 0$ if $j < s$, $\mathrm{adj}_s B = \left(0, \ldots, 0, \frac{\alpha - a_0}{a_s^{\sigma(s)}}\right)$ and $\mathrm{adj}_j B = \mathrm{adj}_j C = 0$ if $j \geq s + 1$. We then return to (1) and use (3), the rank one convexity of f and Theorem 1.5 to get

$$
\begin{aligned}
\lambda g(\alpha) + (1 - \lambda)g(\beta) &= \lambda g(\Phi(B)) + (1 - \lambda)g(\Phi(C)) \\
&= \lambda f(B) + (1 - \lambda)f(C) \\
&\geq f(\lambda B + (1 - \lambda)C) = g(\Phi(\lambda B + (1 - \lambda)C)) \\
&= g(\lambda \alpha + (1 - \lambda)\beta) \ ,
\end{aligned}
$$

which is the desired result.

ii) This part of the theorem has been established by Morrey [1,2]. We give here a slightly different proof based on Dacorogna [2]. Similarly as in the preceding case it is sufficient to show that the rank one convexity of f implies the convexity of g. We therefore want to show that if $\lambda \in (0,1)$, $b, c \in \mathbb{R}^{n+1}$, then

$$
g(\lambda b + (1 - \lambda)c) \leq \lambda g(b) + (1 - \lambda)g(c) \tag{4}
$$

provided f is rank one convex and $f(A) = g(\mathrm{adj}_n A)$. We divide the proof into two cases.

Case 1: $\lambda b + (1 - \lambda)c \neq 0$; we write for simplification $\alpha = \lambda b + (1 - \lambda)c = (\alpha_1, \ldots, \alpha_{n+1}) \in \mathbb{R}^{n+1}$. Since $\alpha \neq 0$, we may assume, without loss of generality, that $\alpha_1 \neq 0$. We then let

$$
A = \begin{pmatrix} A_1^1 & \cdots & A_n^1 \\ A_1^2 & \cdots & A_n^2 \\ \vdots & & \vdots \\ A_1^{n+1} & \cdots & A_n^{n+1} \end{pmatrix} = \begin{pmatrix} -\alpha_2 & -\dfrac{\alpha_3}{\alpha_1} & \cdots & -\dfrac{\alpha_{n+1}}{\alpha_1} \\ \alpha_1 & 0 & \cdots & 0 \\ & 1 & & \\ & & & 0 \\ 0 & & & 1 \end{pmatrix} .
$$

It is then easy to see that

$$
\mathrm{adj}_n A = \alpha = \lambda b + (1 - \lambda)c \neq 0. \tag{6}
$$

We may now apply Lemma 1.11 to get $B, C \in \mathbb{R}^{n(n+1)}$ such that

$$
\begin{cases} A = \lambda B + (1 - \lambda)C \\ \mathrm{adj}_n B = b, \ \mathrm{adj}_n C = c \ . \\ \mathrm{rank}\{B - C\} \leq 1 \end{cases}
$$

Returning to (4), using the rank one convexity of f, we obtain

$$\lambda g(b) + (1 - \lambda)g(c) = \lambda f(B) + (1 - \lambda)f(C)$$
$$\geq f(\lambda B + (1 - \lambda)C) = f(A)$$
$$= g(\mathrm{adj}_n A) = g(\lambda b + (1 - \lambda)c) \, ,$$

which is precisely the result.

Case 2: $\lambda b + (1 - \lambda)c = 0$; observe first that the rank one convexity of f implies that f is continuous (cf. Theorem 1.1), thus from $f(A) = g(\mathrm{adj}_n A)$ we deduce that g is continuous. Therefore using Case 1 for $\beta = b + (\varepsilon, 0, \dots, 0)$ and $\gamma = c + (\varepsilon, 0, \dots, 0)$ where $\varepsilon > 0$ is arbitrary, we deduce (4) by continuity of g.

iii) This result has been established in Dacorogna [7]. Let $A \in \mathbb{R}^{nm}$ and

$$f(A) = g(|A|) \, .$$

In view of Theorem 1.1, it remains to show that 1) f rank one convex $\Rightarrow$ 2) g convex and $g(0) = \inf\{g(x) : x \geq 0\} \Rightarrow$ 3) f convex.

1) $\Rightarrow$ 2): First define g by symmetry, i.e. $g(x) = g(-x)$ if $x \leq 0$. Let $\lambda \in [0, 1]$, $\alpha, \beta \in \mathbb{R}$. We want to show

$$g(\lambda \alpha + (1 - \lambda)\beta) \leq \lambda g(\alpha) + (1 - \lambda)g(\beta) \, . \tag{7}$$

Since g is symmetric we can assume that $\lambda \alpha + (1 - \lambda)\beta \geq 0$. Define $F = (F_j^i)_{1 \leq i \leq m, 1 \leq j \leq n}$ and $G = (G_j^i)$, such that $F_1^1 = \alpha$, $G_1^1 = \beta$ and $F_j^i = G_j^i = 0$ if $(i, j) \neq (1, 1)$. Observing that $\mathrm{rank}\{F - G\} \leq 1$ and using the rank one convexity of f we get

$$g(\lambda \alpha + (1 - \lambda)\beta) = f(\lambda F + (1 - \lambda)G) \leq \lambda f(F) + (1 - \lambda)f(G)$$
$$= \lambda g(|\alpha|) + (1 - \lambda)g(|\beta|) \leq \lambda g(\alpha) + (1 - \lambda)g(\beta) \, ,$$

which is indeed (7). The fact that $g(0) = \inf\{g(x) : x \geq 0\}$ results immediately from the convexity and symmetry of g.

2) $\Rightarrow$ 3): Observe also that the convexity and symmetry of g imply immediately that g is increasing on $\mathbb{R}_+$. We now want to show that g convex $\Rightarrow f$ convex. This is immediate since

$$f(\lambda F + (1 - \lambda)G) = g(|\lambda F + (1 - \lambda)G|) \leq g(\lambda|F| + (1 - \lambda)|G|)$$
$$\leq \lambda g(|F|) + (1 - \lambda)g(|G|)$$
$$= \lambda f(F) + (1 - \lambda)f(G) \, ,$$

and this achieves the proof of the third part of the theorem.

iv) This result has been established by Ball and Murat [1]. Let $n = m$, $A \in \mathbb{R}^{n^2}$, $1 \leq \alpha < 2n$ and

$$f(A) = |A|^\alpha + h(\det A) .$$

It follows from Theorem 1.1 that it remains to prove that f rank one convex $\Rightarrow h$ convex. Let $\lambda \in (0,1)$, $a, b \in \mathbb{R}$, we want to show

$$h(\lambda a + (1 - \lambda)b) \leq \lambda h(a) + (1 - \lambda)h(b) ; \tag{8}$$

there is no loss of generality if we choose $a \neq b$ and $a \neq 0$. Let $e_1 = (1, 0, \ldots, 0) \in \mathbb{R}^n$ and $\varepsilon \neq 0$ with $\varepsilon(b - a) > 0$. Let $v(\varepsilon) = \frac{a\varepsilon}{b-a}$ and let $A(\varepsilon) = (A_j^i(\varepsilon))_{1 \leq i, j \leq n}$ be the diagonal matrix

$$A_1^1(\varepsilon) = v(\varepsilon), \ A_2^2(\varepsilon) = \ldots = A_n^n(\varepsilon) = \left(\frac{a}{v(\varepsilon)} \right)^{1/n-1},$$

$$A_j^i(\varepsilon) = 0 \text{ if } i \neq j .$$

It is then easy to see that

$$\begin{cases} \det A(\varepsilon) = a, \ \det(A(\varepsilon) + \varepsilon \, e_1 \otimes e_1) = b \\ \det(A(\varepsilon) + (1 - \lambda)\varepsilon \, e_1 \otimes e_1) = \lambda a + (1 - \lambda)b \end{cases} .$$

Since f is rank one convex we have

$$f(\lambda A(\varepsilon) + (1 - \lambda)(A(\varepsilon) + \varepsilon \, e_1 \otimes e_1))$$
$$= |A(\varepsilon) + (1 - \lambda)\varepsilon \, e_1 \otimes e_1|^\alpha + h(\lambda a + (1 - \lambda)b)$$
$$\leq \lambda f(A(\varepsilon)) + (1 - \lambda)f(A(\varepsilon) + \varepsilon \, e_1 \otimes e_1)$$
$$\leq \lambda |A(\varepsilon)|^\alpha + (1 - \lambda)|A(\varepsilon) + \varepsilon \, e_1 \otimes e_1|^\alpha$$
$$+ \lambda h(a) + (1 - \lambda)h(b) , \tag{9}$$

Observe that

$$\lambda |A(\varepsilon)|^\alpha + (1 - \lambda)|A(\varepsilon) + \varepsilon(e_1 \otimes e_1)|^\alpha - |A(\varepsilon) + (1 - \lambda)\varepsilon(e_1 \otimes e_1)|^\alpha$$

$$= \lambda \left[(v(\varepsilon))^2 + (n-1) \left(\frac{a}{v(\varepsilon)} \right)^{2/n-1} \right]^{\alpha/2}$$

$$+ (1 - \lambda) \left[(v(\varepsilon) + \varepsilon)^2 + (n-1) \left(\frac{a}{v(\varepsilon)} \right)^{2/n-1} \right]^{\alpha/2}$$

$$- \left[(v(\varepsilon) + (1-\lambda)\varepsilon)^2 + (n-1) \left(\frac{a}{v(\varepsilon)} \right)^{2/n-1} \right]^{\alpha/2} . \qquad (10)$$

It is clear that if $1 \leq \alpha < 2n$, then the right hand side in (10) tends to zero as $\varepsilon \to 0$. Thus combining (9) and (10) we have indeed obtained (8), i.e. that h is convex.

v) Let $m = n = 2$, $A = \begin{pmatrix} A_1^1 & A_2^1 \\ A_1^2 & A_2^2 \end{pmatrix}$ and

$$f(A) = g_1 \left(A_1^1, A_2^1 \right) + g_2 \left(A_1^2 \right) + h(\det A) .$$

As usual, it is sufficient to show that f rank one convex implies g_1, g_2 and h convex. We first show that given $\lambda \in [0,1]$, $(\alpha, \beta), (a, b) \in \mathbb{R}^2$ then

$$g_1(\lambda(\alpha, \beta) + (1 - \lambda)(a, b)) \leq \lambda g_1(\alpha, \beta) + (1 - \lambda)g_1(a, b) . \qquad (11)$$

Choosing

$$A = \begin{pmatrix} \alpha & \beta \\ 0 & 0 \end{pmatrix}, \quad B = \begin{pmatrix} a & b \\ 0 & 0 \end{pmatrix}$$

we have immediately (11) if we use the rank one convexity of f. Similarly one can show the convexity of g_2. We next show the convexity of h, i.e.

$$h(\lambda s + (1 - \lambda)t) \leq \lambda h(s) + (1 - \lambda)h(t) \qquad (12)$$

for every $s, t \in \mathbb{R}$, $\lambda \in [0,1]$. It is sufficient to observe that if $s = t = 0$ then (12) is trivial and if, for example, $s \neq 0$, one chooses

$$A = \begin{pmatrix} s & 0 \\ 0 & 1 \end{pmatrix}, \quad B = \begin{pmatrix} s & 0 \\ 0 & t/s \end{pmatrix} .$$

The rank one convexity of f implies then (12).

vi) The last part of the theorem has been established in Charrier-Dacorogna-Hanouzet-Laborde [1]. We decompose the proof into two steps: 1) $p \geq \frac{s}{n-s}$ $\Rightarrow$ f polyconvex, 2) f rank one convex $\Rightarrow p \geq \frac{s}{n-s}$.

Step 1: $p \geq \frac{s}{n-s} \Rightarrow f$ polyconvex. Define first $h : \mathbb{R}_+ \times \mathbb{R}_+ \to \mathbb{R}$ by

$$h(x, \delta) = x^{np/s} \delta^{-p} . \qquad (13)$$

It is then easy to see that h is convex if and only if $p \geq \frac{s}{n-s}$. We then let $g : \mathbb{R}^{\sigma(s)} \times \mathbb{R}_+ \to \mathbb{R}$ be defined by

$$g(B,\delta) = h(|B|,\delta) = \left(\frac{|B|^{n/s}}{\delta} \right)^p \tag{14}$$

where $\sigma(s) = \binom{n}{s}\binom{n}{s}$, $1 \leq s \leq n-1$. Then from the convexity of h and from the fact that $x \to h(x,\delta)$ is a non decreasing function of x, we deduce that g is convex. Observing that

$$f(A) = g(\mathrm{adj}_s A, \det A)$$

we deduce immediately the polyconvexity of f from the fact that $p \geq \frac{s}{n-s}$.

Step 2: f rank one convex $\Rightarrow p \geq \frac{s}{n-s}$ Let $A \in \mathbb{R}^{n^2}$, $a, b \in \mathbb{R}^n$ be such that

$$\det(A + ta \otimes b) > 0 \text{ for every } t > 0 . \tag{15}$$

Then the rank one convexity of f implies that

$$\varphi(t) = f(A + ta \otimes b) = \left(\frac{|\mathrm{adj}_s(A + ta \otimes b)|^{n/s}}{\det(A + ta \otimes b)} \right)^p \tag{16}$$

is convex in t. We next simplify the notations by letting $\lambda_1, \ldots, \lambda_5$ be such that

$$\begin{cases} |\mathrm{adj}_s(A + ta \otimes b)|^2 = \lambda_1^2 t^2 + \lambda_2 t + \lambda_3^2 \\ \det(A + ta \otimes b) = \lambda_4 t + \lambda_5 \end{cases} . \tag{17}$$

Such $\lambda_1, \ldots, \lambda_5$ exist since $\mathrm{adj}_s(A + t a \otimes b)$ as well as $\det(A + t a \otimes b)$ are linear in t (cf. Proposition 3.1 below). Summarizing (16) and (17) we have

$$\varphi(t) = \left(\lambda_1^2 t^2 + \lambda_2 t + \lambda_3^2 \right)^{np/2s} \left(\lambda_4 t + \lambda_5 \right)^{-p} .$$

After an elementary computation we obtain

$$\varphi''(t) = \left(\lambda_1^2 t^2 + \lambda_2 t + \lambda_3^2 \right)^{\frac{np}{2s}-2} \left(\lambda_4 t + \lambda_5 \right)^{-p-2}$$

$$\times \left[\lambda_1^4 \lambda_4^2 t^4 \frac{p}{s^2} (n-s)^2 \left(p - \frac{s}{n-s} \right) + O(t^3) \right] .$$

Since φ is convex for $t \geq 0$ we must have $p \geq \frac{s}{n-s}$. $\qquad \square$

We now conclude this section by proving Lemma 1.11 which has been used for the minimal surfaces case in the proof of Theorem 1.10.

PROOF of Lemma 1.11. This lemma has been established in Dacorogna [2]. We decompose the proof into three steps. Recall that $0 < \lambda < 1$, $b, c \in \mathbb{R}^{n+1}$ and $A \in \mathbb{R}^{n(n+1)}$ are such that

$$\mathrm{adj}_n A = \lambda b + (1 - \lambda)c \neq 0 \ . \tag{1}$$

We wish to find B and $C \in \mathbb{R}^{n(n+1)}$ such that

$$\begin{cases} A = \lambda B + (1 - \lambda)C \\ \mathrm{adj}_n B = b, \ \mathrm{adj}_n C = c \ . \\ \mathrm{rank}\{B - C\} \leq 1 \end{cases} \tag{2}$$

Step 1: We first introduce the notations and recall some elementary algebraic facts (see Section 4.3 for more details). Let $A \in \mathbb{R}^{n(n+1)}$

$$A = \begin{pmatrix} A_1^1 & \cdots & A_n^1 \\ \vdots & & \vdots \\ A_1^{n+1} & \cdots & A_n^{n+1} \end{pmatrix} = \begin{pmatrix} A^1 \\ \vdots \\ A^{n+1} \end{pmatrix} = (A_1, \ldots, A_n) \ . \tag{3}$$

As seen earlier

$$\mathrm{adj}_n A \equiv (D_1(A), \ldots, D_s(A), \ldots, D_{n+1}(A))$$

$$= \left(\det \begin{pmatrix} A^2 \\ \vdots \\ A^{n+1} \end{pmatrix}, \ldots, (-1)^{s+1} \det \begin{pmatrix} A^1 \\ \vdots \\ A^{s-1} \\ A^{s+1} \\ \vdots \\ A^{n+1} \end{pmatrix}, \ldots, (-1)^{n+2} \det \begin{pmatrix} A^1 \\ \vdots \\ A^n \end{pmatrix} \right)$$

$$= \left(\det \hat{A}^1, \ldots, (-1)^{s+1} \det \hat{A}^s, \ldots, (-1)^{n+2} \det \hat{A}^{n+1} \right) \tag{4}$$

where we denote by $\hat{A}^s$ the $n \times n$ matrix composed with the lines $A^1, \ldots, A^{s-1}$, $A^{s+1}, \ldots, A^{n+1}$. More generally we write $\hat{A}_{j_1, \ldots, j_{l'}}^{i_1, \ldots, i_l}$ for the $(n - l') \times (n + 1 - l)$ matrix obtained from A by suppressing the l' columns $j_1, \ldots, j_{l'}$, and the l lines $i_1, \ldots, i_l$. We now give some elementary properties of $\mathrm{adj}_n A$. The first one is simple (cf. Proposition 3.1 of Chapter 4). Let $A \in \mathbb{R}^{n(n+1)}$, then

$$\langle A_s; \mathrm{adj}_n A \rangle = \sum_{j=1}^{n+1} (-1)^{j+1} A_s^j \det \hat{A}^j = 0, \ s = 1, \ldots, n \ . \tag{5}$$

((5) is easily seen, geometrically, if one takes $u : \mathbf{R}^n \to \mathbf{R}^{n+1}$ and $A = \nabla u$; then as observed above $\mathrm{adj}_n \nabla u$ is the normal to the surface, while $(\nabla u)_s$, $1 \leq s \leq n$, are the tangent vectors to the surface.). We next study the properties of the derivatives of $\mathrm{adj}_n A$ (cf. Proposition 3.1). Since $D_s(A) = (-1)^{s+1} \det \hat{A}^s$ we have

$$\frac{\partial}{\partial \hat{A}^s} D_s(A) = \left(\frac{\partial}{\partial A^1} D_s(A), ..., \frac{\partial}{\partial A^{s-1}} D_s(A), \frac{\partial}{\partial A^{s+1}} D_s(A), ..., \frac{\partial}{\partial A^{n+1}} D_s(A) \right)$$

$$= \mathrm{adj}_{n-1} \hat{A}^s \in \mathbf{R}^{n^2}, \quad s = 1, 2, \ldots, n+1 \ . \tag{6}$$

Using Theorem 1.5 above, we deduce immediately that

$$D_s(A + \alpha \otimes \beta) = D_s(A) + \langle \mathrm{adj}_{n-1} \hat{A}^s; (\alpha \otimes \beta)^s \rangle, \quad s = 1, 2, \ldots, n+1 \ , \tag{7}$$

for every $\alpha \in \mathbf{R}^{n+1}$, $\beta \in \mathbf{R}^n$ and where $\langle \bullet; \bullet \rangle$ denotes the scalar product in $\mathbf{R}^{n^2}$. We may also write (7) in a simpler way, if we let

$$\langle \mathrm{adj}_{n-1} A; \alpha \otimes \beta \rangle = \left(\langle \mathrm{adj}_{n-1} \hat{A}^1; (\alpha \otimes \beta)^1 \rangle, ..., \langle \mathrm{adj}_{n-1} \hat{A}^{n+1}; (\alpha \otimes \beta)^{n+1} \rangle \right)$$

$$\tag{8}$$

then (7) may be rewritten as

$$\mathrm{adj}_n(A + \alpha \otimes \beta) = \mathrm{adj}_n A + \langle \mathrm{adj}_{n-1} A; \alpha \otimes \beta \rangle \ . \tag{9}$$

We finally give two more properties which can be deduced from (6). We let

$$m_{\nu j} = \left(\frac{\partial}{\partial A_\alpha^j} D_\nu(A) \right)_{1 \leq \alpha \leq n} = (-1)^{\nu+1} \left(\frac{\partial}{\partial A_\alpha^j} \det \hat{A}^\nu \right)_{1 \leq \alpha \leq n}$$

$$= \frac{\partial}{\partial A^j} D_\nu(A) = (-1)^{\nu+1} \frac{\partial}{\partial A^j} \det \hat{A}^\nu \in \mathbf{R}^n \ . \tag{10}$$

Observe that if $j < \nu$, then

$$D_\nu(A) = (-1)^{\nu+1} \det \hat{A}^\nu = (-1)^{\nu+1} ((-1)^{j+1} \langle A^j; \mathrm{adj}_{n-1} \hat{A}^{j\nu} \rangle) \tag{11}$$

while

$$D_j(A) = (-1)^{j+1} \det \hat{A}^j = (-1)^{j+1} ((-1)^\nu \langle A^\nu; \mathrm{adj}_{n-1} \hat{A}^{j\nu} \rangle) \ .$$

Hence, using (11), we have for $j < \nu$,

$$m_{\nu j} = \frac{\partial}{\partial A^j} D_\nu(A) = (-1)^{j+\nu} \mathrm{adj}_{n-1}(\hat{A}^{j\nu})$$

$$= -(-1)^{j+\nu+1} \mathrm{adj}_{n-1}(\hat{A}^{j\nu}) = -\frac{\partial}{\partial A^\nu} D_j(A) = -m_{j\nu} \ , \tag{12}$$

and trivially if $j = \nu$, then

$$m_{jj} = \frac{\partial}{\partial A^j} D_j(A) = (-1)^{j+1}\frac{\partial}{\partial A^j}(\det \hat{A}^j) = 0 \ . \tag{13}$$

The last property of the $m_{\nu j}$ that we mention is the following

$$m_{12}\, D_\nu(A) + m_{2\nu}\, D_1(A) - m_{1\nu}\, D_2(A) = 0, \ \ \nu = 3,\ldots,n+1 \tag{14}$$

for every $A \in \mathbb{R}^{n(n+1)}$. To show (14) we first observe that the left hand side can be rewritten, using (4), (11) and (12) as

$$(-1)^{\nu+1} \det \hat{A}^\nu \cdot (-1)^3 \mathrm{adj}_{n-1}\hat{A}^{12}$$

$$+ \det \hat{A}^1 \cdot (-1)^{\nu+2}\mathrm{adj}_{n-1}\hat{A}^{2\nu} + \det \hat{A}^2 \cdot (-1)^{\nu+1}\mathrm{adj}_{n-1}\hat{A}^{1\nu} = 0 \ ,$$

i.e.

$$\det \hat{A}^\nu \mathrm{adj}_{n-1}\hat{A}^{12} + \det \hat{A}^1 \mathrm{adj}_{n-1}\hat{A}^{2\nu} - \det \hat{A}^2 \mathrm{adj}_{n-1}\hat{A}^{1\nu} = 0 \ . \tag{15}$$

Interchanging the lines from $\hat{A}^{12}$ we have that (15) and thus (14) can be replaced by

$$(-1)^{\nu+1} \det \hat{A}^\nu\, \mathrm{adj}_{n-1}\begin{pmatrix} A^\nu \\ \hat{A}^{12\nu} \end{pmatrix} + \det \hat{A}^1 \mathrm{adj}_{n-1}\begin{pmatrix} A^1 \\ \hat{A}^{12\nu} \end{pmatrix}$$

$$- \det \hat{A}^2 \mathrm{adj}_{n-1}\begin{pmatrix} A^2 \\ \hat{A}^{12\nu} \end{pmatrix} = 0 \ ,$$

or equivalently

$$\mathrm{adj}_{n-1}\begin{pmatrix} (-1)^{\nu+1}A^\nu \det \hat{A}^\nu + A^1 \det \hat{A}^1 - A^2 \det \hat{A}^2 \\ \hat{A}^{12\nu} \end{pmatrix} = 0 \ . \tag{16}$$

We now use (5) to get that

$$A^1 \det \hat{A}^1 - A^2 \det \hat{A}^2 + (-1)^{\nu+1}A^\nu \det \hat{A}^\nu = \sum_{j\neq 1,2,\nu}^{n+1} (-1)^j A^j \det \hat{A}^j \ . \tag{17}$$

Using (17) in (16) we have therefore that (14) is equivalent to

$$\mathrm{adj}_{n-1}\begin{pmatrix} \displaystyle\sum_{j\neq 1,2,\nu}^{n+1} (-1)^j A^j \det \hat{A}^j \\ \hat{A}^{12\nu} \end{pmatrix} = 0 \ , \tag{18}$$

and (18) is trivially true since the first line is a linear combination of the $(n-2)$ other lines. We have therefore shown that (14) holds. Before proceeding further we consider the case $n = 2$ for the sake of illustration. Then

1)
$$A = \begin{pmatrix} A_1^1 & A_2^1 \\ A_1^2 & A_2^2 \\ A_1^3 & A_2^3 \end{pmatrix} = \begin{pmatrix} A^1 \\ A^2 \\ A^3 \end{pmatrix} = (A_1, A_2) \ .$$

2)
$$\mathrm{adj}_2 A = \begin{pmatrix} D_1(A) \\ D_2(A) \\ D_3(A) \end{pmatrix} = \begin{pmatrix} \begin{vmatrix} A_1^2 & A_2^2 \\ A_1^3 & A_2^3 \end{vmatrix} = \det \hat{A}^1 \\ -\begin{vmatrix} A_1^1 & A_2^1 \\ A_1^3 & A_2^3 \end{vmatrix} = -\det \hat{A}^2 \\ \begin{vmatrix} A_1^1 & A_2^1 \\ A_1^2 & A_2^2 \end{vmatrix} = \det \hat{A}^3 \end{pmatrix} \ .$$

3)
$$\mathrm{adj}_{n-1}\hat{A}^1 = \mathrm{adj}_1\hat{A}^1 = \begin{pmatrix} A_2^3 & -A_1^3 \\ -A_2^2 & A_1^2 \end{pmatrix}$$

and similarly for $\mathrm{adj}_{n-1}\hat{A}^2$ and $\mathrm{adj}_{n-1}\hat{A}^3$.

4)
$$\begin{cases} m_{21} = \dfrac{\partial}{\partial A^1} D_2(A) = \left(-A_2^3, A_1^3\right) = -m_{12} = -\dfrac{\partial}{\partial A^2} D_1(A) \\[2mm] m_{31} = \dfrac{\partial}{\partial A^1} D_3(A) = \left(A_2^2, -A_1^2\right) = -m_{13} = -\dfrac{\partial}{\partial A^3} D_1(A) \\[2mm] m_{32} = \dfrac{\partial}{\partial A^2} D_3(A) = \left(-A_2^1, A_1^1\right) = -m_{23} = -\dfrac{\partial}{\partial A^3} D_2(A) \ . \end{cases}$$

5) (14) can be read as

$$\begin{cases} A_1^1 \det \hat{A}^1 - A_1^2 \det \hat{A}^2 + A_1^3 \det \hat{A}^3 = 0 \\ A_2^1 \det \hat{A}^1 - A_2^2 \det \hat{A}^2 + A_2^3 \det \hat{A}^3 = 0 \end{cases} \ .$$

With the help of Step 1 we are now in a position to construct B and C satisfying (2).

Step 2: Let $\alpha \in \mathbb{R}^{n+1}$ and $\beta \in \mathbb{R}^n$ to be determined in Step 3, and let

$$\begin{cases} B = A + (1 - \lambda)\alpha \otimes \beta \\ C = A - \lambda\alpha \otimes \beta \end{cases} \ . \tag{19}$$

We then trivially have

$$A = \lambda B + (1 - \lambda)C \quad \text{and} \quad \text{rank}\{B - C\} \leq 1 \ .$$

Furthermore from (19), using (9), we obtain

$$\begin{cases} \text{adj}_n B = \text{adj}_n A + (1 - \lambda)\langle \text{adj}_{n-1} A; \alpha \otimes \beta \rangle \\ \text{adj}_n C = \text{adj}_n A - \lambda \langle \text{adj}_{n-1} A; \alpha \otimes \beta \rangle \end{cases} \ . \qquad (20)$$

Therefore in order to show (1), (2) and thus the lemma, it remains to find $\alpha \in \mathbb{R}^{n+1}$, $\beta \in \mathbb{R}^n$ such that

$$\text{adj}_n B = b, \quad \text{adj}_n C = c$$

i.e. from (20), bearing in mind that $\text{adj}_n A = \lambda b + (1 - \lambda)c$,

$$\langle \text{adj}_{n-1} A; \alpha \otimes \beta \rangle = b - c$$

i.e.

$$\langle \text{adj}_{n-1} \hat{A}^s; (\alpha \otimes \beta)^s \rangle = b^s - c^s \quad s = 1, \ldots, n + 1 \ . \qquad (21)$$

Observe that (21) is composed of $(n + 1)$ non linear equations with $(2n + 1)$ unknowns $(\alpha^1, \ldots, \alpha^{n+1}, \beta_1, \ldots, \beta_n)$.

Step 3: It therefore remains to solve (21), under the hypothesis $\text{adj}_n A \neq 0$, in order to prove the lemma. We first write slightly differently (21), since we have $(n + 1)$ equations and $\alpha \in \mathbb{R}^{n+1}$, we write the left hand side of (21) as $M(\beta)\alpha$ where $M(\beta)$ is an $(n + 1) \times (n + 1)$ matrix whose elements are

$$M(\beta) = (\langle m_{\nu j}; \beta \rangle)_{1 \leq j, \nu \leq n+1} \qquad (22)$$

where $\langle \bullet; \bullet \rangle$ denotes the scalar product in $\mathbb{R}^n$ and $m_{\nu j}$ is defined in (10). Thus (21) has become, upon setting $a = b - c$,

$$M(\beta)\alpha = a \qquad (23)$$

where, using (12) we have

$$M(\beta)^t = -M(\beta) \ . \qquad (24)$$

(For example, if $n = 2$, then

$$M(\beta)\alpha = \begin{pmatrix} 0 & \langle m_{12}; \beta \rangle & \langle m_{13}; \beta \rangle \\ -\langle m_{12}; \beta \rangle & 0 & \langle m_{23}; \beta \rangle \\ -\langle m_{13}; \beta \rangle & -\langle m_{23}; \beta \rangle & 0 \end{pmatrix} \begin{pmatrix} \alpha^1 \\ \alpha^2 \\ \alpha^3 \end{pmatrix} = \begin{pmatrix} a^1 \\ a^2 \\ a^3 \end{pmatrix}$$

and for example

$$\langle m_{12}; \beta \rangle = A_2^3 \beta_1 - A_1^3 \beta_2 \ .)$$

We therefore have to show that given $a \in \mathbf{R}^{n+1}$, there exist $\alpha \in \mathbf{R}^{n+1}$ and $\beta \in \mathbf{R}^n$ satisfying (23). We have to consider two cases.

Case 1: There exist $\nu \in \{2, \ldots, n+1\}$ such that

$$m_{1\nu} \notin \operatorname*{span}_{l \neq 1, \nu} \{ a^\nu m_{1l} - a^1 m_{\nu l} \} \tag{25}$$

where $\operatorname{span}\{-\}$ denotes the vectorial space spanned by the vectors of $\mathbf{R}^n$ $a^\nu m_{1l} - a^1 m_{\nu l}$. Under the hypothesis (25) we may choose $0 \neq \beta \in \mathbf{R}^n$ such that

$$\begin{cases} \langle m_{1\nu}; \beta \rangle \neq 0 \\ \langle -a^\nu m_{1l} + a^1 m_{\nu l}; \beta \rangle = \langle a^l m_{1\nu}; \beta \rangle, \ l \neq 1, \nu, \ 1 \leq l \leq n+1 \end{cases} \tag{26}$$

(Observe that such β exists and it may be constructed as follows. Choose any β orthogonal to the $\operatorname{span}_{l \neq 1, \nu}\{ -a^\nu m_{1l} + a^1 m_{\nu l} - a^l m_{1\nu} \}$. It is then clear that there exists one such β with the property that $\langle m_{1\nu}; \beta \rangle \neq 0$, otherwise this would contradict (25).) So let

$$\alpha^1 = -\frac{a^\nu}{\langle m_{1\nu}; \beta \rangle}, \quad \alpha^\nu = \frac{a^1}{\langle m_{1\nu}; \beta \rangle}, \quad \alpha^l = 0 \text{ if } l \neq \nu \ . \tag{27}$$

It is then easy to check that $M(\beta)\alpha$ is equal to a, indeed
1) if $l = 1$, then

$$(M(\beta)\alpha)^1 = \langle m_{1\nu}; \beta \rangle \alpha^\nu = a^1$$

2) if $l = \nu$, then

$$(M(\beta)\alpha)^\nu = -\langle m_{1\nu}; \beta \rangle \alpha^1 = a^\nu$$

3) if $l \neq 1, \nu$

$$(M(\beta)\alpha)^l = \langle m_{1l}; \beta \rangle \alpha^1 + \langle m_{\nu l}; \beta \rangle \alpha^\nu$$

$$= \frac{1}{\langle m_{1\nu}; \beta \rangle} \left(\langle -a^\nu m_{1l} + a^1 m_{\nu l}; \beta \rangle \right) = a^l$$

where we have used (26) in the last equality.

Case 2: We now assume that for every $\nu \in \{2, \ldots, n+1\}$

$$m_{1\nu} \in \operatorname*{span}_{l \neq 1, \nu} \{ a^\nu m_{1l} - a^1 m_{\nu l} \} \ . \tag{28}$$

This case is slightly more delicate. Assume that we have proved that (28) implies (see below for the proof)

$$(-1)^{\nu+1}a^1 \det \hat{A}^\nu = a^1 D_\nu(A) = a^\nu D_1(A)$$
$$= a^\nu \det \hat{A}^1, \quad \nu = 1,\ldots,n+1 .$$

(29)

Then, since $\mathrm{adj}_n A = (D_1(A),\ldots,D_{n+1}(A)) \neq 0$ (cf. (1)), we may assume, without loss of generality, that $D_1(A) \neq 0$, then this implies in particular that $\frac{\partial}{\partial A^2}D_1(A) = m_{12} \neq 0$. Hence we may choose $\beta \in \mathbf{R}^n$ such that

$$\langle m_{12}; \beta \rangle \neq 0$$

(30)

and $\alpha \in \mathbf{R}^{n+1}$ such that (using (29))

$$\begin{cases} \alpha^1 = -\dfrac{a^2}{\langle m_{12}; \beta \rangle} = -\dfrac{D_2(A)}{D_1(A)}\dfrac{a^1}{\langle m_{12}; \beta \rangle} \\[2ex] \alpha^2 = \dfrac{a^1}{\langle m_{12}; \beta \rangle} \\[2ex] \alpha^l = 0, \quad \text{if } l > 2 \end{cases}.$$

(31)

We then obtain

$$M(\beta)\alpha = \begin{pmatrix} \langle m_{12}; \beta \rangle \alpha^2 \\ -\langle m_{12}; \beta \rangle \alpha^1 \\ \vdots \\ -\langle m_{1,n+1}; \beta \rangle \alpha^1 - \langle m_{2,n+1}; \beta \rangle \alpha^2 \end{pmatrix}.$$

(32)

Therefore using (29), (31), (32) and the identity (14) we have indeed

$$M(\beta)\alpha = \begin{pmatrix} a^1 \\ a^2 \\ \vdots \\ a^{n+1} \end{pmatrix},$$

which is the desired result. Hence in order to conclude the proof it is sufficient to show that (28) implies (29). In order to do that, observe first that, using (12), we have

$$m_{1\nu} = (-1)^\nu \mathrm{adj}_{n-1}\hat{A}^{1\nu}$$

(33)

$$m_{\nu l} = \begin{cases} (-1)^{\nu+l}\mathrm{adj}_{n-1}\hat{A}^{l\nu} = (-1)^{\nu+l}\mathrm{adj}_{n-1}\begin{pmatrix} A^1 \\ \hat{A}^{1\nu l} \end{pmatrix} & \text{if } l < \nu \\[3ex] (-1)^{\nu+l+1}\mathrm{adj}_{n-1}\hat{A}^{l\nu} = (-1)^{\nu+l+1}\mathrm{adj}_{n-1}\begin{pmatrix} A^1 \\ \hat{A}^{1\nu l} \end{pmatrix} & \text{if } l > \nu \end{cases}$$

(34)

$$
m_{1l} = (-1)^l \mathrm{adj}_{n-1} \hat{A}^{1l} =
\begin{cases}
(-1)^{l+\nu+1} \mathrm{adj}_{n-1} \begin{pmatrix} A^\nu \\ \hat{A}^{1l\nu} \end{pmatrix} & \text{if } l < \nu \\[4mm]
(-1)^{l+\nu} \mathrm{adj}_{n-1} \begin{pmatrix} A^\nu \\ \hat{A}^{1l\nu} \end{pmatrix} & \text{if } l > \nu
\end{cases}
\tag{35}
$$

where in the right hand side of (35) we have interchanged the lines of the matrix $\hat{A}^{1l}$, in order to have A^ν in the first line. Therefore using (34) and (35) we have

$$
a^\nu m_{1l} - a^1 m_{\nu l} =
\begin{cases}
(-1)^{l+\nu+1} \mathrm{adj}_{n-1} \begin{pmatrix} a^1 A^1 + a^\nu A^\nu \\ \hat{A}^{1l\nu} \end{pmatrix} & \text{if } l < \nu \\[4mm]
(-1)^{l+\nu} \mathrm{adj}_{n-1} \begin{pmatrix} a^1 A^1 + a^\nu A^\nu \\ \hat{A}^{1l\nu} \end{pmatrix} & \text{if } l > \nu
\end{cases}
\tag{36}
$$

Therefore (28), i.e. $m_{1\nu} \in \mathrm{span}_{l \neq 1,\nu}\{a^\nu m_{1l} - a^1 m_{\nu l}\}$ is equivalent to

$$
\mathrm{adj}_{n-1} \hat{A}^{1\nu} \in \mathop{\mathrm{span}}_{l \neq 1,\nu} \left\{ \mathrm{adj}_{n-1} \begin{pmatrix} a^1 A^1 + a^\nu A^\nu \\ \hat{A}^{1l\nu} \end{pmatrix} \right\} .
\tag{37}
$$

Using now the fact that if $G \in \mathbb{R}^{n(n-1)}$ then $\langle G^1; \mathrm{adj}_{n-1} G \rangle = 0$ (cf. Proposition 3.1) we get from (37) that

$$
\langle a^1 A^1 + a^\nu A^\nu; \mathrm{adj}_{n-1} \hat{A}^{1\nu} \rangle = 0 .
\tag{38}
$$

Recall that we want to show (29), i.e.

$$
a^1 D_\nu(A) = (-1)^{\nu+1} a^1 \det \hat{A}^\nu = a^\nu \det \hat{A}^1 = a^\nu D_1(A) ,
$$

which is equivalent to

$$
a^1 D_\nu(A) = (-1)^{\nu+1} a^1 \langle A^1; \mathrm{adj}_{n-1} \hat{A}^{1\nu} \rangle = a^\nu D_1(A)
$$
$$
= (-1)^\nu a^\nu \langle A^\nu; \mathrm{adj}_{n-1} \hat{A}^{1\nu} \rangle .
\tag{39}
$$

It is then obvious that (38) implies (39) and thus (29). And this completes the proof of the lemma. $\qquad\square$

4.1.2.4 Counterexamples

We now close this section by indicating some counterexamples. We have seen in Theorem 1.1 that

$$f \text{ convex} \Rightarrow f \text{ polyconvex} \Rightarrow f \text{ quasiconvex} \Rightarrow f \text{ rank one convex} .$$

We have also seen that the implication

$$f \text{ polyconvex} \Rightarrow f \text{ convex}$$

is trivially false. In Theorem 1.7 we have seen that if $m, n \geq 3$ then

$$f \text{ rank one convex} \Rightarrow f \text{ polyconvex}$$

does not hold. We show in this subparagraph that the last implication remains false even if $m = n = 2$. Recall also that the implication

$$f \text{ rank one convex} \Rightarrow f \text{ quasiconvex}$$

is still an open question. We give two counterexamples, the first one involves functions f allowed to take the value $+\infty$, the other one involving functions finite everywhere.

We start with the first example given by Dacorogna [8]. We let $A_1, A_2, A_3 \in \mathbb{R}^4$ (2×2 matrices), $\lambda_1, \lambda_2, \lambda_3 \in (0,1)$ such that

$$\begin{cases} \lambda_1 + \lambda_2 + \lambda_3 = 1 \\[2mm] \displaystyle\sum_{i=1}^{3} \lambda_i \det A_i = \det \left(\sum_{i=1}^{3} \lambda_i A_i \right) \\[2mm] \det(A_1 - A_2) \neq 0, \ \det(A_1 - A_3) \neq 0, \ \det(A_2 - A_3) \neq 0 \end{cases} .$$

Such $(\lambda_i, A_i)_{1 \leq i \leq 3}$ exist, e.g. $\lambda_1 = \lambda_2 = \lambda_3 = \frac{1}{3}$ and

$$A_1 = \begin{pmatrix} 1 & 0 \\ 2 & 0 \end{pmatrix}, \ A_2 = \begin{pmatrix} 0 & 1 \\ 0 & 1 \end{pmatrix}, \ A_3 = \begin{pmatrix} -1 & -1 \\ 0 & 0 \end{pmatrix} .$$

Note also that $(\lambda_i, A_i)_{1 \leq i \leq 3}$, as above, do not satisfy property (H_3) of Section 4.1.1.3. We then define $f : \mathbb{R}^4 \rightarrow \bar{\mathbb{R}} = \mathbb{R} \cup \{+\infty\}$ as

$$f(X) = \begin{cases} 0 & \text{if } X = A_1, A_2, A_3 \\ +\infty & \text{otherwise} \end{cases} .$$

Proposition 1.12. *f is rank one convex but not polyconvex.*

PROOF.

Part 1: To show that f is rank one convex, we have to prove that

$$f(\lambda X + (1 - \lambda)Y) \leq \lambda f(X) + (1 - \lambda)f(Y) \tag{1}$$

for every $X, Y \in \mathbf{R}^4$ such that $\mathrm{rank}\{X - Y\} \leq 1$ and for every $\lambda \in [0, 1]$. Three cases may happen.

Case 1: $X \neq A_i$ or $Y \neq A_j$, then $f(X) = +\infty$ or $f(Y) = +\infty$ and therefore (1) is trivially satisfied.

Case 2: $X = A_i$ and $Y = A_j$ with $i \neq j$. This case is impossible since by construction $\mathrm{rank}\{A_i - A_j\} = 2$ if $i \neq j$.

Case 3: $X = Y = A_i$, then (1) is trivially satisfied.

Part 2: It now remains to show that f is not polyconvex. We proceed by contradiction. If f were polyconvex, we should have, using Theorem 1.3 and the construction of $(\lambda_i, A_i)_{1 \leq i \leq 3}$, that

$$f\left(\sum_{i=1}^{3} \lambda_i A_i\right) \leq \sum_{i=1}^{3} \lambda_i f(A_i) .$$

This is however impossible since the left hand side takes the value $+\infty$ while the right hand side is 0. $\qquad\square$

We now turn our attention to a counterexample involving functions f, finite everywhere. The first step in this direction is a counterexample of Aubert [1,2] which is expressed in terms of isotropic functions (cf. Remark iv) following Theorem 1.10) and which is finite only on the set $\{A \in \mathbf{R}^4 : \det A > 0\}$. We here present the extension of this example to the whole of $\mathbf{R}^4$ as proved by Dacorogna-Marcellini [1]. We let

$$A = \begin{pmatrix} A_1^1 & A_2^1 \\ A_1^2 & A_2^2 \end{pmatrix}, \quad |A|^2 = (A_1^1)^2 + (A_2^1)^2 + (A_1^2)^2 + (A_2^2)^2 .$$

Proposition 1.13. *Let $f : \mathbf{R}^4 \to \mathbf{R}$ be such that*

$$f(A) = |A|^4 - a(\det A)^2 - b|A|^2 \det A ,$$

where $|b| \leq 4$ and $8a + 3b^2 < 16$. Then f is rank one convex.

Proposition 1.14. *Let $f : \mathbf{R}^4 \to \mathbf{R}$ be such that*

$$f(A) = |A|^4 - \frac{4}{\sqrt{3}}|A|^2 \det A .$$

Then f is not polyconvex.

REMARK. It is clear that the combination of the two propositions gives the claimed counterexample.

PROOF of Proposition 1.14. We proceed by contradiction. If f were polyconvex, we should have from Theorem 1.3 that there exists $\beta \in \mathbb{R}^4$ and $\gamma \in \mathbb{R}$ such that

$$f(A) \geq \langle \beta; A \rangle + \gamma \det A = \beta_1^1 A_1^1 + \beta_2^1 A_2^1 + \beta_1^2 A_1^2 + \beta_2^2 A_2^2 + \gamma \det A \ . \quad (1)$$

Since the right hand side of (1) is a sum of a quadratic and a linear term, we deduce that there exists a constant $c \in \mathbb{R}$ such that

$$\frac{f(A)}{1 + |A|^2} \geq c, \text{ for every } A \in \mathbb{R}^4 \ . \quad (2)$$

This is however absurd since choosing $A = \begin{pmatrix} t & 0 \\ 0 & t \end{pmatrix}$ we have

$$\frac{f(A)}{1 + |A|^2} = \frac{(2t^2)^2 - 8t^4/\sqrt{3}}{1 + 2t^2} = \frac{4(\sqrt{3} - 2)}{\sqrt{3}} \frac{t^4}{1 + 2t^2} \ .$$

This last expression tends to $-\infty$ as $t \to \infty$ and therefore contradicts (2). Thus the proposition is established. $\qquad\qquad\qquad\qquad\qquad\qquad\qquad\qquad\qquad\square$

PROOF of Proposition 1.13. We decompose the proof into three steps.
Step 1: We first introduce some notations. To every $A \in \mathbb{R}^4$ we associate $\tilde{A} \in \mathbb{R}^4$ in the following way

$$A = \begin{pmatrix} A_1^1 & A_2^1 \\ A_1^2 & A_2^2 \end{pmatrix}, \ \tilde{A} = \begin{pmatrix} A_2^2 & -A_1^2 \\ -A_2^1 & A_1^1 \end{pmatrix} \ .$$

We observe immediately that

$$\begin{cases} |A| = |\tilde{A}| \\ \det A = \det \tilde{A} \\ \langle A; \tilde{B} \rangle = \langle \tilde{A}; B \rangle \\ \langle A; \tilde{A} \rangle = 2 \det A \end{cases}, \quad (1)$$

where $\langle \bullet; \bullet \rangle$ denotes the scalar product in $\mathbb{R}^4$. We have also immediately that

$$\frac{\partial}{\partial A_j^i}(\det A) = \tilde{A}_j^i, \text{ i.e. } \nabla(\det A) = \tilde{A} \ . \quad (2)$$

(Note that in the notations of the appendix $\tilde{A} = \mathrm{adj}_1 A$.)

Step 2: We next compute the following expression

$$\psi(A, B) \equiv \sum_{i,j=1}^{2} \sum_{\alpha,\beta=1}^{2} \frac{\partial^2 f}{\partial A_\alpha^i \, \partial A_\beta^j} B_\alpha^i \, B_\beta^j \; .$$

The aim is to show that $\psi \geq 0$ when restricted to matrices B with $\det B = 0$. This being equivalent to Legendre-Hadamard condition, Theorem 1.1 will then imply that f is rank one convex. We first calculate ∇f; observe that

$$\frac{\partial f}{\partial A_\alpha^i} = 4|A|^2 A_\alpha^i - 2a(\det A)\tilde{A}_\alpha^i - 2b(\det A)A_\alpha^i - b|A|^2 \tilde{A}_\alpha^i \; .$$

We then deduce that, for $1 \leq i, j, \alpha, \beta \leq 2$,

$$\frac{\partial^2 f}{\partial A_\alpha^i \partial A_\beta^j} = 8A_\alpha^i A_\beta^j + 4|A|^2 \delta^{ij}\delta_{\alpha\beta} - 2a\tilde{A}_\alpha^i \tilde{A}_\beta^j - 2a(\det A)\tilde{\delta}^{ij}\tilde{\delta}_{\alpha\beta}$$

$$- 2bA_\alpha^i \tilde{A}_\beta^j - 2b(\det A)\delta^{ij}\delta_{\alpha\beta} - 2b\tilde{A}_\alpha^i A_\beta^j - b|A|^2\tilde{\delta}^{ij}\tilde{\delta}_{\alpha\beta} \; , \quad (3)$$

where

$$\delta^{ij} = \begin{cases} 1 & \text{if } i = j \\ 0 & \text{if } i \neq j \end{cases} , \qquad \tilde{\delta}^{ij} = \begin{cases} (-1)^j & \text{if } i \neq j \\ 0 & \text{otherwise} \end{cases}$$

and similarly for $\delta_{\alpha\beta}$ and $\tilde{\delta}_{\alpha\beta}$. We have therefore that

$$\psi(A, B) = \sum_{i,j,\alpha,\beta=1}^{2} \frac{\partial^2 f}{\partial A_\alpha^i \, \partial A_\beta^j} B_\alpha^i \, B_\beta^j$$

$$= 8(\langle A; B\rangle)^2 + 4|A|^2|B|^2 - 2a(\langle \tilde{A}; B\rangle)^2 - 4a(\det A)(\det B)$$

$$- 2b\langle A; B\rangle\langle \tilde{A}; B\rangle - 2b(\det A)|B|^2$$

$$- 2b\langle \tilde{A}; B\rangle\langle A; B\rangle - 2b(\det B)|A|^2 \; .$$

In order to end the proof of the proposition we therefore only need to show that

$$\psi(A, B) = 8(\langle A; B\rangle)^2 + 4|A|^2|B|^2 - 2a(\langle \tilde{A}; B\rangle)^2 - 4b\langle A; B\rangle\langle \tilde{A}; B\rangle$$

$$- 2b(\det A)|B|^2 \geq 0, \quad \text{whenever } \det B = 0 \; . \quad (4)$$

Step 3: We now prove that (4) holds. Observe first that $\psi(A, B)$ is homogeneous of degree 2 in A (and in B). Therefore in order to show (4) it is sufficient to establish that

$$\min_{A \in \mathbb{R}^4} \{\psi(A, B) : |A| = 1\} \geq 0, \quad \text{for every } B \text{ with } \det B = 0 \; . \quad (5)$$

It is clear that the minimum of $\psi(\bullet, B)$ is attained on the manifold $|A| = 1$. Let α be a Lagrange multiplier we have therefore to prove that $\psi(A_0, B) \geq 0$ for every A_0 which is a critical point of the function of A, $\varphi(A, B) = \psi(A, B) - \alpha(|A|^2 - 1)$. The gradient, with respect to A, of φ is therefore equal to zero when

$$8\langle A_0; B\rangle B + 4|B|^2 A_0 - 2a\langle \tilde{A}_0; B\rangle \tilde{B} - 2b\langle \tilde{A}_0; B\rangle B$$
$$- 2b\langle A_0; B\rangle \tilde{B} - b|B|^2 \tilde{A}_0 = \alpha A_0 \ . \tag{6}$$

Upon multiplication by A_0, bearing in mind that $|A_0| = 1$ and (1), we obtain

$$\psi(A_0, B) = \alpha \ . \tag{7}$$

Multiplying (6) first by B, then by $\tilde{B}$, bearing in mind that $\langle B; \tilde{B}\rangle = 2\det B = 0$, we have

$$\begin{cases} (12|B|^2 - \alpha)\langle A_0; B\rangle - 3b|B|^2\langle \tilde{A}_0; B\rangle = 0 \\ -3b|B|^2\langle A_0; B\rangle + ((4 - 2a)|B|^2 - \alpha)\langle \tilde{A}_0; B\rangle = 0 \end{cases} \ . \tag{8}$$

Two cases may then happen in (8).

Case 1: The first possibility is that

$$\langle A_0; B\rangle = \langle \tilde{A}_0; B\rangle = 0 \ .$$

We have then that

$$\psi(A_0, B) = 4|A_0|^2|B|^2 - 2b(\det A_0)|B|^2 \ .$$

Since $|b| \leq 4$ and since for every $A \in \mathbf{R}^4$ we have $2\det A \leq |A|^2$ we deduce immediately that $\psi(A_0, B) \geq 0$ and hence (5) and (4) hold.

Case 2: The other possibility is that

$$(12|B|^2 - \alpha)((4 - 2a)|B|^2 - \alpha) - 9b^2|B|^4 = 0 \ ,$$

i.e.

$$\alpha^2 - 2\alpha|B|^2(8 - a) + 3|B|^4(16 - 8a - 3b^2) = 0 \ . \tag{9}$$

It is clear that since $8a + 3b^2 \leq 16$ (and hence $a \leq 2 \leq 8$), we must have that the solutions α of (9), if they are real, are both positive. Hence using (7) we have indeed obtained (5) and (4). This concludes the proof of the proposition.

$\square$

4.2 Weak Continuity, Weak Lower Semicontinuity and Existence Theorems

Let

$$I(u, \Omega) \equiv \int_\Omega f(x, u(x), \nabla u(x))\, dx$$

where $u : \Omega \subset \mathbb{R}^n \to \mathbb{R}^m$, $\nabla u \in \mathbb{R}^{nm}$. We now use the different notions introduced in Section 4.1 in order to study the behaviour of the functional I with respect to weak convergence in $W^{1,p}(\Omega; \mathbb{R}^m)$ and then deduce existence theorems.

The main result of the first part will be:

I is weakly lower semicontinuous

if and only if

$f(x, u, \bullet)$ is quasiconvex

(i.e. quasiconvex in the last variable), provided f satisfies some continuity and growth conditions.

We shall then turn over our attention in the second part to the weak continuity of the functional I, i.e. both I and $-I$ are weakly lower semicontinuous and we shall show (dropping, for simplicity, the dependence on x and u) that

$$I(u_\nu) \to I(u) \text{ for every sequence } u_\nu \rightharpoonup u \text{ in } W^{1,p}$$

(or in other words, since Ω is arbitrary,

$$f(\nabla u_\nu) \rightharpoonup f(\nabla u) \text{ in } \mathcal{D}'(\Omega) \text{ for every sequence } u_\nu \rightharpoonup u \text{ in } W^{1,p})$$

if and only if

f is quasiaffine ,

i.e. (cf. Theorem 1.5) f is a linear combination of minors of the matrix ∇u.

It is interesting to note that both results are typical of the vectorial case. In the scalar case there is no *non linear* weakly continuous function.

Finally in Section 4.2.3 we use both results in order to show two existence theorems; one for quasiconvex (and finite) functions, one for polyconvex functions, taking possibly the value $+\infty$.

4.2.1 Weak Lower Semicontinuity

4.2.1.1 Necessity of Quasiconvexity

The main theorem of this paragraph is

Theorem 2.1. *Let Ω be a bounded open set of $\mathbf{R}^n$, let $u : \Omega \subset \mathbf{R}^n \to \mathbf{R}^m$ and let $f : \Omega \times \mathbf{R}^m \times \mathbf{R}^{nm} \to \mathbf{R}$ be continuous and such that*

$$|f(x, u, A)| \leq a(x, |u|, |A|)$$

for every $(x, u, A) \in \Omega \times \mathbf{R}^m \times \mathbf{R}^{nm}$ and where a is increasing with respect to $|s|$, $|A|$ and locally summable in x. Finally let

$$I(u, \Omega) = \int_\Omega f(x, u(x), \nabla u(x)) \, dx \ .$$

If I is weak $*$ lower semicontinuous in $W^{1,\infty}(\Omega, \mathbf{R}^m)$, for every bounded open set Ω, i.e.

$$\liminf_{u_\nu \overset{*}{\rightharpoonup} \bar{u}} I(u_\nu, \Omega) \geq I(\bar{u}, \Omega)$$

then $f(x, u, \bullet)$ is quasiconvex, i.e.

$$\frac{1}{\operatorname{meas} D} \int_D f(x_0, u_0, A_0 + \nabla \varphi(x)) \, dx \geq f(x_0, u_0, A_0)$$

for every $(x_0, u_0, A_0) \in \Omega \times \mathbf{R}^m \times \mathbf{R}^{nm}$, for every bounded open set $D \subset \mathbf{R}^n$ and for every $\varphi \in W_0^{1,\infty}(D; \mathbf{R}^n)$.

REMARKS.

i) The above theorem has already been proved, cf. Lemma 3.3 of Chapter 3;
ii) it is clear that if I is weakly lower semicontinuous in $W^{1,p}$, then I is weak $*$ lower semicontinuous in $W^{1,\infty}$ and therefore the quasiconvexity of f is also necessary for the weak lower semicontinuity of I in $W^{1,p}$.

4.2.1.2 Sufficiency of Quasiconvexity

We now turn our attention to the sufficiency of the quasiconvexity to obtain weak lower semicontinuity in $W^{1,p}$. We consider two cases: first the case $f(x, u, \nabla u) \equiv f(\nabla u)$ and then the general case.

We first start with an elementary lemma concerning convex functions (see Morrey [2], Fusco [1], Marcellini [5]).

Lemma 2.2. *Let $f : \mathbf{R}^N \to \mathbf{R}$ be convex in each variable and let*

$$|f(x)| \leq \alpha(1 + |x|^p)$$

for every $x \in \mathbf{R}^N$ and where $\alpha \geq 0$, $p \geq 1$. Then there exists $\beta \geq 0$ such that

$$|f(x) - f(y)| \leq \beta(1 + |x|^{p-1} + |y|^{p-1})|x - y| ,$$

for every $x, y \in \mathbf{R}^N$.

REMARK. In particular if f is convex, polyconvex, quasiconvex or rank one convex, then f is convex in each variable and therefore Lemma 2.2 applies to such functions.

PROOF. We present here the proof of Marcellini [5]. Consider the function

$$g(x_1) = f(x_1, x_2, \ldots, x_N)$$

where $(x_2, \ldots, x_N) \in \mathbf{R}^{N-1}$ are fixed. Since g is convex, $g'(x_1)$ is defined at almost all points and

$$g(x_1 + h) - g(x_1) \geq g'(x_1)h$$

for every $h \in \mathbf{R}$ and for almost all $x_1 \in \mathbf{R}$. Therefore for $h > 0$, we have

$$|g'(x_1)| = \left|\frac{\partial f}{\partial x_1}(x)\right| \leq \max\left\{\frac{g(x_1 + h) - g(x_1)}{h}, \frac{g(x_1 - h) - g(x_1)}{h}\right\} .$$

Using the growth condition on f and choosing $h = 1 + |x|$ we find

$$|g'(x_1)| \leq \gamma(1 + |x|^{p-1})$$

where $\gamma \geq 0$. Proceeding similarly with the other components, we have indeed established the lemma. $\square$

We now proceed with the proof of the lower semicontinuity when $f(x, u, \nabla u) \equiv f(\nabla u)$. This case is contained in the more general case studied below, but we do it here for the sake of illustration. We first introduce a growth condition on f.

Definition. Let $f : \mathbf{R}^{nm} \to \mathbf{R}$ and $1 \leq p \leq \infty$, then f is said to satisfy growth condition (C_p) if

1) when $p = \infty$

$$(C_\infty) \qquad |f(A)| \leq \eta(|A|), \text{ for every } A \in \mathbf{R}^{nm} ,$$

where η is a continuous and increasing function;

2) when $1 < p < \infty$

$$(C_p) \qquad -\alpha(1 + |A|^q) \leq f(A) \leq \alpha(1 + |A|^p), \text{ for every } A \in \mathbf{R}^{nm} ,$$

where $\alpha \geq 0, 1 \leq q < p$;

3) when $p = 1$

$$(C_1) \qquad\qquad |f(A)| \leq \alpha(1 + |A|), \text{ for every } A \in \mathbf{R}^{nm} \ ,$$

where $\alpha \geq 0$.

We may now state the theorem.

Theorem 2.3. *Let* $f : \mathbf{R}^{nm} \to \mathbf{R}$ *be quasiconvex and satisfying growth condition* (C_p). *Let* $\Omega \subset \mathbf{R}^n$ *be a bounded open set and*

$$I(u, \Omega) = \int_\Omega f(\nabla u(x))\, dx \ .$$

Then I *is weakly lower semicontinuous in* $W^{1,p}(\Omega; \mathbf{R}^m)$ *(weak $*$ lower semi-continuous if* $p = \infty$*), i.e.*

$$\liminf_{u_\nu \rightharpoonup u} I(u_\nu, \Omega) \geq I(u, \Omega) \ .$$

REMARKS.

i) The above theorem is essentially due to Morrey [1,2] and has been refined
   by Meyers [1] and several authors since then. With the hypothesis (C_p),
   the result is due to Marcellini [5] and we shall follow his proof;
ii) if f is convex, instead of quasiconvex, then there exists $A^* \in \mathbf{R}^{nm}$ such
   that

$$f(0) + \langle A^*; A \rangle \leq f(A) \qquad\qquad (*)$$

for every $A \in \mathbf{R}^{nm}$. Therefore, in the convex case, we impose only the
above natural growth condition on f. As seen in the preceding section
there is no known equivalent to $(*)$ for quasiconvex functions, therefore
one need to impose conditions of the type (C_p) below and above;

iii) one should also note that the condition (C_p) if $1 < p < +\infty$ is optimal
   in the sense that one cannot allow the lower bound in (C_p) to be of the
   form $-\alpha(1 + |A|^p)$ with the same p as in the upper bound $\alpha(1 + |A|^p)$,
   as for the case $p = 1$.

PROOF of Remark iii). We give here an example which is essentially due to
Tartar (see Ball-Murat [1]). Let $m = n = p = 2$ and

$$f(\nabla u) = \det \nabla u$$

then (C_2) is satisfied only if $q = 2$. We shall show that if $0 < a < 1$, $\Omega = (0, a)^2$ and

$$u_\nu(x, y) = \frac{1}{\sqrt{\nu}}(1 - y)^\nu (\sin \nu x, \cos \nu x)$$

then

$$u_\nu \rightharpoonup 0 = \bar{u} \text{ in } W^{1,2}(\Omega; \mathbb{R}^2)$$

while

$$\liminf_{\nu \to \infty} \int_\Omega f(\nabla u_\nu)\, dx dy < 0 = \int_\Omega f(\nabla \bar{u})\, dx dy \ .$$

The fact that $u_\nu \to 0$ in $L^\infty(\Omega; \mathbb{R}^2)$ is obvious. We also have

$$\nabla u_\nu = \begin{pmatrix} \sqrt{\nu}(1-y)^\nu \cos \nu x & -\sqrt{\nu}(1-y)^{\nu-1} \sin \nu x \\ -\sqrt{\nu}(1-y)^\nu \sin \nu x & -\sqrt{\nu}(1-y)^{\nu-1} \cos \nu x \end{pmatrix}$$

therefore

$$\|\nabla u_\nu\|_{L^2} = \int_0^a \int_0^a \nu[(1-y)^{2\nu} + (1-y)^{2\nu-2}]\, dx dy$$

$$= a\nu \left[\frac{1}{2\nu+1} + \frac{1}{2\nu-1} - \frac{(1-a)^{2\nu+1}}{2\nu+1} - \frac{(1-a)^{2\nu-1}}{2\nu-1} \right] < 2a$$

if $\nu \geq 1$; and hence $u_\nu \rightharpoonup 0 = \bar{u}$ in $W^{1,2}(\Omega; \mathbb{R}^2)$. However

$$\int_\Omega f(\nabla u_\nu(x,y))\, dx dy = \int_0^a \int_0^a (-\nu(1-y)^{2\nu-1})\, dx dy$$

$$= -\nu a \left[\frac{1}{2\nu} - \frac{(1-a)^{2\nu}}{2\nu} \right]$$

and therefore

$$\liminf_{\nu \to \infty} \int_\Omega f(\nabla u_\nu) = -\frac{a}{2} < \int_\Omega f(\nabla \bar{u}) = 0 \ . \qquad \qquad \square$$

PROOF of Theorem 2.3. We divide the proof into two steps.

Step 1: We first approximate Ω by a union of cubes in $\mathbb{R}^n$ and we take the average of ∇u over each of these cubes (so that ∇u is constant on each cube).
Step 2: We then show that we can take the boundary value of u_ν to be u without altering too much the value of $I(u_\nu, \Omega)$. We finally use the quasiconvexity of f to get the result.

Step 1: Let $\varepsilon > 0$ and let N be an integer. We approximate Ω by a union of cubes D_s whose edge length is $\frac{1}{N}$ and we denote this union by H_N. We then choose N large enough so that

$$\begin{cases} \operatorname{meas}(\Omega - H_N) \leq \varepsilon \\ H_N = \bigcup D_s \end{cases} . \tag{1}$$

We then take the average of ∇u over each of the D_s, i.e.

$$A_s = \frac{1}{\operatorname{meas} D_s} \int_{D_s} \nabla u(x)\, dx \ . \tag{2}$$

Choosing N larger if necessary, we have

$$\sum_s \int_{D_s} |\nabla u(x) - A_s|^p\, dx < \varepsilon \ . \tag{3}$$

We recall that

$$u_\nu \rightharpoonup u \ \text{in} \ W^{1,p}(\Omega; \mathbb{R}^m)$$

$(u_\nu \overset{*}{\rightharpoonup} u \ \text{if} \ p = +\infty)$. We then consider

$$I(u_\nu; \Omega) - I(u; \Omega)$$

$$= \int_\Omega [f(\nabla u_\nu(x)) - f(\nabla u(x))]\, dx$$

$$= \int_{\Omega - H_N} [f(\nabla u_\nu(x)) - f(\nabla u(x))]\, dx$$

$$+ \sum_s \int_{D_s} [f(\nabla u + (\nabla u_\nu - \nabla u)) - f(A_s + (\nabla u_\nu - \nabla u))]\, dx$$

$$+ \sum_s \int_{D_s} [f(A_s + (\nabla u_\nu - \nabla u)) - f(A_s)]\, dx$$

$$+ \sum_s \int_{D_s} [f(A_s) - f(\nabla u)]\, dx$$

$$= J_1 + J_2 + J_3 + J_4 \ . \tag{4}$$

The difficult term to estimate is J_3 and this will be done in Step 2. We now estimate J_1, J_2 and J_4.

Estimation of J_1

$$J_1 = \int_{\Omega - H_N} [f(\nabla u_\nu) - f(\nabla u)] \, dx \ .$$

We want to prove that choosing N large enough, then

$$J_1 \geq -\varepsilon \tag{5}$$

uniformly in ν.

Case 1: If $p = +\infty$ then (5) is trivial since $\|\nabla u_\nu\|_{L^\infty}$ is uniformly bounded and since (C_∞) holds.

Case 2: If $1 < p < \infty$, then use (C_p) to get

$$J_1 \geq - \int_{\Omega - H_N} [\alpha(1 + |\nabla u_\nu|^q) + f(\nabla u)] \, dx$$

$$= - \int_{\Omega - H_N} (\alpha + f(\nabla u)) - \alpha \int_{\Omega - H_N} |\nabla u_\nu|^q \ .$$

Using Hölder inequality and the fact that $q < p$ we get

$$J_1 \geq - \int_{\Omega - H_N} (\alpha + f(\nabla u))$$

$$- \alpha \left(\int_{\Omega - H_N} |\nabla u_\nu|^p \right)^{q/p} \text{meas}(\Omega - H_N)^{(p-q)/p} \ .$$

Choosing N large enough we get (5).

Case 3: If $p = 1$ then by (C_1) we have

$$J_1 \geq - \int_{\Omega - H_N} [\alpha(1 + |\nabla u_\nu|) + f(\nabla u)] \, dx \ .$$

Since $\nabla u_\nu \rightharpoonup \nabla u$ in L^1, we may use the equiintegrability of ∇u_ν (cf. Lemma 1.4 of Chapter 2) to get immediately (5).

Estimation of J_2

$$J_2 = \sum_s{}' \int_{D_s} [f(\nabla u + (\nabla u_\nu - \nabla u)) - f(A_s + (\nabla u_\nu - \nabla u))] \, dx \ .$$

Using Lemma 2.2 if $1 \leq p < +\infty$ we find that there exists a constant $\beta > 0$ such that

$$|J_2| \leq \beta \sum_s \int_{D_s} (1 + |\nabla u_\nu|^{p-1} + |\nabla u_\nu + A_s - \nabla u|^{p-1}) |\nabla u - A_s| \, dx \ .$$

Using Hölder inequality, (3) and the fact that $\|\nabla u_\nu\|_{L^p}$ is uniformly bounded we deduce that for N large enough

$$|J_2| \leq \varepsilon \tag{6}$$

uniformly in ν. If $p = +\infty$, (6) is obtained in the same way using (C_∞) and the fact that $\|\nabla u_\nu\|_{L^\infty}$ is uniformly bounded.

Estimation of J_4

$$J_4 = \sum_s \int_{D_s} [f(A_s) - f(\nabla u)]\, dx \ .$$

It is then easy (similarly to the estimation of J_2) to see that for N large enough

$$|J_4| \leq \varepsilon \tag{7}$$

uniformly in ν.

We now return to (4) gathering (5), (6) and (7). We have therefore for N large enough,

$$I(u_\nu; \Omega) - I(u; \Omega) \geq -3\varepsilon + \sum_s \int_{D_s} [f(A_s + \nabla u_\nu - \nabla u) - f(A_s)]\, dx \ .$$

Taking the limit as $\nu \to \infty$ we get

$$\liminf_{\nu \to \infty} I(u_\nu; \Omega) - I(u; \Omega)$$

$$\geq -3\varepsilon + \sum_s \liminf_{\nu \to \infty} \int_{D_s} [f(A_s + (\nabla u_\nu - \nabla u)) - f(A_s)]\, dx \ . \tag{8}$$

If we now show that

$$\liminf_{\nu \to \infty} \int_{D_s} f(A_s + (\nabla u_\nu - \nabla u))\, dx \geq \int_{D_s} f(A_s)\, dx \tag{9}$$

then combining (8), (9) and the arbitrariness of ε we have indeed obtained that I is weakly lower semicontinuous. It is the aim of the next step to show (9).

Step 2: Let $v_\nu = u_\nu - u$, D be a cube in $\mathbb{R}^n$, $A \in \mathbb{R}^{nm}$ and

$$v_\nu \rightharpoonup 0 \text{ in } W^{1,p}(D; \mathbb{R}^m)$$

($v_\nu \overset{*}{\rightharpoonup} 0$ if $p = \infty$). We have then to show:

$$\liminf_{\nu \to \infty} \int_D f(A + \nabla v_\nu(x))\, dx \geq f(A)\, \text{meas}\, D \ . \tag{10}$$

To infer (10) from the quasiconvexity of f, the only problem is to change slightly v_ν in order to have $v_\nu = 0$ on ∂D. This is classical in the calculus of variations, see Chapter 3, see also Morrey [1,2], Meyers [1], De Giorgi [1], Marcellini-Shordone [1], Dacorogna [3], Acerbi-Fusco [1]. We here proceed as in Marcellini [5]. Let $D^\circ \subset\subset D$ be a cube and let

$$R = \frac{1}{2} \text{dist}(D^\circ, \partial D) \ . \tag{11}$$

Let K be an integer and let $D^\circ \subset D^k \subset D$ (see Figure 4.2) $1 \leq k \leq K$ be such that

$$\text{dist}(D^\circ; \partial D^k) = \frac{k}{K} R, \ 1 \leq k \leq K \ . \tag{12}$$

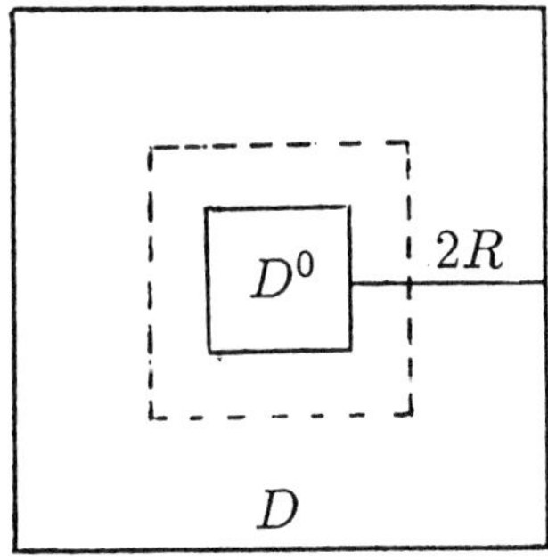

Figure 4.2

We then choose $\varphi^k \in C^\infty(D)$, $1 \le k \le K$ such that

$$
\begin{cases}
0 \le \varphi^k \le 1 \\
\varphi^k(x) = \begin{cases} 1 & \text{if } x \in D^{k-1} \\ 0 & \text{if } x \in D - D^k \end{cases} \\
|\operatorname{grad} \varphi^k| \le a\dfrac{K}{R}
\end{cases}
\tag{13}
$$

where $a > 0$ is a constant. Let

$$
v_\nu^k = \varphi^k\, v_\nu \ ,
\tag{14}
$$

then $v_\nu^k = 0$ on ∂D. We may therefore use the quasiconvexity of f to get

$$
\int_D f(A)\, dx \le \int_D f(A + \nabla v_\nu^k(x))\, dx
$$

$$
\le \int_{D-D^k} f(A)\, dx + \int_{D^k - D^{k-1}} f(A + \nabla v_\nu^k(x))\, dx
$$

$$
+ \int_{D^{k-1}} f(A + \nabla v_\nu(x))\, dx \ .
$$

We then deduce that

$$
\int_{D^k} f(A)\, dx \le \int_{D^k - D^{k-1}} f(A + \nabla v_\nu^k(x))\, dx + \int_{D^{k-1}} f(A + \nabla v_\nu(x))\, dx \ .
$$

We may also rewrite the above inequality in the following way

$$
\int_{D^k} f(A)\, dx \le \int_D f(A + \nabla v_\nu(x))\, dx - \int_{D - D^{k-1}} f(A + \nabla v_\nu(x))\, dx
$$

$$
+ \int_{D^k - D^{k-1}} f(A + \nabla v_\nu^k(x))\, dx
$$

$$
= \int_D f(A + \nabla v_\nu(x))\, dx + \alpha_1 + \alpha_2 \ .
\tag{15}
$$

We now estimate α_1 and α_2.

Estimation of α_1

$$
\alpha_1 = - \int_{D - D^{k-1}} f(A + \nabla v_\nu(x))\, dx
$$

We want to show that by choosing R sufficiently small (cf. (11)) we have

$$\alpha_1 \leq \varepsilon \tag{16}$$

uniformly in ν.

Case 1: If $p = +\infty$, then (16) is trivial since $\|\nabla v_\nu\|_{L^\infty}$ is bounded uniformly.

Case 2: If $1 < p < +\infty$, then use (C_p) to get

$$\alpha_1 \leq \alpha \int_{D-D^{k-1}} (1 + |A + \nabla v_\nu|^q)\, dx$$

$$\leq \alpha \int_{D-D^\circ} (1 + |A|^q + |\nabla v_\nu|^q)\, dx \ .$$

Since $q < p$, we use Hölder inequality (as in the estimation of J_1 above) to get (16).

Case 3: If $p = 1$, using (C_1) we obtain

$$\alpha_1 \leq \alpha \int_{D-D^\circ} (1 + |A| + |\nabla v_\nu|)\, dx \ .$$

Using the fact that $\nabla v_\nu \rightharpoonup 0$ in L^1 and hence the equiintegrability of ∇v_ν (as in the estimation of J_1 above) we get (16).

Estimation of α_2

$$\alpha_2 = \int_{D^k-D^{k-1}} f(A + \nabla v_\nu^k(x))\, dx \ .$$

If $1 \leq p < +\infty$ (the case $p = +\infty$ is simpler and proceeds similarly) we have, using (C_p) and denoting by α a generic constant,

$$\alpha_2 \leq \alpha \int_{D^k-D^{k-1}} (1 + |A + \nabla v_\nu^k|^p)\, dx$$

$$\leq \alpha \int_{D^k-D^{k-1}} (1 + |A|^p + |\varphi^k \nabla v_\nu + v_\nu \operatorname{grad} \varphi^k|^p)\, dx$$

$$\leq \alpha \int_{D^k-D^{k-1}} \left(1 + |A|^p + |\nabla v_\nu|^p + \left(\frac{aK}{R}\right)^p |v_\nu|^p\right) dx \ , \tag{17}$$

where we have used (13) in the last inequality.

Returning to (15), using (16) and (17) and summing the left and right hand side of (15) from $k = 1$ to K we have

$$K \int_D f(A + \nabla v_\nu(x))\, dx - f(A) \left(\sum_{k=1}^{K} \operatorname{meas} D^k \right)$$

$$\geq -K\varepsilon - \alpha \int_{D^K - D^\circ} \left(1 + |A|^p + |\nabla v_\nu|^p + \left(\frac{aK}{R} \right)^p |v_\nu|^p \right) dx \ .$$

Dividing the above inequality by K and letting $\nu \to +\infty$ we get (recalling that $v_\nu \rightharpoonup 0$ in $W^{1,p}(\Omega; \mathbf{R}^m)$)

$$\liminf_{\nu \to \infty} \int_D f(A + \nabla v_\nu(x))\, dx - f(A) \left(\frac{1}{K} \sum_{k=1}^{K} \operatorname{meas} D^k \right) \geq -\varepsilon - \frac{\gamma}{K} \ , \quad (18)$$

where γ is a constant. Letting $K \to \infty$, taking into account the arbitrariness of D° (see (11)) and of ε we have indeed obtained from (18) that

$$\liminf_{\nu \to \infty} \int_D f(A + \nabla v_\nu(x))\, dx \geq f(A) \operatorname{meas} D$$

which is the desired result. $\square$

We now turn our attention to general integrands of the type $f(x, u, \nabla u)$.

Theorem 2.4. *Let $\Omega \subset \mathbf{R}^n$ be a bounded open set and let $f : \Omega \times \mathbf{R}^m \times \mathbf{R}^{nm} \to \mathbf{R}$ be continuous and quasiconvex, i.e.*

$$\frac{1}{\operatorname{meas} D} \int_D f(x_0, u_0, A_0 + \nabla\varphi(x))\, dx \geq f(x_0, u_0, A_0) \tag{1}$$

for every bounded domain $D \subset \mathbf{R}^n$, for every $(x_0, u_0, A_0) \in \Omega \times \mathbf{R}^m \times \mathbf{R}^{nm}$ and for every $\varphi \in W_0^{1,\infty}(D; \mathbf{R}^m)$. Assume furthermore that f satisfies for $1 \leq p < \infty$

$$0 \leq f(x, u, A) \leq \alpha(1 + |u|^p + |A|^p) \tag{2}$$

where $\alpha > 0$,

$$|f(x, u, A) - f(x, v, B)| \leq \beta(1 + |u|^{p-1} + |v|^{p-1} + |A|^{p-1} + |B|^{p-1})$$
$$\times (|u - v| + |A - B|) \tag{3}$$

where $\beta > 0$ and

$$|f(x, u, A) - f(y, u, A)| \leq \eta(|x - y|)(1 + |u|^p + |A|^p) \ , \tag{4}$$

where η is a continuous increasing function with $\eta(0) = 0$. Let

$$I(u, \Omega) = \int_\Omega f(x, u(x), \nabla u(x)) \, dx \; , \tag{5}$$

then I is weakly lower semicontinuous in $W^{1,p}(\Omega; \mathbf{R}^m)$.

REMARKS.

i) The above theorem is due to Morrey [1,2] and has been slightly refined by Meyers [1];

ii) one can replace easily (2) by the following conditions: if $p \leq n$, then (2) can be replaced by

$$-\alpha(|A|^q + |u|^r) - \beta(x) \leq f(x, u, A) \leq \alpha(|A|^p + |u|^r) + \gamma(x) \tag{2'}$$

where $\alpha \geq 0$, $\beta, \gamma \in L^1$, $1 \leq q < p$ (if $p = 1$, $q = 1$) and $1 \leq r < \frac{np}{n-p}$ ($1 \leq r < \infty$ if $p = n$) (r is just the Sobolev exponent which ensures compactness of the imbedding of $W^{1,p} \to L^r$). If $p > n$ then (2) can be replaced by

$$-\alpha|A|^q - g(x, u) \leq f(x, u, A) \leq \alpha|A|^p + g(x, u) \tag{2''}$$

where $\alpha \geq 0$, $1 \leq q < p$ and $g \geq 0$ is a continuous function. The proof is then almost identical as the one of Theorem 2.4 using similar devices as those of Theorem 2.3;

iii) the case $p = +\infty$ is, as usual, simpler. One can replace (2) by

$$|f(x, u, A)| \leq \eta(x, |u|, |A|) \tag{2'''}$$

where η is an increasing function in each of its arguments. One also does not need conditions (3) and (4) of the theorem in order to ensure that I is weak $*$ lower semicontinuous in $W^{1,\infty}$;

iv) recently Acerbi-Fusco [1] and Marcellini [5] have improved Morrey's theorem in two ways. Firstly, one can assume that f is a Carathéodory function, instead of continuous as above. Secondly, one has the lower semicontinuity even without assuming (3) and (4), exactly as in the case $p = +\infty$;

v) hypotheses (2) (or (2'), (2'')), (3) and (4) are satisfied by functions of the type

$$f(x, u, A) = g(x, A) + h(x, u)$$

with the appropriate growth conditions, but not by functions of the type

$$f(x, u, A) = g(u)h(A) \; ;$$

vi) finally if one assumes that f is quasiconvex and satisfies

$$|f(x,u,A)| \leq \alpha(1 + |u|^p + |A|^p) ,$$

then Theorem 2.4, as well as Theorem 2.3, is proved in a much simpler way if one wants to show that I is weakly lower semicontinuous in $W^{1,p+\varepsilon}$, where $\varepsilon > 0$, instead of $W^{1,p}$. This observation is useful since often, for minimization problems, it is possible to see that some minimizing sequences are bounded uniformly in $W^{1,p+\varepsilon}$, instead of $W^{1,p}$ (see for example Ekeland and Témam [1], Chapter IX and X, or Marcellini-Sbordone [2]).

PROOF of Theorem 2.4. Let

$$u_\nu \rightharpoonup u \text{ in } W^{1,p}(\Omega; \mathbb{R}^m) .$$

Let $\varepsilon > 0$ and let N be an integer. We approximate Ω by a union of cubes D_s of edge length $\frac{1}{N}$ and we denote this union by H_N. We then choose N large enough so that

$$\begin{cases} \text{meas}\,(\Omega - H_N) \leq \varepsilon \\ H_N = \bigcup D_s \end{cases} \tag{6}$$

We then define

$$\begin{cases} \bar{x}_s = \dfrac{1}{\text{meas}\,D_s} \displaystyle\int_{D_s} x\,dx \\ \bar{u}_s = \dfrac{1}{\text{meas}\,D_s} \displaystyle\int_{D_s} u(x)\,dx \end{cases} \tag{7}$$

We now estimate

$$\begin{aligned}
I(u_\nu, \Omega) - I(u, \Omega) &= \int_{\Omega-H_N} [f(x,u_\nu,\nabla u_\nu) - f(x,u,\nabla u)]\,dx \\
&\quad + \int_{H_N} [f(x,u_\nu,\nabla u_\nu) - f(x,u,\nabla u_\nu)]\,dx \\
&\quad + \int_{H_N} [f(x,u,\nabla u_\nu) - f(x,u,\nabla u)]\,dx \\
&= J_1 + J_2 + \int_{H_N} [f(x,u,\nabla u_\nu) - f(x,u,\nabla u)]\,dx
\end{aligned}$$

$$= J_1 + J_2 + \sum_s \int_{D_s} [f(x, u, \nabla u_\nu) - f(\bar{x}_s, \bar{u}_s, \nabla u_\nu)]\, dx$$

$$+ \sum_s \int_{D_s} [f(\bar{x}_s, \bar{u}_s, \nabla u_\nu) - f(\bar{x}_s, \bar{u}_s, \nabla u)]\, dx$$

$$+ \sum_s \int_{D_s} [f(\bar{x}_s, \bar{u}_s, \nabla u) - f(x, u, \nabla u)]\, dx$$

$$= J_1 + J_2 + J_3 + J_4 + J_5 \; . \tag{8}$$

We now show that the hypotheses (2), (3) and (4) ensure that J_1, J_2, J_3 and J_5 are small, the term J_4 will be shown to be positive by Theorem 2.3.

Estimation of J_1

$$J_1 = \int_{\Omega - H_N} [f(x, u_\nu, \nabla u_\nu) - f(x, u, \nabla u)]\, dx \; .$$

Since $f \geq 0$ we have

$$J_1 \geq - \int_{\Omega - H_N} f(x, u, \nabla u)\, dx \; .$$

(If f satisfies $(2')$ or $(2'')$ instead of $f \geq 0$, then one can estimate J_1 exactly as in Theorem 2.3). Choosing N larger, if necessary, we have

$$J_1 \geq -\varepsilon \; . \tag{9}$$

Estimation of J_2

$$J_2 = \int_{H_N} [f(x, u_\nu, \nabla u_\nu) - f(x, u, \nabla u_\nu)]\, dx \; .$$

Using (3) we get

$$J_2 \geq -\beta \int_{H_N} (1 + |u|^{p-1} + |u_\nu|^{p-1} + |\nabla u_\nu|^{p-1})|u - u_\nu|\, dx \; .$$

Using Hölder inequality and the fact that $u_\nu \to u$ in $L^p(H_N)$ we find that by choosing ν large enough

$$J_2 \geq -\varepsilon \; . \tag{10}$$

Estimation of J_3 and J_5

$$J_3 = \sum_s \int_{D_s} [f(x, u, \nabla u_\nu) - f(\bar{x}_s, \bar{u}_s, \nabla u_\nu)] \, dx$$

$$= \sum_s \int_{D_s} [f(x, u, \nabla u_\nu) - f(x, \bar{u}_s, \nabla u_\nu)] \, dx$$

$$+ \sum_s \int_{D_s} [f(x, \bar{u}_s, \nabla u_\nu) - f(\bar{x}_s, \bar{u}_s, \nabla u_\nu)] \, dx \ .$$

Using (3), (4), the uniform convergence of $\bar{x}_s \to x$ as $N \to \infty$, the convergence of $\bar{u}_s \to u$ as $N \to \infty$ in L^p and Hölder inequality, we find that for N large enough

$$J_3 \geq -\varepsilon \ . \tag{11}$$

And similarly we find

$$J_5 \geq -\varepsilon \ . \tag{12}$$

We now return to (8), using (9), (10), (11) and (12) and letting $\nu \to +\infty$, we find

$$\liminf_{\nu \to \infty} I(u_\nu, \Omega) - I(u, \Omega)$$

$$\geq -4\varepsilon + \liminf_{\nu \to \infty} \sum_s \int_{D_s} [f(\bar{x}_s, \bar{u}_s, \nabla u_\nu) - f(\bar{x}_s, \bar{u}_s, \nabla u)] \, dx$$

$$\geq -4\varepsilon + \sum_s \liminf_{\nu \to \infty} \int_{D_s} [f(\bar{x}_s, \bar{u}_s, \nabla u_\nu) - f(\bar{x}_s, \bar{u}_s, \nabla u)] \, dx \ .$$

Using now Theorem 2.3, since $\bar{x}_s, \bar{u}_s$ are constant on D_s we obtain

$$\liminf_{\nu \to \infty} I(u_\nu, \Omega) \geq I(u, \Omega) - 4\varepsilon \ .$$

Since $\varepsilon > 0$ is arbitrary, we have indeed obtained the result. $\qquad\square$

4.2.2 Weak Continuity

We now turn our attention to results on weak continuity of nonlinear functions. Let $f : \mathbb{R}^{nm} \to \mathbb{R}$ be continuous, we shall show that

$$f(\nabla u_\nu) \rightharpoonup f(\nabla u) \text{ in } \mathcal{D}'(\Omega)$$

for every sequence $u_\nu \rightharpoonup u$ in $W^{1,p}(\Omega; \mathbb{R}^m)$ if and only if f is quasiaffine (i.e. from Theorem 1.5, f is a linear combination of minors of the matrix ∇u).

Plainly the existence of nonlinear weakly continuous functions is purely due to the vectorial nature of the problem, since if $m = 1$ (or $n = 1$), the only minors of the matrix ∇u are just the linear terms $\partial u/\partial x_i$, $1 \leq i \leq n$ (or if $n = 1$, the linear terms du_i/dx, $1 \leq i \leq m$).

It is also clear that Theorem 2.1 and Theorem 2.3 applied to f, I and $-f$, $-I$, added to the arbitrariness of the domain Ω give immediately the result if p is large enough. We shall use Theorem 2.1 for the necessary condition; however, for reasons explained below, we shall not use Theorem 2.3 for the sufficiency result and we shall give a new proof of the weak continuity of the minors.

The results of this section are essentially due to Reshetnyak [2,3] and Ball [1-3]. Considerations on weak continuity have been developed in a more general context, called *compensated compactness*, by Murat [1,2], Tartar [2], Murat-Tartar [1] (for a presentation of this theory see also Dacorogna [4]). More recently Hanouzet-Joly [1], Hanouzet [1], Bachelot [1] have also obtained results on weak continuity, considering such a property as particular case of products of functions belonging to Sobolev spaces. Both approaches contain the theory presented here.

4.2.2.1 Necessary Condition

Theorem 2.5. *Let $1 \leq p \leq \infty$, let $\Omega \subset \mathbb{R}^n$ be a bounded open set and let $f : \mathbb{R}^{nm} \to \mathbb{R}$ be continuous. If, for every sequence $u_\nu \rightharpoonup u$ in $W^{1,p}(\Omega; \mathbb{R}^m)$ $(u_\nu \overset{*}{\rightharpoonup} u$ if $p = \infty)$*

$$f(\nabla u_\nu) \rightharpoonup f(\nabla u) \text{ in } \mathcal{D}'(\Omega) \ ,$$

i.e.

$$\int_\Omega f(\nabla u_\nu(x))\varphi(x)\,dx \rightarrow \int_\Omega f(\nabla u(x))\varphi(x)\,dx \tag{1}$$

for every $\varphi \in \mathcal{D}(\Omega)$ (the set of C^∞ functions with compact support), then f is quasiaffine, i.e. there exist $\alpha \in \mathbb{R}$, $\beta \in \mathbb{R}^{\tau(n,m)}$ such that

$$f(A) = \alpha + \langle \beta; T(A) \rangle \tag{2}$$

for every $A \in \mathbb{R}^{nm}$, where

$$\begin{cases} n \wedge m = \min\{n, m\} \\ T(A) = (A, \mathrm{adj}_2 A, \ldots, \mathrm{adj}_{n \wedge m} A) \\ \tau(n,m) = \sum_{s=1}^{n \wedge m} \sigma(s), \ \ \sigma(s) = \binom{m}{s}\binom{n}{s} \end{cases},$$

and $\langle \bullet; \bullet \rangle$ denotes the scalar product in $\mathbb{R}^{\tau(n,m)}$.

PROOF. Let $\varphi \in \mathcal{D}(\Omega)$ and let

$$I(u, \Omega) = \int_\Omega \varphi(x) f(\nabla u(x))\, dx \ ,$$

then (1) is equivalent to

$$\lim_{\nu \to \infty} I(u_\nu, \Omega) = I(u, \Omega) \ .$$

We may therefore apply Theorem 2.1 to I and $-I$ and get that f and $-f$ are quasiconvex, i.e. f is quasiaffine. Theorem 1.5 implies then (2) and the theorem. $\square$

4.2.2.2 Sufficient Condition

For the clarity of the exposition we shall always give the results for the cases $m = n = 2$, $m = n = 3$ and then $m = n$, before giving the general result when $m, n \geq 2$.

If $m = n = 2$, the only nonlinear quasiaffine function is $\det \nabla u$; while, if $m = n = 3$, the only nonlinear quasiaffine functions are linear combination of elements of the matrix $\mathrm{adj}_2 \nabla u$ and of $\det \nabla u$. We now give the main theorem which shows that these functions are actually weakly continuous.

Theorem 2.6. *Let $\Omega \subset \mathbb{R}^n$ be a bounded open set, $1 < p < \infty$, and let*

$$u_\nu \rightharpoonup u \text{ in } W^{1,p}(\Omega; \mathbb{R}^m) \tag{1}$$

(if $p = +\infty$, $u_\nu \overset{*}{\rightharpoonup} u$).
Part 1: Let $m = n = 2$ and $p \geq 2$, then

$$\det \nabla u_\nu \rightharpoonup \det \nabla u \text{ in } \mathcal{D}'(\Omega) \ . \tag{2}$$

Part 2: Let $m = n = 3$.
1) Let $p \geq 2$, then

$$\mathrm{adj}_2 \nabla u_\nu \rightharpoonup \mathrm{adj}_2 \nabla u \text{ in } (\mathcal{D}'(\Omega))^9 \ . \tag{3}$$

2) Let $p \geq 3$, then
$$\det \nabla u_\nu \rightharpoonup \det \nabla u \text{ in } \mathcal{D}'(\Omega) \ . \tag{4}$$

Part 3: Let $m = n$ and $p \geq n$, then

$$\det \nabla u_\nu \rightharpoonup \det \nabla u \text{ in } \mathcal{D}'(\Omega) \ . \tag{5}$$

Part 4: Let $m, n \geq 2$, $2 \leq s \leq n \wedge m = \min\{n, m\}$ and $p \geq s$, then

$$\mathrm{adj}_s \nabla u_\nu \rightharpoonup \mathrm{adj}_s \nabla u \text{ in } (\mathcal{D}'(\Omega))^{\sigma(s)} . \tag{6}$$

where $\sigma(s) = \binom{m}{s}\binom{n}{s} = \frac{m! n!}{(s!)^2 (m-s)! (n-s)!}$.
Part 5: Let $m, n \geq 2$, $2 \leq s \leq n \wedge m$ and assume that

$$\mathrm{adj}_{s-1} \nabla u_\nu \rightharpoonup \mathrm{adj}_{s-1} \nabla u \text{ in } (L^r(\Omega))^{\sigma(s-1)} . \tag{7}$$

where $r > 1$ with $\frac{1}{p} + \frac{1}{r} \leq 1$, then

$$\mathrm{adj}_s \nabla u_\nu \rightharpoonup \mathrm{adj}_s \nabla u \text{ in } (\mathcal{D}'(\Omega))^{\sigma(s)} . \tag{8}$$

REMARKS.

i) Let $m = n = 2$. Note that if $p \geq 2$, then $\det \nabla u \in L^{p/2}(\Omega)$. Therefore
if $p > 2$, (2) is equivalent to

$$\det \nabla u_\nu \rightharpoonup \det \nabla u \text{ in } L^{p/2}(\Omega) , \tag{2'}$$

since $L^{p/2}(\Omega)$ is reflexive if $p > 2$. Similarly, if $p = \infty$, then

$$\det \nabla u_\nu \overset{*}{\rightharpoonup} \det \nabla u \text{ in } L^\infty(\Omega) ; \tag{2''}$$

ii) let $m = n = 2$. If $p > 2$, (2) (or equivalently (2'), (2'')) results immediately from Theorem 2.3, since, trivially

$$-(1 + |\nabla u|^2) \leq -|\nabla u|^2 \leq \det \nabla u \leq |\nabla u|^2 \leq 1 + |\nabla u|^p ;$$

iii) let $m = n = 2$. If $p = 2$, Theorem 2.3 cannot be applied and indeed as
seen in Remark iii) following Theorem 2.3, there are examples of sequences
$u_\nu \rightharpoonup u$ in $W^{1,2}(\Omega; \mathbb{R}^2)$ such that $\det \nabla u_\nu \not\rightharpoonup \det \nabla u$ in $L^1(\Omega)$. Theorem
2.6 ensures, however, that $\det \nabla u_\nu \rightharpoonup \det \nabla u$ in $\mathcal{D}'(\Omega)$;
iv) all the above remarks can be made for the general case $m, n \geq 2$.

The main tool in proving Theorem 2.6 is the observation that any minor
of ∇u can be expressed as a divergence of a vector field.

Lemma 2.7. *Let $\Omega \subset \mathbb{R}^n$ be a bounded open set and let $u \in W^{1,p}(\Omega; \mathbb{R}^m)$,
$1 < p \leq \infty$.*
Part 1: Let $m = n = 2$ and define

$$\mathrm{Det}\, \nabla u \equiv \frac{\partial}{\partial x_1}\left(u^1 \frac{\partial u^2}{\partial x_2}\right) - \frac{\partial}{\partial x_2}\left(u^1 \frac{\partial u^2}{\partial x_1}\right) . \tag{1}$$

i) If $p \geq \frac{4}{3}$, then $\mathrm{Det}\, \nabla u \in \mathcal{D}'(\Omega)$.
ii) If $p \geq 2$, then

$$\mathrm{Det}\, \nabla u = \det \nabla u \text{ in } \mathcal{D}'(\Omega) \ , \tag{2}$$

in particular if $u \in C^2(\Omega; \mathbf{R}^2)$, then (2) holds in the usual sense.

Part 2: Let $m = n = 3$.

1) Recall that

$$\mathrm{adj}_2 \nabla u = \big((\mathrm{adj}_2 \nabla u)^i_\alpha\big)_{1 \leq i,\, \alpha \leq 3} \tag{3}$$

where

$$(\mathrm{adj}_2 \nabla u)^i_\alpha = (-1)^{i+\alpha} \frac{\partial(u^\beta, u^\gamma)}{\partial(x_j, x_k)}$$

$$= (-1)^{i+\alpha} \left[\frac{\partial u^\beta}{\partial x_j} \frac{\partial u^\gamma}{\partial x_k} - \frac{\partial u^\gamma}{\partial x_j} \frac{\partial u^\beta}{\partial x_k} \right] \ ,$$

where $j < k$ and $j, k \neq i$, $\beta < \gamma$ and $\beta, \gamma \neq \alpha$. Define

$$\mathrm{Adj}_2 \nabla u = \big((\mathrm{Adj}_2 \nabla u)^i_\alpha\big)_{1 \leq i,\, \alpha \leq 3} \tag{4}$$

where

$$(\mathrm{Adj}_2 \nabla u)^i_\alpha = (-1)^{i+\alpha} \left[\frac{\partial}{\partial x_j} \left(u^\beta \frac{\partial u^\gamma}{\partial x_k} \right) - \frac{\partial}{\partial x_k} \left(u^\beta \frac{\partial u^\gamma}{\partial x_j} \right) \right]$$

where $j < k$ and $j, k \neq i$, $\beta < \gamma$ and $\beta, \gamma \neq \alpha$.
i) If $p \geq \frac{3}{2}$, then $\mathrm{Adj}_2 \nabla u \in (\mathcal{D}'(\Omega))^9$.
ii) If $p \geq 2$, then

$$\mathrm{Adj}_2 \nabla u = \mathrm{adj}_2 \nabla u \text{ in } (\mathcal{D}'(\Omega))^9 \ , \tag{5}$$

in particular if $u \in C^2(\Omega; \mathbf{R}^3)$, then (5) holds in the usual sense.

2) Recall that

$$\det \nabla u = \sum_{i=1}^{3} \frac{\partial u^1}{\partial x_i} (\mathrm{adj}_2 \nabla u)^1_i \ . \tag{6}$$

Define

$$\mathrm{Det}\, \nabla u = \sum_{i=1}^{3} \frac{\partial}{\partial x_i} \big(u^1 (\mathrm{adj}_2 \nabla u)^1_i \big) \ . \tag{7}$$

i) If $p \geq \frac{9}{4}$, then $\mathrm{Det}\, \nabla u \in \mathcal{D}'(\Omega)$.
ii) If $p \geq 3$, then

$$\mathrm{Det}\, \nabla u = \det \nabla u \text{ in } \mathcal{D}'(\Omega) \ , \tag{8}$$

in particular if $u \in C^2(\Omega; \mathbf{R}^3)$, then (8) holds in the usual sense.

Part 3: Let $m = n$, recall that

$$\det \nabla u = \frac{\partial(u^1, \ldots, u^n)}{\partial(x_1, \ldots, x_n)} = \sum_{\alpha=1}^{n} (-1)^{\alpha+1} \frac{\partial u^1}{\partial x_\alpha} \frac{\partial(u^2, \ldots, u^n)}{\partial(x_1, \ldots, x_{\alpha-1}, x_{\alpha+1}, \ldots, x_n)} \ . \tag{9}$$

Define

$$\mathrm{Det}\, \nabla u = \sum_{\alpha=1}^{n} (-1)^{\alpha+1} \frac{\partial}{\partial x_\alpha} \left(u^1 \frac{\partial(u^2, \ldots, u^n)}{\partial(x_1, \ldots, x_{\alpha-1}, x_{\alpha+1}, \ldots, x_n)} \right) \ . \tag{10}$$

i) If $p \geq \frac{n^2}{n+1}$, then $\mathrm{Det}\, \nabla u \in \mathcal{D}'(\Omega)$.
ii) If $p \geq n$, then

$$\mathrm{Det}\, \nabla u = \det \nabla u \ \text{in} \ \mathcal{D}'(\Omega) \ , \tag{11}$$

in particular if $u \in C^2(\Omega; \mathbf{R}^n)$, then (11) holds in the usual sense.

Part 4: Let $m, n \geq 2$ and $2 \leq s \leq n \wedge m = \min\{n, m\}$. Recall that

$$\mathrm{adj}_s \nabla u = \left((\mathrm{adj}_s \nabla u)^i_\alpha \right)_{1 \leq \alpha \leq \binom{n}{s}, 1 \leq i \leq \binom{m}{s}} \tag{12}$$

where

$$(\mathrm{adj}_s \nabla u)^i_\alpha = (-1)^{i+\alpha} \frac{\partial(u^{i_1}, \ldots, u^{i_s})}{\partial(x_{\alpha_1}, \ldots, x_{\alpha_s})}$$

for the precise relation between $i, i_1, \ldots, i_s$ and $\alpha, \alpha_1, \ldots, \alpha_s$ see the appendix.
Define

$$\mathrm{Adj}_s \nabla u = \left((\mathrm{Adj}_s \nabla u)^i_\alpha \right)_{1 \leq \alpha \leq \binom{n}{s}, 1 \leq i \leq \binom{m}{s}} \tag{13}$$

where

$$(\mathrm{Adj}_s \nabla u)^i_\alpha$$
$$= (-1)^{i+\alpha} \sum_{t=1}^{s} (-1)^{t+1} \frac{\partial}{\partial x_{\alpha_t}} \left(u^{i_1} \frac{\partial(u^{i_2}, \ldots, u^{i_s})}{\partial(x_{\alpha_1}, \ldots, x_{\alpha_{t-1}}, x_{\alpha_{t+1}}, \ldots, x_{\alpha_s})} \right) \ . \tag{14}$$

i) If $p \geq \frac{sn}{n+1}$, then $\mathrm{Adj}_s \nabla u \in (\mathcal{D}'(\Omega))^{\sigma(s)}$, where $\sigma(s) = \binom{m}{s} \binom{n}{s}$.
ii) If $p \geq s$, then

$$\mathrm{Adj}_s \nabla u = \mathrm{adj}_s \nabla u \ \text{in} \ (\mathcal{D}'(\Omega))^{\sigma(s)} \ , \tag{15}$$

in particular if $u \in C^2(\Omega; \mathbf{R}^m)$, then (15) holds in the usual sense.
Part 5: Let $m, n \geq 2$ and $2 \leq s \leq n \wedge m$. Assume that

$$\mathrm{adj}_{s-1} \nabla u \in (L^r(\Omega))^{\sigma(s-1)}, \ r > 1 \ . \tag{16}$$

i) If $\frac{1}{p} + \frac{1}{r} \leq 1 + \frac{1}{n}$, then $\mathrm{Adj}_s \nabla u \in (\mathcal{D}'(\Omega))^{\sigma(s)}$, where $\mathrm{Adj}_s \nabla u$ is defined by (13).

ii) If $\frac{1}{p} + \frac{1}{r} \leq 1$, then

$$\mathrm{Adj}_s \nabla u = \mathrm{adj}_s \nabla u \quad \text{in} \quad (\mathcal{D}'(\Omega))^{\sigma(s)} , \tag{17}$$

REMARKS.

i) Let $m = n = 2$ and $\frac{4}{3} \leq p < 2$. If $\det \nabla u$ is defined in the usual way, i.e.

$$\det \nabla u = \frac{\partial u^1}{\partial x_1} \frac{\partial u^2}{\partial x_2} - \frac{\partial u^2}{\partial x_1} \frac{\partial u^1}{\partial x_2} ,$$

then $\det \nabla u$ is not necessarily a distribution; while $\mathrm{Det}\, \nabla u$ defined by (1) is. In fact $\mathrm{Det}\, \nabla u$ is the (unique) extension by continuity as a distribution of $\det \nabla u$ when $\frac{4}{3} \leq p < 2$ (cf. for more general results of this type, Hanouzet-Jolly [1], Hanouzet [1], Bachelot [1]). Note also that if $p < \frac{4}{3}$, then even $\mathrm{Det}\, \nabla u$ need not be a distribution;

ii) one should observe that if $m = n = 3$, Ball [2] defines

$$\mathrm{Det}\, \nabla u = \sum_{i=1}^{3} \frac{\partial}{\partial x_i}(u^1 (\mathrm{Adj}_2 \nabla u)_i^1)$$

which does not correspond to our definition (7) (note the change from adj in (7) to Adj above). The two definitions need not be the same if $p < 2$;

iii) similar remarks apply to the general case $m, n \geq 1$.

PROOF of Lemma 2.7. We prove Part 1 for illustration and then prove Part 4 and 5.

Part 1: Let $m = n = 2$.

i) If $p \geq \frac{4}{3}$ and since $u \in W^{1,p}(\Omega; \mathbb{R}^2)$, we have by the Sobolev imbedding theorem that $u \in (L^4_{\mathrm{loc}}(\Omega))^2$. Using Hölder inequality we deduce that

$$u^1 \frac{\partial u^2}{\partial x_2}, \quad u^1 \frac{\partial u^2}{\partial x_1} \in L^1_{\mathrm{loc}}(\Omega) ,$$

and thus $\mathrm{Det}\, \nabla u \in \mathcal{D}'(\Omega)$;

ii) If $p \geq 2$, then $\det \nabla u \in L^1(\Omega)$. Observe that if $u \in C^2$, then

$$\det \nabla u = \frac{\partial u^1}{\partial x_1} \frac{\partial u^2}{\partial x_2} - \frac{\partial u^2}{\partial x_1} \frac{\partial u^1}{\partial x_2}$$

$$= \frac{\partial}{\partial x_1}\left(u^1 \frac{\partial u^2}{\partial x_2}\right) - \frac{\partial}{\partial x_2}\left(u^1 \frac{\partial u^2}{\partial x_1}\right) = \mathrm{Det}\, \nabla u . \tag{18}$$

Therefore multiplying (18) by $\varphi \in \mathcal{D}(\Omega)$ and integrating by parts we find

$$\int_\Omega \det \nabla u \cdot \varphi \, dx = -\int_\Omega \left(u^1 \frac{\partial u^2}{\partial x_2} \frac{\partial \varphi}{\partial x_1} - u^1 \frac{\partial u^2}{\partial x_1} \frac{\partial \varphi}{\partial x_2} \right) dx \ . \qquad (19)$$

Since C^2 is dense in $W^{1,2}(\Omega; \mathbb{R}^2)$, then (19) holds for every $u \in W^{1,2}$; and this concludes Part 1.

Part 4: Let $m, n \geq 2$.

i) If $p \geq \frac{sn}{n+1}$ and since $u \in W^{1,p}(\Omega; \mathbb{R}^m)$, we have by the Sobolev imbedding theorem that $u \in (L_{\mathrm{loc}}^{sn/(n+1-s)}(\Omega))^m$. Using Hölder inequality we have also that

$$\frac{\partial(u^{i_2}, \ldots, u^{i_s})}{\partial(x_{\alpha_1}, \ldots, x_{\alpha_{t-1}}, x_{\alpha_{t+1}}, \ldots, x_{\alpha_s})} \in L^{sn/(n+1)(s-1)}(\Omega) \ .$$

Therefore, using (13), (14) we get that $\mathrm{Adj}_s \nabla u \in (\mathcal{D}'(\Omega))^{\sigma(s)}$.

ii) If $p \geq s$, then $\mathrm{adj}_s \nabla u \in (L^1(\Omega))^{\sigma(s)}$. Observe that if $u \in C^2(\Omega; \mathbb{R}^m)$, then (cf. Theorem 3.2 in the appendix)

$$\mathrm{Adj}_s \nabla u = \mathrm{adj}_s \nabla u \ . \qquad (20)$$

Multiplying (20) by $\varphi \in \mathcal{D}(\Omega)$ and integrating by parts the left hand side we find that

$$\int_\Omega \left((\mathrm{adj}_s \nabla u)_\alpha^i \right) \varphi \, dx$$

$$= \sum_{t=1}^s (-1)^t \int_\Omega u^{i_1} \frac{\partial(u^{i_2}, \ldots, u^{i_s})}{\partial(x_{\alpha_1}, \ldots, x_{\alpha_{t-1}}, x_{\alpha_{t+1}}, \ldots, x_{\alpha_s})} \frac{\partial \varphi}{\partial x_{\alpha_t}} \, dx \ . \quad (21)$$

Since C^2 is dense in $W^{1,s}(\Omega; \mathbb{R}^m)$, we deduce that (21) holds for every $u \in W^{1,s}$ and this concludes Part 4 of the lemma.

Part 5: Let $m, n \geq 2$.

i) Using Sobolev imbedding theorem, we have that $u \in (L_{\mathrm{loc}}^{np/(n-p)}(\Omega))^m$, since $u \in W^{1,p}$. We now combine (13), (14), (16), Holder inequality and the fact that $\frac{1}{p} + \frac{1}{r} - \frac{1}{n} \leq 1$, to deduce that

$$u^{i_1}(\mathrm{adj}_{s-1} \nabla u)_\alpha^i \in L_{\mathrm{loc}}^1(\Omega) \ .$$

Therefore, using (13), we have $\mathrm{Adj}_s \nabla u \in (\mathcal{D}'(\Omega))^{\sigma(s)}$.

ii) Since $\frac{1}{p} + \frac{1}{r} \leq 1$ and (16) holds we have that $\operatorname{adj}_s \nabla u \in L^1$. Since (17) holds (cf. Theorem 3.2 in the appendix) for every $u \in C^2$, we deduce (17) by density. $\square$

We are now in a position to show Theorem 2.6.

PROOF of Theorem 2.6. We prove Part 1 for illustration, then Part 4 and Part 5.

Part 1: Let $m = n = 2$ and $p \geq 2$. Let $\varphi \in \mathcal{D}(\Omega)$, then by Lemma 2.7 we have

$$\int_\Omega \det \nabla u_\nu \cdot \varphi \, dx = -\int_\Omega \left(u_\nu^1 \frac{\partial u_\nu^2}{\partial x_2} \frac{\partial \varphi}{\partial x_1} - u_\nu^1 \frac{\partial u_\nu^2}{\partial x_1} \frac{\partial \varphi}{\partial x_2} \right) dx \ .$$

Since $u_\nu \rightharpoonup u$ in $W^{1,p}(\Omega; \mathbb{R}^2)$, $u_\nu \to u$ in $(L^q_{\text{loc}}(\Omega))^2$ with $q < \infty$, and therefore

$$\left(u_\nu^1 \frac{\partial u_\nu^2}{\partial x_2}, u_\nu^1 \frac{\partial u_\nu^2}{\partial x_1} \right) \rightharpoonup \left(u^1 \frac{\partial u^2}{\partial x_2}, u^1 \frac{\partial u^2}{\partial x_1} \right) \ \text{in } L^r_{\text{loc}}(\Omega), \ 1 \leq r < 2 \ .$$

We therefore deduce that

$$\int_\Omega \det \nabla u_\nu \cdot \varphi \, dx \to \int_\Omega \det \nabla u \cdot \varphi \, dx \ .$$

Part 4: Let $m, n \geq 2$, $2 \leq s \leq n \wedge m$ and $p \geq s$. In order to show the theorem it is sufficient to establish that for every $\varphi \in \mathcal{D}(\Omega)$ we have

$$\int_\Omega \frac{\partial(u_\nu^{i_1}, \ldots, u_\nu^{i_s})}{\partial(x_{\alpha_1}, \ldots, x_{\alpha_s})} \varphi \, dx \to \int_\Omega \frac{\partial(u^{i_1}, \ldots, u^{i_s})}{\partial(x_{\alpha_1}, \ldots, x_{\alpha_s})} \varphi \, dx \ . \tag{9}$$

To show (9), we proceed by induction on s. Suppose that the theorem has been established up to the order $s - 1$ (the case $s = 2$ has been dealt with in Part 1). Since $u_\nu \rightharpoonup u$ in $W^{1,p}(\Omega; \mathbb{R}^m)$ and $p \geq s$ we deduce that

$$u_\nu \to u \ \text{in } (L^q_{\text{loc}}(\Omega))^m \ \text{with } 1 \leq q < \frac{ns}{n - s} \ . \tag{10}$$

By hypothesis of induction

$$\frac{\partial(u_\nu^{i_2}, \ldots, u_\nu^{i_s})}{\partial(x_{\alpha_1}, \ldots, x_{\alpha_{t-1}}, x_{\alpha_{t+1}}, \ldots, x_{\alpha_s})}$$

$$\rightharpoonup \frac{\partial(u^{i_2}, \ldots, u^{i_s})}{\partial(x_{\alpha_1}, \ldots, x_{\alpha_{t-1}}, x_{\alpha_{t+1}}, \ldots, x_{\alpha_s})} \ \text{in } \mathcal{D}'(\Omega) \ , \tag{11}$$

for every $1 \leq t \leq s$. Since the above $(s-1)$ minor is in $L^{p/(s-1)}(\Omega)$ and $p \geq s$, we get that the convergence in (11) is actually in $L^{p/(s-1)}(\Omega)$. Combining (10) and (11) we obtain that

$$u_\nu^{i_1} \frac{\partial(u_\nu^{i_2},\ldots,u_\nu^{i_s})}{\partial(x_{\alpha_1},\ldots,x_{\alpha_{t-1}},x_{\alpha_{t+1}},\ldots,x_{\alpha_s})}$$

$$\rightharpoonup u^{i_1} \frac{\partial(u^{i_2},\ldots,u^{i_s})}{\partial(x_{\alpha_1},\ldots,x_{\alpha_{t-1}},x_{\alpha_{t+1}},\ldots,x_{\alpha_s})}$$

$$\text{in } L_{\text{loc}}^r(\Omega) \text{ with } 1 \leq r < \frac{n}{n-1} \;, \tag{12}$$

for every $1 \leq t \leq s$. We finally use Lemma 2.7 and (12) to get for every $\varphi \in \mathcal{D}(\Omega)$

$$\int_\Omega \frac{\partial(u_\nu^{i_1},\ldots,u_\nu^{i_s})}{\partial(x_{\alpha_1},\ldots,x_{\alpha_s})} \varphi \, dx$$

$$= \sum_{t=1}^s (-1)^t \int_\Omega u_\nu^{i_1} \frac{\partial(u_\nu^{i_2},\ldots,u_\nu^{i_s})}{\partial(x_{\alpha_1},\ldots,x_{\alpha_{t-1}},x_{\alpha_{t+1}},\ldots,x_{\alpha_s})} \frac{\partial\varphi}{\partial x_t} \, dx$$

$$\rightarrow \sum_{t=1}^s (-1)^t \int_\Omega u^{i_1} \frac{\partial(u^{i_2},\ldots,u^{i_s})}{\partial(x_{\alpha_1},\ldots,x_{\alpha_{t-1}},x_{\alpha_{t+1}},\ldots,x_{\alpha_s})} \frac{\partial\varphi}{\partial x_t} \, dx \; .$$

Using again Lemma 2.7 on the right hand side, we obtain (9) and thus Part 4.

Part 5: Let $m,n \geq 2$, $2 \leq s \leq n \wedge m$ and $\frac{1}{p}+\frac{1}{r} \leq 1$. We then have that $\text{adj}_s \nabla u \in (L^1(\Omega))^{\sigma(s)}$. Proceeding exactly as in Part 4 we obtain the result and hence the theorem. $\qquad \square$

It is clear that from Theorem 2.6 and Lemma 2.7 (with the same notations as in the lemma) we get immediately (cf. Ball [2,3]).

Corollary 2.8. *Let $\Omega \subset \mathbf{R}^n$ be a bounded open set and let*

$$u_\nu \rightharpoonup u \text{ in } W^{1,p}(\Omega;\mathbf{R}^m) \;.$$

Part 1: Let $m = n$ and $p > \frac{n^2}{n+1}$, then

$$\text{Det } \nabla u_\nu \rightharpoonup \text{Det } \nabla u \text{ in } \mathcal{D}'(\Omega) \;.$$

Part 2: Let $m,n \geq 2$, $2 \leq s \leq n \wedge m$ and $p > \frac{sn}{n+1}$, then

$$\text{Adj}_s \nabla u_\nu \rightharpoonup \text{Adj}_s \nabla u \text{ in } (\mathcal{D}'(\Omega))^{\sigma(s)} \;.$$

REMARKS.

i) The proof of corollary 2.8 is almost identical to that of Theorem 2.6, using Lemma 2.7;

ii) note that, for example in the case $m = n$, although $\mathrm{Det}\, \nabla u \in \mathcal{D}'(\Omega)$ if $p \geq \frac{n^2}{n+1}$, the weak continuity holds only if $p > \frac{n^2}{n+1}$. Since the imbedding $W^{1,p} \to L^q_{\mathrm{loc}}$, which holds if $1 \leq q \leq \frac{np}{n-p}$, is compact only if $1 \leq q < \frac{np}{n-p}$.

4.2.3 Existence Theorems

We now collect the results of the two preceding sections to get existence theorems in the classical way. There will be two results:

1) The first one involving quasiconvex functions which are finite everywhere,
2) the second one using polyconvex functions which are allowed to take the value $+\infty$ in a subset of $\mathbf{R}^{nm}$.

4.2.3.1 Existence Theorem for Quasiconvex Functions

We now combine the lower semicontinuity result obtained in Theorem 2.4 with a coercivity condition to get the first existence theorem.

Theorem 2.9. *Let $\Omega \subset \mathbf{R}^n$ be a bounded open set. Let $f : \Omega \times \mathbf{R}^m \times \mathbf{R}^{nm} \to \mathbf{R}$ be continuous and quasiconvex and satisfying*

$$(H1) \quad \begin{cases} \alpha|A|^p \leq f(x,u,A) \leq \gamma(1 + |u|^p + |A|^p) & (1) \\[2ex] |f(x,u,A) - f(x,v,B)| \leq \beta(1 + |u|^{p-1} + |v|^{p-1} + |A|^{p-1} \\ \qquad\qquad\qquad\qquad + |B|^{p-1})(|u-v| + |A-B|) & (2) \\[2ex] |f(x,u,A) - f(y,u,A)| \leq \eta(|x-y|)(1 + |u|^p + |A|^p) & (3) \end{cases}$$

where $p > 1$, $\alpha, \beta, \gamma > 0$ and η is a continuous increasing function with $\eta(0) = 0$. Let

$$(P) \quad \inf\left\{ I(u) = \int_\Omega f(x, u(x), \nabla u(x))\, dx : u \in u_0 + W^{1,p}_0(\Omega; \mathbf{R}^m) \right\},$$

then (P) admits at least one solution.

REMARKS.

i) The above theorem is due to Morrey [1,2] (cf. also Meyers [1]);
ii) the hypothesis $(H1)$ can be weakened in several ways:

1) First the coercivity and growth conditions (1) can be replaced (the proof being almost unaltered) by
   - if $p \leq n$

$$\alpha_1 |A|^p + \alpha_2(x) \leq f(x, u, A) \leq \alpha_3 |A|^p + \alpha_4 |u|^q + \alpha_5(x) \qquad (1')$$

where $\alpha_2, \alpha_5 \in L^1(\Omega)$, $\alpha_3 \geq \alpha_1 > 0$, $\alpha_4 \geq 0$, $p > 1$ and $1 \leq q < \frac{np}{n-p}$ (if $p = n$ then $1 \leq q < \infty$);
   - if $p > n$

$$\alpha_1 |A|^p \leq f(x, u, A) \leq \alpha_2 |A|^p + g(x, u) \qquad (1'')$$

where $\alpha_2 \geq \alpha_1 > 0$, $p > 1$, $g \geq 0$ and continuous;

2) if $f(x, u, A) \equiv f(A)$ then (2) is implied by the quasiconvexity of f and (1) (cf. Lemma 2.2);

3) as mentioned in the remarks following Theorem 2.4, Acerbi-Fusco [1] and Marcellini [5] have obtained similar results with f a Carathéodory function satisfying (1) without the hypotheses (2), (3). Marcellini [6] has also shown that, under some more restrictive hypotheses on f, one can show existence of solutions even if the coercivity condition and the growth condition in (1) do not have the same power p.

PROOF of Theorem 2.9. Observe first that $\inf(P)$ is finite, since for example $I(u_0) < +\infty$, by (1). So let u_ν be a minimizing sequence, i.e.

$$I(u_\nu) \to \inf(P) \ .$$

The coercivity condition (1) ensures that $\|\nabla u_\nu\|_{L^p}$ is uniformly bounded. Poincaré inequality ensures then that $\|u_\nu\|_{W^{1,p}}$ is also uniformly bounded. Since $p > 1$, we then deduce that, up to the extraction of a subsequence,

$$u_\nu \rightharpoonup \bar{u} \text{ in } W^{1,p}(\Omega; \mathbf{R}^m) \ .$$

Using Theorem 2.4, we immediately get that

$$I(\bar{u}) = \inf(P) \ . \qquad \qquad \square$$

4.2.3.2 Existence Theorem for Polyconvex Functions

We now give a theorem which is applicable to functions in a smaller class than the previous one from the point of view of convexity (since f polyconvex $\Rightarrow f$ quasiconvex), but in a larger class from the point of view of growth and coercivity conditions. More precisely the previous theorem excludes two important cases:

1) functions f allowed to take the value $+\infty$

2) functions f of the type (if for example $m = n = 2$)

$$f(A) = |A|^2 + |\det A|^2 \ .$$

These two cases are important for applications. For example the first one is useful when one deals with minimization problems with constraints, as it is the case for example in elasticity where a natural constraint is $\det A > 0$.

We now state the theorem for polyconvex functions.

Theorem 2.10. *Let $\Omega \subset \mathbf{R}^n$ be a bounded open set, $f : \Omega \times \mathbf{R}^m \times \mathbf{R}^{nm} \to \bar{\mathbf{R}} = \mathbf{R} \cup \{+\infty\}$. Let $g : \Omega \times \mathbf{R}^m \times \mathbf{R}^{\tau(n,m)} \to \bar{\mathbf{R}} = \mathbf{R} \cup \{+\infty\}$ be a Carathéodory function which is such that*

$$\begin{cases} g(x,u,\bullet) \text{ is convex for every } u \in \mathbf{R}^m \text{ and almost every } x \in \Omega \\ f(x,u,A) \equiv g(x,u,T(A)) \geq \alpha(x) + \displaystyle\sum_{s=1}^{n\wedge m} \beta_s |\text{adj}_s A|^{p_s} \end{cases} \tag{1}$$

where $T(A) = (A, \text{adj}_2 A, \ldots, \text{adj}_{n\wedge m} A)$, $\alpha \in L^1(\Omega)$, $\beta_s > 0$ and $p_1 \geq 2$, $p_s \geq \frac{p_1}{p_1 - 1}$ if $2 \leq s < n \wedge m$ and $p_{n\wedge m} > 1$. Let

$$(P) \quad \inf\left\{ I(u) = \int_\Omega f(x,u(x),\nabla u(x))\,dx \ : \ u \in u_0 + W_0^{1,p}(\Omega;\mathbf{R}^m) \right\} \ .$$

Assume that there exists $\tilde{u} \in u_0 + W_0^{1,p}(\Omega;\mathbf{R}^m)$ such that

$$I(\tilde{u}) < +\infty \ . \tag{2}$$

Then (P) admits at least one solution.

EXAMPLE. If $m = n = 2$ then (1) is read

$$f(x,u,A) \geq \alpha(x) + \beta_1 |A|^{p_1} + \beta_2 |\det A|^{p_2}$$

with $p_1 \geq 2$ and $p_2 > 1$, therefore $f(A) = |A|^2 + (\det A)^2$ satisfies (1).

REMARKS.

i) The above theorem is due to Ball [2,3] and has been applied to find minima in nonlinear elasticity (c.f. Theorem 1.3 in the appendix);

ii) the hypothesis (2) is important to ensure that $I(u) \not\equiv +\infty$ over the whole of $u_0 + W_0^{1,p}$. A way of satisfying (2) would be to impose a growth condition of the same type as the coercivity condition (1), as it was done in Theorem 2.9 and then u_0 would trivially satisfy (2);

iii) often in applications $f(x, u, \nabla u)$ is given as a function of the eigenvalues of the matrix $(\nabla u^t \nabla u)^{1/2}$. For example if $m = n = 3$ and v_1, v_2, v_3 are the eigenvalues of $(\nabla u^t \nabla u)^{1/2}$ and if

$$f(x, u, \nabla u) = \psi(x, u, v_1, v_2, v_3, v_1 v_2, v_1 v_3, v_2 v_3, v_1 v_2 v_3)$$

with ψ satisfying some convexity and cercivity conditions, it is then possible to infer from Theorem 2.10, an existence theorem for such functions, cf. Ball [2,3]; see the appendix below.

PROOF of Theorem 2.10.

Step 1: Let u_ν be a minimizing sequence for (P) then by (1) and (2) we have

$$\int_\Omega \alpha(x)\, dx + \beta_1 \int_\Omega |\nabla u_\nu|^{p_1}\, dx$$

$$+ \sum_{s=2}^{n \wedge m} \beta_s \int_\Omega |\mathrm{adj}_s \nabla u_\nu|^{p_s}\, dx \leq I(u_\nu) < +\infty\ .$$

Using Poincaré inequality and denoting by γ a positive generic constant, we find

$$\gamma \left(1 + \|u_\nu\|_{W^{1,p_1}} + \sum_{s=2}^{n \wedge m} \|\mathrm{adj}_s \nabla u_\nu\|_{L^{p_s}} \right) \leq I(u_\nu) < +\infty\ .$$

Note that since $p_s > 1, 1 \leq s \leq n \wedge m$, we can extract a weakly convergent subsequence (still denoted u_ν) such that

$$\begin{cases} u_\nu \rightharpoonup u \ \text{in}\ W^{1,p_1}(\Omega; \mathbf{R}^m) \\ \mathrm{adj}_s \nabla u_\nu \rightharpoonup A_s \ \text{in}\ (L^{p_s}(\Omega))^{\sigma(s)} \quad s = 2, \ldots, n \wedge m \end{cases} \tag{3}$$

Step 2: We now show, by induction, that in fact (3) implies

$$\mathrm{adj}_s \nabla u_\nu \rightharpoonup \mathrm{adj}_s \nabla u \ \text{in}\ (\mathcal{D}'(\Omega))^{\sigma(s)}\ . \tag{4}$$

If $s = 2$, then Theorem 2.6 combined with the fact that $p_1 \geq 2$ gives immediately (3). Assume that we have proved (4) up to $s - 1 \leq (m \wedge n) - 2$, then we have using (3) and (4) that

$$\mathrm{adj}_{s-1} \nabla u - A_{s-1} \in L^{p_s-1}\ .$$

We have therefore

$$\begin{cases} u_\nu \rightharpoonup u \ \text{in}\ W^{1,p_1} \\ \mathrm{adj}_{s-1} \nabla u_\nu \rightharpoonup \mathrm{adj}_{s-1} \nabla u \ \text{in}\ L^{p_s-1} \end{cases}\ .$$

Since $\frac{1}{p_1} + \frac{1}{p_s} \leq 1$, we have immediately (4) by Part 5 of Theorem 2.6.

Step 3: Therefore summarizing (3) and (4) we have that, up to a subsequence,

$$\begin{cases} u_\nu \rightarrow u \text{ in } (L^{p_1}(\Omega))^m \\ T(\nabla u_\nu) \rightharpoonup T(\nabla u) \text{ in } (L^1(\Omega))^{\tau(n,m)} \end{cases} .$$

Hence we may apply Theorem 3.4 of Chapter 3 to g and obtain that

$$\liminf_{\nu \to \infty} I(u_\nu) = \liminf_{\nu \to \infty} \int_\Omega g(x, u_\nu(x), T(\nabla u_\nu(x))) \, dx$$

$$\geq \int_\Omega g(x, u(x), T(\nabla u(x))) \, dx = I(u) \ ,$$

and therefore u is a minimum for (P). $\square$

4.2.3.3 Some Remarks on Regularity Results

We give here some regularity results. We do not intend to give neither a complete view of all known results nor any proofs. We refer for a recent presentation to the book of Giaquinta [1]. We here only want to stress again the difference between the scalar case (cf. Chapter 3) and the vectorial case.

We now quote a theorem which is essentially due to Evans [1] and has been improved by Giaquinta-Modica [1], Fusco-Hutchinson [1,2], Acerbi-Fusco [2].

Theorem 2.11. *Let $\Omega \subset \mathbb{R}^n$ be an open set. Let $f : \Omega \times \mathbb{R}^m \times \mathbb{R}^{nm} \to \mathbb{R}$ be such that*

$$(H1) \qquad \begin{cases} c_1(|A|^k - 1) \leq f(x, u, A) \leq c_2(|A|^k + 1), \\ \text{for every } (x, u, A) \in \Omega \times \mathbb{R}^m \times \mathbb{R}^{nm} \end{cases}$$

where $k \geq 2$, $c_1, c_2 > 0$;

$$(H2) \qquad |f(x, u, A) - f(y, v, A)| \leq c_3(1 + |A|^k)\omega(|x - y|^2 + |u - v|^2) \ ,$$

where $\omega(t) \leq t^\sigma$, $0 < \sigma \leq \frac{1}{2}$ and ω is bounded, concave, nonnegative and increasing;

$$(H3) \qquad \begin{cases} f_{AA}(x, u, A) = \left(\dfrac{\partial^2 f}{\partial A_\alpha^i \, \partial A_\beta^j} \right)_{\substack{1 \leq i,j \leq m \\ 1 \leq \alpha,\beta \leq n}} \quad \text{exists and is continuous} \\ |f_{AA}(x, u, A)| \leq c_4(|A|^{k-2} + 1) \end{cases} \ ;$$

$$(H4) \qquad \begin{aligned} \int_\Omega [f(x_0, u_0, A_0) + \gamma(|\nabla\varphi|^2 + |\nabla\varphi|^k)]\, dx \\ \leq \int_\Omega f(x_0, u_0, A_0 + \nabla\varphi(x))\, dx \end{aligned}$$

for every $(x_0, u_0, A_0) \in \Omega \times \mathbf{R}^m \times \mathbf{R}^{nm}$, for every $\varphi \in C_0^1(\Omega; \mathbf{R}^m)$ and for some $\gamma > 0$, $k \geq 2$ (as in $(H1)$, $(H2)$, $(H3)$).
   Let $u \in u_0 + W_0^{1,k}(\Omega; \mathbf{R}^m)$ be a minimum of (P)

$$(P) \quad \inf\left\{ I(u) = \int_\Omega f(x, u(x), \nabla u(x))\, dx \; : \; u \in u_0 + W_0^{1,k}(\Omega; \mathbf{R}^m) \right\},$$

then there exists an open set $\Omega_0 \subset \Omega$ such that

$$\begin{cases} \operatorname{meas}(\Omega - \Omega_0) = 0 \\ u \in C^{1,\alpha}(\Omega_0; \mathbf{R}^m) \end{cases},$$

for some $\alpha < 1$.

REMARKS.

i) The above theorem as stated can be found in Fusco-Hutchinson [1]. There are some improvements of this result for example with a weakened version of $(H3)$, see Acerbi-Fusco [2];

ii) the hypothesis $(H4)$ is a kind of *uniform quasiconvexity*;

iii) a function f satisfying $(H1)$-$(H4)$ is for example

$$f(x, u, A) = a|A|^2 + b|A|^k + g(A)$$

where $a, b > 0$, g is quasiconvex with $|g_{AA}(A)| \leq c(1 + |A|^{k-2})$;

iv) it is well known that in general the set $\Omega - \Omega_0$ may be non empty. We give here some examples which are taken from Giaquinta [1] and we refer to his book for the exact references;

v) finally one should compare the above theorem to the corresponding theorems for the scalar case (Theorem 4.2, 4.5 of Chapter 3).

EXAMPLES.

i) (Necas) Let Ω be the unit ball of $\mathbf{R}^n$, let $u : \Omega \subset \mathbf{R}^n \to \mathbf{R}^{n^2}$ and $f : \mathbf{R}^{nm} = \mathbf{R}^{n^3} \to \mathbf{R}$ strictly convex and satisfying $(H1)$-$(H4)$. Then Necas produces for n large enough a function f which is such that

$$I(u) = \int_\Omega f(\nabla u(x))^2\, dx \geq I(\bar{u})$$

where $u \in \bar{u} + W_0^{1,p}(\Omega; \mathbf{R}^{n^2})$ and

$$\bar{u}_{ij}(x) = \frac{x_i x_j}{|x|} \ .$$

Observe that $\bar{u} \in W^{1,\infty}$ but $\bar{u} \notin C^1$.

ii) (Giusti-Miranda) Let Ω be the unit ball of $\mathbf{R}^n$, $u : \Omega \subset \mathbf{R}^n \to \mathbf{R}^n$ with n sufficiently large and let

$$I(u) = \int_\Omega f(u(x), \nabla u(x)) \, dx$$

$$= \int_\Omega \left\{ \sum_{i,j=1}^n \left(\frac{\partial u_j}{\partial x_i} \right)^2 + \left[\sum_{i,j=1}^n \left(\delta_{ij} + \frac{4}{n-2} \frac{u_i u_j}{1+|u|^2} \right) \frac{\partial u_j}{\partial x_i} \right]^2 \right\} dx$$

where $\delta_{ij} = 0$ if $i \neq j$ and 1 if $i = j$. Then $u_0(x) = \frac{x}{|x|}$ is the unique minimum of I in the class of functions $x + W_0^{1,2}(\Omega; \mathbf{R}^n)$.

iii) (Hildebrandt-Widmann) Let Ω be the unit ball of $\mathbf{R}^3$, $u : \Omega \subset \mathbf{R}^3 \to \mathbf{R}^3$, $a \in C^\infty(\mathbf{R})$ with $a'(1) = -2a(1)$, and let

$$I(u) = \int_\Omega f(u(x), \nabla u(x)) \, dx = \int_\Omega a(|u|)|\nabla u|^2 \, dx \ .$$

Then $u_0(x) = \frac{x}{|x|}$ is a minimum for I among the class of functions $u \in W^{1,2}(\Omega; \mathbf{R}^3) \cap L^\infty(\Omega)$ which agrees with x on $\partial\Omega$.

4.3 Appendix: Some Elementary Properties of Determinants

In the whole of Chapter 4 we have seen the importance of *determinants* in the vectorial calculus of variations. We gather in this appendix some well known algebraic and analytic properties of determinants. In the first part, we introduce carefully the notation for the minors $\text{adj}_s A$ of a given matrix A. In the second part, we shall give some properties of Jacobians of the form $\text{adj}_s \nabla u$.

4.3.1 Definitions and Properties of Determinants

We first introduce some notations. Let $n \in \mathbb{N}$ (the set of positive integers) and let $1 \leq s \leq n$. We define

$$I_s^n = \{(\alpha_1, \ldots, \alpha_s) \in \mathbb{N}^s : 1 \leq \alpha_1 < \alpha_2 < \ldots < \alpha_s \leq n\} \ .$$

We call the elements of I_s^n increasing s-tuples. The number of elements of I_s^n is then

$$\mathrm{Card}\, I_s^n = \binom{n}{s} = \frac{n!}{s!(n-s)!} \ .$$

We next endow I_s^n with the following ordering relation:

$$\alpha = (\alpha_1, \ldots, \alpha_s) > (\beta_1, \ldots, \beta_s) = \beta$$

if and only if

$$\alpha_k < \beta_k$$

where k is the largest integer less than or equal to s such that $\alpha_k \neq \beta_k$ and $\alpha_l = \beta_l$ for every $l > k$. (This is the inverse of the lexicographical order when read backward).

EXAMPLES.

i) $n = 4$, $s = 2$, then

$$(1,2) > (1,3) > (2,3) > (1,4) > (2,4) > (3,4) \ ;$$

ii) $n = 5$, $s = 3$, then

$$(1,2,3) > (1,2,4) > (1,3,4) > (2,3,4) > (1,2,5)$$
$$> (1,3,5) > (2,3,5) > (1,4,5) > (2,4,5) > (3,4,5) \ .$$

We then define the map φ_s^n

$$\varphi_s^n : \{1, 2, 3, \ldots, \binom{n}{s}\} \to I_s^n$$

as the only bijection which respects the order defined above.

EXAMPLE. $n = 4$, $s = 2$, then

$$\varphi_2^4(1) = (3,4), \ \varphi_2^4(2) = (2,4), \ \varphi_2^4(3) = (1,4),$$
$$\varphi_2^4(4) = (2,3), \ \varphi_2^4(5) = (1,3), \ \varphi_2^4(6) = (1,2) \ .$$

We are now in a position to define for a given matrix $A \in \mathbf{R}^{nm}$, the adjugate matrix of order s, $\mathrm{adj}_s A$, $1 \leq s \leq n \wedge m = \min\{n,m\}$. Let $A \in \mathbf{R}^{nm}$ be such that

$$A = \begin{pmatrix} A_1^1 & \cdots & A_n^1 \\ \vdots & & \vdots \\ A_1^m & \cdots & A_n^m \end{pmatrix} = \begin{pmatrix} A^1 \\ \vdots \\ A^m \end{pmatrix} = (A_1, \ldots, A_n) \ .$$

We define $\mathrm{adj}_s A$ to be the following matrix in $\mathbf{R}^{\sigma(s)}$, where $\sigma(s) = \binom{n}{s}\binom{m}{s}$,

$$\mathrm{adj}_s A = \begin{pmatrix} (\mathrm{adj}_s A)_1^1 & \cdots & (\mathrm{adj}_s A)_{\binom{n}{s}}^1 \\ \vdots & & \vdots \\ (\mathrm{adj}_s A)_1^{\binom{m}{s}} & \cdots & (\mathrm{adj}_s A)_{\binom{n}{s}}^{\binom{m}{s}} \end{pmatrix}$$

$$= \begin{pmatrix} (\mathrm{adj}_s A)^1 \\ \vdots \\ (\mathrm{adj}_s A)^{\binom{m}{s}} \end{pmatrix} = \left((\mathrm{adj}_s A)_1, \ldots, (\mathrm{adj}_s A)_{\binom{n}{s}} \right) \ ,$$

where

$$(\mathrm{adj}_s A)_\alpha^i = (-1)^{i+\alpha} \det \begin{pmatrix} A_{\alpha_1}^{i_1} & \cdots & A_{\alpha_s}^{i_1} \\ \vdots & & \vdots \\ A_{\alpha_1}^{i_s} & \cdots & A_{\alpha_s}^{i_s} \end{pmatrix}$$

and $(i_1, \ldots, i_s)$, $(\alpha_1, \ldots, \alpha_s)$ are the s-tuples corresponding to i and α by the bijections φ_s^m and φ_s^n.

EXAMPLES.

i) $m = n = 2$, $s = 1$: let

$$A = \begin{pmatrix} A_1^1 & A_2^1 \\ A_1^2 & A_2^2 \end{pmatrix} \ .$$

Then $I_s^n = I_s^m = \{1,2\}$ and the bijection $\varphi : \{1,2\} \to \{2,1\}$. Hence

$$\mathrm{adj}_1 A = \begin{pmatrix} (\mathrm{adj}_1 A)_1^1 & (\mathrm{adj}_1 A)_2^1 \\ (\mathrm{adj}_1 A)_1^2 & (\mathrm{adj}_1 A)_2^2 \end{pmatrix} = \begin{pmatrix} A_2^2 & -A_1^2 \\ -A_2^1 & A_1^1 \end{pmatrix} \ .$$

(note that $\mathrm{adj}_1 A$ is exactly $\tilde{A}$ defined in Proposition 1.13 above);

ii) $m = n = s = 2$, then $I_s^n = I_s^m = \{(1,2)\}$ and $\varphi_2^2(1) = (1,2)$. Hence

$$\mathrm{adj}_2 A = \det \begin{pmatrix} A_1^1 & A_2^1 \\ A_1^2 & A_2^2 \end{pmatrix} = \det A \ .$$

iii) $m = 3$, $s = n = 2$, then $I_s^n = I_2^2 = \{(1,2)\}$ and $\varphi_2^2(1) = (1,2)$ while $I_s^m = I_2^3 = \{(1,2);(1,3);(2,3)\}$ and $\varphi_2^3(1) = (2,3)$, $\varphi_2^3(2) = (1,3)$, $\varphi_2^3(3) = (1,2)$. Therefore if

$$A = \begin{pmatrix} A_1^1 & A_2^1 \\ A_1^2 & A_2^2 \\ A_1^3 & A_2^3 \end{pmatrix} = \begin{pmatrix} A^1 \\ A^2 \\ A^3 \end{pmatrix} = (A_1, A_2) \ ,$$

then

$$\operatorname{adj}_2 A = \begin{pmatrix} (\operatorname{adj}_2 A)_1^1 \\ (\operatorname{adj}_2 A)_1^2 \\ (\operatorname{adj}_2 A)_1^3 \end{pmatrix} = \begin{pmatrix} \det \begin{pmatrix} A_1^2 & A_2^2 \\ A_1^3 & A_2^3 \end{pmatrix} \\ -\det \begin{pmatrix} A_1^1 & A_2^1 \\ A_1^3 & A_2^3 \end{pmatrix} \\ \det \begin{pmatrix} A_1^1 & A_2^1 \\ A_1^2 & A_2^2 \end{pmatrix} \end{pmatrix} \ .$$

iv) $m = n + 1$, $s = n$, we let

$$A = \begin{pmatrix} A_1^1 & \cdots & A_n^1 \\ \vdots & & \vdots \\ A_1^{n+1} & \cdots & A_n^{n+1} \end{pmatrix} = \begin{pmatrix} A^1 \\ \vdots \\ A^{n+1} \end{pmatrix} = (A_1, \ldots, A_n) \ .$$

Then

$$\operatorname{adj}_n A = \begin{pmatrix} (\operatorname{adj}_n A)_1^1 \\ \vdots \\ (\operatorname{adj}_n A)_1^{n+1} \end{pmatrix} = \begin{pmatrix} \det \begin{pmatrix} A_1^2 & \cdots & A_n^2 \\ \vdots & & \vdots \\ A_1^{n+1} & \cdots & A_n^{n+1} \end{pmatrix} \\ \vdots \\ (-1)^{n+2} \det \begin{pmatrix} A_1^1 & \cdots & A_n^1 \\ \vdots & & \vdots \\ A_1^n & \cdots & A_n^n \end{pmatrix} \end{pmatrix}$$

$$= \begin{pmatrix} \det \hat{A}^1 \\ \vdots \\ (-1)^{n+2} \det \hat{A}^{n+1} \end{pmatrix}$$

where $\hat{A}^k$ denotes the $n \times n$ matrix obtained by suppressing the k-th row in the matrix A;

v) $m = n = s = 3$, then $I_3^3 = \{(1,2,3)\}$ and therefore

$$\operatorname{adj}_3 A = \det A \ .$$

vi) $m = n = 3$, $s = 2$, then

$$\mathrm{adj}_2 A = \begin{pmatrix} \det\begin{pmatrix} A_2^2 & A_3^2 \\ A_2^3 & A_3^3 \end{pmatrix} & -\det\begin{pmatrix} A_1^2 & A_3^2 \\ A_1^3 & A_3^3 \end{pmatrix} & \det\begin{pmatrix} A_1^2 & A_2^2 \\ A_1^3 & A_2^3 \end{pmatrix} \\ -\det\begin{pmatrix} A_2^1 & A_3^1 \\ A_2^3 & A_3^3 \end{pmatrix} & \det\begin{pmatrix} A_1^1 & A_3^1 \\ A_1^3 & A_3^3 \end{pmatrix} & -\det\begin{pmatrix} A_1^1 & A_2^1 \\ A_1^3 & A_2^3 \end{pmatrix} \\ \det\begin{pmatrix} A_2^1 & A_3^1 \\ A_2^2 & A_3^2 \end{pmatrix} & -\det\begin{pmatrix} A_1^1 & A_3^1 \\ A_1^2 & A_3^2 \end{pmatrix} & \det\begin{pmatrix} A_1^1 & A_2^1 \\ A_1^2 & A_2^2 \end{pmatrix} \end{pmatrix}.$$

The above expression is the usual transpose of the matrix of *cofactors*.

REMARK. Note that one can write the rows of $\mathrm{adj}_s A$ in the following way

$$(\mathrm{adj}_s A)^i = (-1)^{i+1} \mathrm{adj}_s \begin{pmatrix} A^{i_1} \\ \vdots \\ A^{i_s} \end{pmatrix}, \quad 1 \leq i \leq \binom{m}{s},$$

where $(i_1, \ldots, i_s) = \varphi_s^m(i)$ is the s-tuple associated to the integer i. So in particular

$$(\mathrm{adj}_s A)^1 = \mathrm{adj}_s \begin{pmatrix} A^{m-s+1} \\ A^{m-s+2} \\ \vdots \\ A^{m-1} \\ A^m \end{pmatrix};$$

$$(\mathrm{adj}_s A)^{\binom{m}{s}} = (-1)^{\binom{m}{s}+1} \mathrm{adj}_s \begin{pmatrix} A^1 \\ A^2 \\ \vdots \\ A^{s-1} \\ A^s \end{pmatrix}.$$

A similar remark applies to the columns of $\mathrm{adj}_s A$.

Notations. We sometimes, as in the example iv) above, denote by $\hat{A}^{i_1,\ldots,i_k}_{\alpha_1,\ldots,\alpha_l}$ the $(n-l) \times (m-k)$ matrix obtained from $A \in \mathbf{R}^{nm}$ by suppressing the k rows $i_1, \ldots, i_k$ and the l columns $\alpha_1, \ldots, \alpha_l$.

We now give some elementary properties of determinants whose proofs are standard.

Proposition 3.1. *Let $A \in \mathbb{R}^{nm}$.*

i) If $m = n$, then

$$\det A = \langle A^\nu; (\mathrm{adj}_{n-1} A)^\nu \rangle = \langle A_\nu; (\mathrm{adj}_{n-1} A)_\nu \rangle, \quad \nu = 1, 2, \ldots, n \ ,$$

where $\langle \bullet; \bullet \rangle$ denotes the scalar product in $\mathbb{R}^n$.

ii) If $m = n$, then

$$A(\mathrm{adj}_{n-1} A)^t = \det A \cdot I$$

where I is the identity matrix in $\mathbb{R}^{n^2}$ and A^t denotes the transpose of the matrix A. In particular if $\det A \neq 0$, then

$$A^{-1} = \frac{1}{\det A} (\mathrm{adj}_{n-1} A)^t \ .$$

iii) If $m = n + 1$, then

$$\langle A_\nu; \mathrm{adj}_n A \rangle = 0, \quad \nu = 1, \ldots, n \ ,$$

where $\langle \bullet; \bullet \rangle$ denotes the scalar product in $\mathbb{R}^{n+1}$.

iv) If $m = n - 1$, then

$$\langle A^\nu; \mathrm{adj}_{n-1} A \rangle = 0, \quad \nu = 1, \ldots, n - 1 \ ,$$

where $\langle \bullet; \bullet \rangle$ denotes the scalar product in $\mathbb{R}^n$.

v) If $m = n$, then

$$\frac{\partial}{\partial A^i_\alpha}(\det A) = (\mathrm{adj}_{n-1} A)^i_\alpha, \quad 1 \leq i, \alpha \leq n = m \ .$$

vi) Denote by

$$T(A) = (A, \mathrm{adj}_2 A, \ldots, \mathrm{adj}_{n \wedge m} A) \in \mathbb{R}^{\tau(n,m)}$$

where $n \wedge m = \min\{n, m\}$ and

$$\tau(n, m) = \sum_{s=1}^{n \wedge m} \sigma(s) = \sum_{s=1}^{n \wedge m} \binom{n}{s}\binom{m}{s} \ .$$

Let $a \in \mathbb{R}^n$, $b \in \mathbb{R}^m$. Define $a \otimes b = (a^i b_\alpha)_{1 \leq \alpha \leq n, \, 1 \leq i \leq m} \in \mathbb{R}^{nm}$. Let $\lambda \in [0, 1]$, then

$$T(A + (1 - \lambda)a \otimes b) = \lambda T(A) + (1 - \lambda)T(A + a \otimes b) \ .$$

PROOF.

 i) and ii) are standard.

iii) Let $m = n + 1$ and $\nu \in \{1, \ldots, n\}$. We have to show that

$$\langle A_\nu; \operatorname{adj}_n A \rangle = 0 . \tag{1}$$

Define the matrix $B = [A_\nu; A] \in \mathbb{R}^{(n+1)(n+1)}$ (recall that $A \in \mathbb{R}^{n(n+1)}$). Then $B_1 = B_{\nu+1}$ and therefore $\det B = 0$. Using i) we obtain

$$0 = \det B = \langle B_1; (\operatorname{adj}_n B)_1 \rangle = \langle A_\nu; \operatorname{adj}_n A \rangle .$$

 iv) This is established exactly as above.

  v) This is a direct consequence of i).

 vi) The result is equivalent to

$$\operatorname{adj}_s(A + (1 - \lambda)a \otimes b) = \lambda \operatorname{adj}_s A + (1 - \lambda)\operatorname{adj}_s(A + a \otimes b) , \tag{2}$$

for every $1 \leq s \leq n \wedge m$. In terms of components this is equivalent to

$$\begin{aligned}
&(\operatorname{adj}_s(A + (1 - \lambda)a \otimes b))^i_\alpha \\
&\qquad = \lambda(\operatorname{adj}_s A)^i_\alpha + (1 - \lambda)(\operatorname{adj}_s(A + a \otimes b))^i_\alpha ,
\end{aligned} \tag{3}$$

$1 \leq i \leq \binom{m}{s}, 1 \leq \alpha \leq \binom{n}{s}$. Recall that

$$(\operatorname{adj}_s A)^i_\alpha = (-1)^{i+\alpha} \det \begin{pmatrix} A^{i_1}_{\alpha_1} & \cdots & A^{i_1}_{\alpha_s} \\ \vdots & & \vdots \\ A^{i_s}_{\alpha_1} & \cdots & A^{i_s}_{\alpha_s} \end{pmatrix} .$$

Let, by abuse of notation,

$$A = \begin{pmatrix} A^{i_1}_{\alpha_1} & \cdots & A^{i_1}_{\alpha_s} \\ \vdots & & \vdots \\ A^{i_s}_{\alpha_1} & \cdots & A^{i_s}_{\alpha_s} \end{pmatrix}, \quad a \otimes b = \begin{pmatrix} a^{i_1} b_{\alpha_1} & \cdots & a^{i_1} b_{\alpha_s} \\ \vdots & & \vdots \\ a^{i_s} b_{\alpha_1} & \cdots & a^{i_s} b_{\alpha_s} \end{pmatrix} .$$

Therefore (3) is equivalent to show that for every $A \in \mathbb{R}^{s^2}$, $a, b \in \mathbb{R}^s$, $\lambda \in [0, 1]$

$$\det(A + (1 - \lambda)a \otimes b) = \lambda \det A + (1 - \lambda)\det(A + a \otimes b) . \tag{4}$$

This is however a standard property of determinants. $\qquad\square$

4.3.2 Some Properties of Jacobians

We now let $u : \mathbf{R}^n \to \mathbf{R}^m$ be a C^2 function (hence $\nabla u \in \mathbf{R}^{nm}$) and study the analytical properties of $\mathrm{adj}_s \nabla u$. The main result is that they may be written in divergence form.

Theorem 3.2. *Let $u \in C^2(\mathbf{R}^n; \mathbf{R}^m)$.*

i) Let $2 \leq s \leq n \wedge m$, $1 \leq i_1 < \ldots < i_s \leq n$ and $1 \leq \alpha_1 < \ldots < \alpha_s \leq m$, then

$$\sum_{t=1}^{s} (-1)^{t+1} \frac{\partial}{\partial x_{\alpha_t}} \left[\frac{\partial(u_{i_2}, \ldots, u_{i_s})}{\partial(x_{\alpha_1}, \ldots, x_{\alpha_{t-1}}, x_{\alpha_{t+1}}, \ldots, x_{\alpha_s})} \right] = 0 \ , \qquad (1)$$

$$
\begin{aligned}
&\frac{\partial(u_{i_1}, \ldots, u_{i_s})}{\partial(x_{\alpha_1}, \ldots, x_{\alpha_s})} \\
&= \sum_{t=1}^{s} (-1)^{t+1} \frac{\partial}{\partial x_{\alpha_t}} \left[u_{i_1} \frac{\partial(u_{i_2}, \ldots, u_{i_s})}{\partial(x_{\alpha_1}, \ldots, x_{\alpha_{t-1}}, x_{\alpha_{t+1}}, \ldots, x_{\alpha_s})} \right] .
\end{aligned}
\qquad (2)
$$

ii) Let $\Omega \subset \mathbf{R}^n$ be a bounded open set and let

$$T(A) = (A, \mathrm{adj}_2 A, \ldots, \mathrm{adj}_{n \wedge m} A)$$

then

$$\int_{\Omega} T(A + \nabla\varphi(x)) \, dx = T(A) \cdot \mathrm{meas} \, \Omega \qquad (3)$$

for every $A \in \mathbf{R}^{nm}$ and for every $\varphi \in C_0^2(\Omega; \mathbf{R}^m)$.

EXAMPLE. Let $m = n = 2$, $u(x_1, x_2) = (u_1, u_2)$ and $\nabla u = \left(\begin{smallmatrix} \mathrm{grad}\, u_1 \\ \mathrm{grad}\, u_2 \end{smallmatrix} \right)$. Then (1) expresses just the fact that $\mathrm{curl}(\mathrm{grad}\, u_1) = \mathrm{curl}(\mathrm{grad}\, u_2) = 0$. Equation (2) is then

$$
\begin{aligned}
\det \nabla u &= \frac{\partial}{\partial x_1} \left(u_1 \frac{\partial u_2}{\partial x_2} \right) - \frac{\partial}{\partial x_2} \left(u_1 \frac{\partial u_2}{\partial x_1} \right) \\
&= \frac{\partial}{\partial x_2} \left(u_2 \frac{\partial u_1}{\partial x_1} \right) - \frac{\partial}{\partial x_1} \left(u_2 \frac{\partial u_1}{\partial x_2} \right) .
\end{aligned}
$$

Finally note that

$$T(A) = (A, \det A)$$

and hence (3) becomes

$$
\begin{aligned}
T(A) \cdot \mathrm{meas} \, \Omega &= (A, \det A) \cdot \mathrm{meas} \, \Omega \\
&= \int_{\Omega} (A + \nabla\varphi(x), \det(A + \nabla\varphi(x))) \, dx .
\end{aligned}
$$

REMARKS.

i) It is obvious that in (1) and (2) one can interchange the role of u_{i_1} with any u_{i_t};

ii) it is also simple to see that (3) holds not only for every $\varphi \in C_0^2(\Omega; \mathbf{R}^m)$ but also for every $u \in W_0^{1,\infty}(\Omega; \mathbf{R}^m)$, by a density argument.

PROOF.

i) Note that (2) is a direct consequence of (1). To show (1) we proceed by induction. For simplification we take $(i_1, \ldots, i_s) = (1, \ldots, s)$ and $(\alpha_1, \ldots, \alpha_s) = (1, \ldots, s)$. Assume that (1) and (2) have been established up to the order $(s-1)$, the case $s = 1$ being trivial. Using the hypothesis of induction we have

$$
\frac{\partial(u_2, \ldots, u_s)}{\partial(x_1, \ldots, \hat{x}_t, \ldots, x_s)}
$$

$$
= \sum_{\alpha=1}^{t-1} (-1)^{\alpha+1} \frac{\partial}{\partial x_\alpha} \left(u_2 \frac{\partial(u_3, \ldots, u_s)}{\partial(x_1, \ldots, \hat{x}_\alpha, \ldots, \hat{x}_t, \ldots, x_s)} \right)
$$

$$
+ \sum_{\alpha=t+1}^{s} (-1)^{\alpha} \frac{\partial}{\partial x_\alpha} \left(u_2 \frac{\partial(u_3, \ldots, u_s)}{\partial(x_1, \ldots, \hat{x}_t, \ldots, \hat{x}_\alpha, \ldots, x_s)} \right)
$$

where we have denoted by $(x_1, \ldots, \hat{x}_t, \ldots, x_s) = (x_1, \ldots, x_{t-1}, x_{t+1}, \ldots, x_s)$ and similarly for $(x_1, \ldots, \hat{x}_t, \ldots, \hat{x}_\alpha, \ldots, x_s)$. Returning to (1) and using the above identity we have

$$
\sum_{t=1}^{s} (-1)^{t+1} \frac{\partial}{\partial x_t} \left[\frac{\partial(u_2, \ldots, u_s)}{\partial(x_1, \ldots, \hat{x}_t, \ldots, x_s)} \right]
$$

$$
= \sum_{t=1}^{s} \left[\sum_{\alpha=1}^{t-1} (-1)^{\alpha+t} \frac{\partial^2}{\partial x_t\, \partial x_\alpha} \left(u_2 \frac{\partial(u_3, \ldots, u_s)}{\partial(x_1, \ldots, \hat{x}_\alpha, \ldots, \hat{x}_t, \ldots, x_s)} \right) \right.
$$

$$
\left. + \sum_{\alpha=t+1}^{s} (-1)^{\alpha+t+1} \frac{\partial^2}{\partial x_t\, \partial x_\alpha} \left(u_2 \frac{\partial(u_3, \ldots, u_s)}{\partial(x_1, \ldots, \hat{x}_t, \ldots, \hat{x}_\alpha, \ldots, x_s)} \right) \right] . \tag{4}
$$

Now observe that for any $r < \beta$

$$
\frac{\partial^2}{\partial x_r\, \partial x_\beta} \left(u_2 \frac{\partial(u_3, \ldots, u_s)}{\partial(x_1, \ldots, \hat{x}_r, \ldots, \hat{x}_\beta, \ldots, x_s)} \right) [(-1)^{r+\beta} + (-1)^{r+\beta+1}] \equiv 0 .
$$

$$
\tag{5}
$$

We have therefore established (1) by combining (4) and (5).

ii) In order to establish (3) we only need to show that

$$\int_\Omega \mathrm{adj}_s(A + \nabla\varphi(x))\, dx = \mathrm{adj}_s A \cdot \mathrm{meas}\,\Omega \tag{6}$$

for every $1 \le s \le n \wedge m$, for every $A \in \mathbf{R}^{nm}$ and for every $\varphi \in C_0^2(\Omega; \mathbf{R}^m)$. Recall that for $1 \le i \le \binom{m}{s}$, $1 \le \alpha \le \binom{n}{s}$, we have

$$(\mathrm{adj}_s A)_\alpha^i = (-1)^{i+\alpha} \det \begin{pmatrix} A_{\alpha_1}^{i_1} & \cdots & A_{\alpha_s}^{i_1} \\ \vdots & & \vdots \\ A_{\alpha_1}^{i_s} & \cdots & A_{\alpha_s}^{i_s} \end{pmatrix}.$$

Let, by abuse of notation,

$$A = \begin{pmatrix} A_{\alpha_1}^{i_1} & \cdots & A_{\alpha_s}^{i_1} \\ \vdots & & \vdots \\ A_{\alpha_1}^{i_s} & \cdots & A_{\alpha_s}^{i_s} \end{pmatrix}, \quad \nabla\varphi = \begin{pmatrix} \dfrac{\partial \varphi_{i_1}}{\partial x_{\alpha_1}} & \cdots & \dfrac{\partial \varphi_{i_1}}{\partial x_{\alpha_s}} \\ \vdots & & \vdots \\ \dfrac{\partial \varphi_{i_s}}{\partial x_{\alpha_1}} & \cdots & \dfrac{\partial \varphi_{i_s}}{\partial x_{\alpha_s}} \end{pmatrix}.$$

Therefore (3) or (6) are equivalent to show that

$$\int_\Omega \det(A + \nabla\varphi(x))\, dx = \det A \cdot \mathrm{meas}\,\Omega \tag{7}$$

for every bounded domain $\Omega \subset \mathbf{R}^s$, for every $A \in \mathbf{R}^{s^2}$, for every $\varphi \in C_0^2(\Omega; \mathbf{R}^s)$. To show (7) we proceed by induction on s. The result is trivial if $s = 1$. Assume therefore that (7) has been established up to the order $s - 1$. Using Proposition 3.1, we have that

$$\det(A + \nabla\varphi)$$
$$= \langle (A + \nabla\varphi)^1; (\mathrm{adj}_{s-1}(A + \nabla\varphi))^1 \rangle$$
$$= \langle A^1; (\mathrm{adj}_{s-1}(A + \nabla\varphi))^1 \rangle + \langle (\nabla\varphi)^1; (\mathrm{adj}_{s-1}(A + \nabla\varphi))^1 \rangle$$
$$= \sum_{\alpha=1}^{s} \left[A_\alpha^1 (\mathrm{adj}_{s-1}(A + \nabla\varphi))_\alpha^1 + \frac{\partial \varphi_1}{\partial x_\alpha}(\mathrm{adj}_{s-1}(A + \nabla\varphi))_\alpha^1 \right]. \tag{8}$$

Integrating (8), using the hypothesis of induction on the first part, an integration by part in the second term and (1) we have indeed obtained (7) and thus the theorem. $\qquad\square$

CHAPTER 5

Non-Convex Integrands

5.0 Introduction

In Chapter 3 and 4 we have seen that in order to get existence theorems for

$$(P) \qquad \inf \left\{ I(u) = \int_\Omega f(x, u(x), \nabla u(x)) \, dx \; : \; u \in u_0 + W_0^{1,p}(\Omega; \mathbf{R}^m) \right\}$$

the convexity (or quasiconvexity in the vectorial case) of f, with respect to the last variable, plays a central role. In this chapter we shall study the case where f fails to be convex (quasiconvex in the vectorial case).

In Section 5.1 we shall consider functions $f = f(\nabla u) : \mathbf{R}^{nm} \to \bar{\mathbf{R}} = \mathbf{R} \cup \{+\infty\}$ and characterize, with the help of the Hahn-Banach and Carathéodory theorem the *convex, polyconvex, quasiconvex* and *rank one convex envelopes*, i.e.

$$Cf = \sup\{g \le f : g \text{ convex}\}$$

$$Pf = \sup\{g \le f : g \text{ polyconvex}\}$$

$$Qf = \sup\{g \le f : g \text{ quasiconvex}\}$$

$$Rf = \sup\{g \le f : g \text{ rank one convex}\} \; .$$

In view of Theorem 1.1 of Chapter 4 one has

$$Cf \le Pf \le Qf \le Rf \le f \; .$$

At the end of Section 5.1 we shall give some examples where one can explicitly compute such envelopes. Note also that in the scalar case, i.e. if $m = 1$ or

$n = 1$ then $Cf = Pf = Qf = Rf$. In Section 5.2 we shall turn our attention to *relaxation theorems*. We define

$$(QP) \quad \inf\left\{ \bar{I}(u) = \int_\Omega Qf(x, u(x), \nabla u(x))\, dx \; : \; u \in u_0 + W_0^{1,p}(\Omega; \mathbb{R}^m) \right\}$$

where Qf is the quasiconvex envelope of f (with respect to the last variable ∇u). We show that, even though the original f is not quasiconvex (not convex in the scalar case) and therefore in general the infimum of (P) is not attained, one has

$$\inf(P) = \inf(QP)$$

and with some extra coercivity condition the infimum of (QP) is attained. More precisely if $\bar{u}$ is a solution of (QP), then there exists a minimizing sequence $\{u_\nu\}$ of (P) such that

$$\begin{cases} u_\nu \rightharpoonup \bar{u} \text{ in } W_0^{1,p}(\Omega; \mathbb{R}^m) \\ I(u_\nu) \to \bar{I}(\bar{u}) = \inf(P) = \inf(QP) \\ I'(u_\nu) \to 0, \text{ in the sense of distributions} \end{cases} \quad .$$

In other words even if (P), as well as the associated Euler equations $I'(u) = 0$, has no solution in $W^{1,p}(\Omega; \mathbb{R}^m)$, one can consider the solutions of (QP), respectively the solutions of $\bar{I}'(u) = 0$, as generalized solutions of (P), respectively $I'(u) = 0$, in the sense of weak convergence.

Weak convergence measuring some kind of averages (Chapter 2), this is a particularly interesting result for applications to non linear elasticity (cf. Dacorogna [1], Gurtin-Témam [1], Témam-Strang [1], Knowles-Sternberg [2], de Campos-Oden [1], Fonseca [1], [2]), to *optimal design* (cf. Kohn-Strang [1,2], Kohn-Vogelius [1], Strang [1], Lurie-Cherkaev [1]), since in these applications, usually, only averages of physical quantities are actually measured (cf. Appendix).

Finally at the end of Section 5.2 we shall give some examples of existence and non existence of solutions of (P) when f fails to be convex.

5.1 Convex, Polyconvex, Quasiconvex, Rank One Convex Envelopes

We now proceed with the characterization of Cf, Pf, Qf and Rf. As mentioned before there will be two types of characterization, one via duality and Hahn-Banach theorem for Cf and Pf and one via Carathéodory theorem for Cf, Pf, Qf and Rf. These two approaches are classical for Cf in the context of convex analysis.

5.1.1 Definitions and Properties

5.1.1.1 Duality for Convex and Polyconvex Functions

We first recall (cf. Chapter 2) some facts about duality in convex analysis.

Definition. Let $f : \mathbb{R}^{nm} \to \mathbb{R} \cup \{\pm\infty\}$ and let $f^* : \mathbb{R}^{nm} \to \mathbb{R} \cup \{\pm\infty\}$ be defined as follows

$$f^*(x^*) = \sup_{x \in \mathbb{R}^{nm}} \{\langle x; x^* \rangle - f(x)\}$$

where $\langle \bullet; \bullet \rangle$ denotes the scalar product in $\mathbb{R}^{nm}$. And let

$$f^{**} = (f^*)^* \, ,$$

i.e.

$$f^{**}(x) = \sup_{x^* \in \mathbb{R}^{nm}} \{\langle x; x^* \rangle - f^*(x^*)\} \, .$$

REMARKS. As seen in Chapter 2 we always have

i) $f^{***} = f^*$

ii) f^{**} is convex and lower semicontinuous and therefore

$$f^{**} \leq Cf \leq f \, .$$

We now proceed in an analogous way for polyconvex functions and we follow here the idea of Kohn and Strang [1,2]. We also adopt the notations of Chapter 4.

Definition. Let $f : \mathbb{R}^{nm} \to \mathbb{R} \cup \{\pm\infty\}$. One defines

$$f^P : \mathbb{R}^{\tau(n,m)} \to \mathbb{R} \cup \{\pm\infty\}$$

where $\tau(n,m) = \sum_{s=1}^{n \wedge m} \binom{m}{s}\binom{n}{s}$, by

$$f^P(X) = \sup_{A \in \mathbb{R}^{nm}} \{\langle X; T(A) \rangle - f(A)\}$$

where $\langle \bullet; \bullet \rangle$ denotes the scalar product in $\mathbb{R}^{\tau(n,m)}$. Let $(f^P)^* : \mathbb{R}^{\tau(n,m)} \to \mathbb{R} \cup \{\pm\infty\}$ be defined by

$$(f^P)^*(Y) = \sup_{X \in \mathbb{R}^{\tau(n,m)}} \{\langle X; Y \rangle - f^P(X)\} \, .$$

Finally let for $A \in \mathbb{R}^{nm}$

$$f^{PP}(A) = (f^P)^*(T(A)) \, .$$

REMARKS.

i) It is clear that if $m = 1$ or $n = 1$

$$f^P = f^* \quad \text{and} \quad f^{PP} = f^{**} \ ;$$

ii) it is also simple to see that

$$f^{PPP} = f^P$$

and that f^{PP} is polyconvex, lower semicontinuous and less than f and therefore

$$f^{PP} \leq Pf \leq f \ ;$$

iii) if $m = n = 2$, then $\tau(n,m) = 5$ and we can write

$$f^P(X,\gamma) = \sup_{A \in \mathbb{R}^4} \{\langle X; A\rangle + \gamma \det A - f(A)\}$$

where $X \in \mathbb{R}^4$, $\gamma \in \mathbb{R}$ and $\langle \bullet; \bullet \rangle$ denotes the scalar product in $\mathbb{R}^4$. Similarly

$$(f^P)^*(Y,\delta) = \sup_{\substack{X \in \mathbb{R}^4 \\ \gamma \in \mathbb{R}}} \{\langle X; Y\rangle + \gamma\delta - f^P(X,\gamma)\}$$

and therefore

$$f^{PP}(A) = (f^P)^*(A, \det A) \ .$$

We now give a simple example where one can compute explicitly f^*, f^{**}, f^P, f^{PP}.

EXAMPLE. Let $m = n = 2$ and

$$f(A) = \det A \ .$$

i) It is easy to show that

$$f^*(X) = \sup_{A \in \mathbb{R}^4} \{\langle X; A\rangle - \det A\} \equiv +\infty$$

and therefore

$$f^{**}(A) \equiv -\infty \ .$$

ii) Similarly

$$f^P(A,\gamma) = \sup_{X \in \mathbb{R}^4} \{\langle A; X\rangle + (\gamma - 1)\det X\}$$

$$= \begin{cases} 0 & \text{if } \gamma = 1 \text{ and } A = 0 \\ +\infty & \text{elsewhere} \end{cases} \ .$$

We therefore obtain

$$(f^P)^*(B, \delta) = \sup_{\substack{A \in \mathbf{R}^4 \\ \gamma \in \mathbf{R}}} \{\langle A; B \rangle + \gamma\delta - f^P(A, \gamma)\} = \delta \ ,$$

and hence

$$f^{PP}(A) = \det A \ .$$

5.1.1.2 The Main Theorem

We are now in position to state the main theorem of this section. Recall that

$$Cf = \sup\{g \leq f : g \text{ convex}\}$$

$$Pf = \sup\{g \leq f : g \text{ polyconvex}\}$$

$$Qf = \sup\{g \leq f : g \text{ quasiconvex}\}$$

$$Rf = \sup\{g \leq f : g \text{ rank one convex}\}$$

then

Theorem 1.1. *Let* $f : \mathbf{R}^{nm} \rightarrow \bar{\mathbf{R}} = \mathbf{R} \cup \{+\infty\}$ *and let* $c : \mathbf{R}^{\tau(n,m)} \rightarrow \bar{\mathbf{R}} = \mathbf{R} \cup \{+\infty\}$ *be convex and such that* $f(A) \geq c(T(A))$.
Part 1: For every $A \in \mathbf{R}^{nm}$,

$$Cf(A) = \inf\left\{ \sum_{i=1}^{nm+1} \lambda_i f(A_i) : \sum_{i=1}^{nm+1} \lambda_i A_i = A \right\} \tag{1}$$

$$Pf(A) = \inf\left\{ \sum_{i=1}^{\tau(n,m)+1} \lambda_i f(A_i) : \sum_{i=1}^{\tau(n,m)+1} \lambda_i T(A_i) = T(A) \right\} \tag{2}$$

$$Rf(A) = \inf\left\{ \sum_{i=1}^{I} \lambda_i f(A_i) : \sum_{i=1}^{I} \lambda_i A_i = A, \ (\lambda_i, A_i) \text{ satisfy } (H_I) \right\} \tag{3}$$

where (H_I) *is defined as in Section 4.1.1.3. Furthermore if* $f : \mathbf{R}^{nm} \rightarrow \mathbf{R}$ *is locally bounded and Borel measurable then*

$$Qf(A) = \inf\left\{ \frac{1}{\text{meas } D} \int_D f(A + \nabla\varphi(x))\, dx : \varphi \in W_0^{1,\infty}(D; \mathbf{R}^m) \right\} \tag{4}$$

where $D \subset \mathbf{R}^n$ *is a bounded domain. In particular the infimum in (4) is independent of the choice of* D.

Part 2: Let $f : \mathbb{R}^{nm} \to \mathbb{R}$, i.e. f is finite, and let f^{**} and f^{PP} be defined as in the preceding section, then

$$f^{**} = Cf \tag{5}$$

$$f^{PP} = Pf . \tag{6}$$

REMARKS.

i) Part 1 should be seen as the equivalent of Carathéodory theorem, while Part 2 should be interpreted as the counterpart of Hahn-Banach theorem. One should also note that there is no equivalent to Part 2 for Qf and Rf;

ii) one can also rewrite (5) in the following way (if f takes only finite values)

$$Cf = \sup\{g \le f : g \text{ affine}\} .$$

Similarly for (6)

$$Pf = \sup\{g \le f : g \text{ quasiaffine}\} ;$$

iii) it is also interesting to note that (5) and (6) do not hold if f is allowed to take the value $+\infty$. For example if $m = n = 1$ and

$$f(x) = \chi_{(0,1)}(x) = \begin{cases} 0 & \text{if } x \in (0,1) \\ +\infty & \text{otherwise} \end{cases} ,$$

then $f = Cf$ and $f^{**} = \chi_{[0,1]}$;

iv) the identities (1), (5) are standard in convex analysis. The identities (2), (3), (4) have been established by Dacorogna [3,7,8], while (6) was proved in Kohn-Strang [1,2];

v) in Kohn-Strang [1,2], there is a very similar construction to (3) in order to characterize Rf, namely they define

$$\begin{cases} R_0 f = f \\ R_{k+1}f(A) = \inf\{\lambda R_k f(A_1) + (1-\lambda)R_k f(A_2) \\ \qquad\qquad \lambda A_1 + (1-\lambda)A_2 = A \\ \qquad\qquad \text{and } \operatorname{rank}\{A_1 - A_2\} \le 1\} \\ Rf = \lim_{k \to \infty} R_k f \end{cases} \tag{7}$$

The two approaches (3) or (7) present a serious defect in the sense that in (3) one cannot prescribe *a priori* the value of the integer I, while in (7) one cannot prescribe that the limit Rf is attained after a given finite number of steps. Therefore such formulas are useful for computing Rf only when there is a hint on what is the number of steps required in order to get Rf.

We now proceed with the proof of the theorem.

PROOF. Observe first that (1) and (5) are direct consequences of (2) and (6).

Formula of Pf. We write when there is no ambiguity $\tau = \tau(n,m)$. We first define for $I \geq \tau + 1$

$$P'f(A) = \inf \left\{ \sum_{i=1}^{I} \lambda_i f(A_i) : \sum_{i=1}^{I} \lambda_i T(A_i) = T(A) \right\} . \qquad (8)$$

We decompose the proof into three steps.
Step 1: We first show that $P'f$ is polyconvex.
Step 2: We next prove that $P'f = Pf$.
Step 3: Finally we show that we can take $I = \tau + 1$ in (8).

Step 1: In view of Theorem 1.3 of Chapter 4, to show the polyconvexity of $P'f$, it is sufficient to see that

$$\sum_{\nu=1}^{\tau+1} \lambda_\nu P'f(B_\nu) \geq P'f\left(\sum_{\nu=1}^{\tau+1} \lambda_\nu B_\nu\right) \qquad (9)$$

whenever

$$\sum_{\nu=1}^{\tau+1} \lambda_\nu T(B_\nu) = T\left(\sum_{\nu=1}^{\tau+1} \lambda_\nu B_\nu\right) . \qquad (10)$$

Fix $\varepsilon > 0$. From (8) we have that there exist $I_\nu \geq \tau + 1$, $1 \leq \nu \leq \tau + 1$, $\alpha_i^\nu \geq 0$, with $\sum_{i=1}^{I_\nu} \alpha_i^\nu = 1$ and $A_i^\nu \in \mathbb{R}^{nm}$ such that

$$\begin{cases} \varepsilon + P'f(B_\nu) \geq \sum_{i=1}^{I_\nu} \alpha_i^\nu f(A_i^\nu), \ 1 \leq \nu \leq \tau + 1 \\ \\ \sum_{i=1}^{I_\nu} \alpha_i^\nu T(A_i^\nu) = T(B_\nu), \ 1 \leq \nu \leq \tau + 1 \end{cases} .$$

Relabelling α_i^ν and A_i^ν in the following way

$$\begin{cases} \beta_i = \lambda_1 \alpha_i^1 & C_i = A_i^1 & 1 \leq i \leq I_1 \\ \beta_{I_1+i} = \lambda_2 \alpha_i^2 & C_{I_1+1} = A_i^2 & 1 \leq i \leq I_2 \\ \quad \vdots \\ \beta_{I_1+...+I_\tau+i} = \lambda_{\tau+1}\alpha_i^{\tau+1} & C_{I_1+...+I_\tau+i} = A_i^{\tau+1} & 1 \leq i \leq I_{\tau+1} \end{cases} ;$$

we get that

$$
\begin{cases}
\varepsilon + \displaystyle\sum_{\nu=1}^{\tau+1} \lambda_\nu P' f(B_\nu) \geq \sum_{i=1}^{I_1+\dots+I_{\tau+1}} \beta_i f(C_i) \\[2ex]
\displaystyle\sum_{\nu=1}^{\tau+1} \lambda_\nu T(B_\nu) = \sum_{i=1}^{I_1+\dots+I_{\tau+1}} \beta_i T(C_i) \\[3ex]
\qquad\qquad = T\left(\displaystyle\sum_{i=1}^{I_1+\dots+I_{\tau+1}} \beta_i C_i \right) = T\left(\sum_{\nu=1}^{\tau+1} \lambda_\nu B_\nu \right)
\end{cases}
\tag{11}
$$

Using (8) in the right hand side of the inequality in (11) and the arbitrariness of ε we have indeed obtained (9), and therefore shown that $P'f$ is polyconvex.

Step 2: We next want to prove that $P'f = Pf$. We first observe that $P'f \leq f$ and, using Theorem 1.3 of Chapter 4 and Step 1, that $P'(P'f) = P'f$. Hence if $g \leq f$ is any polyconvex function then

$$
g = P'g \leq P'(P'f) = P'f \leq f \;.
$$

Thus $P'f \geq Pf$ and since $P'f$ is polyconvex we have indeed $P'f = Pf$.

Step 3: It now remains to show that in (8) we can choose $I = \tau + 1$. The proof is almost identical to that of Step 2 of Theorem 1.3 of Chapter 4 and we shall not reproduce it here.

Formula for Rf. We first define

$$
R'f(A) = \inf \left\{ \sum_{i=1}^{I} \lambda_i f(A_i) : \sum_{i=1}^{I} \lambda_i A_i = A, (\lambda_i, A_i) \text{ satisfy } (H_I) \right\} \;.
\tag{12}
$$

We decompose the proof into three steps.
Step 1: We first establish a preliminary result on combination of matrices which satisfy (H_I).
Step 2: We then establish that $R'f$ is rank one convex.
Step 3: We finally conclude by showing that $R'f = Rf$.

Step 1: We want to show that if

$$
\begin{cases}
\lambda_n \geq 0 \text{ with } \displaystyle\sum_{n=1}^{N} \lambda_n = 1, (\lambda_n, A_n) \text{ satisfy } (H_N) \\[2ex]
\mu_m \geq 0 \text{ with } \displaystyle\sum_{m=1}^{M} \mu_m = 1, (\mu_m, B_m) \text{ satisfy } (H_M)
\end{cases}
\tag{13}
$$

and if

$$\text{rank}\left\{\sum_{n=1}^{N}\lambda_n A_n - \sum_{m=1}^{M}\mu_m B_m\right\} \leq 1 \tag{14}$$

then, for every $\alpha \in [0,1]$, we have

$$((\alpha\lambda_n, A_n)_{1\leq n\leq N}, ((1-\alpha)\mu_m, B_m)_{1\leq m\leq M}) \text{ satisfy } (H_{N+M}) . \tag{15}$$

To show (15) we proceed by induction over $N + M$. The case $N + M = 2$ is trivial since this implies that $N = M = 1$ and therefore (14) is equivalent, by definition, to (15). Assume therefore that (15) has been established for $N + M - 1$; we may also assume without loss of generality that $N \geq 2$. Since (λ_n, A_n) satisfy (H_N) we have, up to a permutation, that

$$\text{rank}\{A_1 - A_2\} \leq 1 \tag{16}$$

and if

$$\begin{cases} \tilde{\lambda}_1 = \lambda_1 + \lambda_2, \ \tilde{A}_1 = \dfrac{\lambda_1 A_1 + \lambda_2 A_2}{\lambda_1 + \lambda_2} \\ \tilde{\lambda}_i = \lambda_{i+1}, \ \tilde{A}_i = A_{i+1}, \ i \geq 2 \end{cases}$$

then

$$(\tilde{\lambda}_i, \tilde{A}_i)_{1\leq i\leq N-1} \text{ satisfy } (H_{N-1}) . \tag{17}$$

Note that (14) implies then that

$$\text{rank}\left\{\sum_{n=1}^{N-1}\tilde{\lambda}_n \tilde{A}_n - \sum_{m=1}^{M}\mu_m B_m\right\} \leq 1 .$$

The hypothesis of induction therefore ensures that

$$\left((\alpha\tilde{\lambda}_n, \tilde{A}_n)_{1\leq n\leq N-1}, ((1-\alpha)\mu_m, B_m)_{1\leq m\leq M}\right) \text{ satisfy } (H_{N+M-1}) . \tag{18}$$

Coupling (16) and (18) we have indeed obtained (15).

Step 2: We now show that $R'f$ is rank one convex, i.e.

$$\alpha R'f(A) + (1-\alpha)R'f(B) \geq R'f(\alpha A + (1-\alpha)B) \tag{19}$$

for every A and B such that

$$\text{rank}\{A - B\} \leq 1 . \tag{20}$$

Fix $\varepsilon > 0$ and use (12) to get

$$\begin{cases} \varepsilon + R'f(A) \geq \displaystyle\sum_{n=1}^{N} \lambda_n f(A_n) \\ \varepsilon + R'f(B) \geq \displaystyle\sum_{m=1}^{M} \mu_m f(B_m) \\ \displaystyle\sum_{n=1}^{N} \lambda_n A_n = A, \ (\lambda_n, A_n) \text{ satisfy } (H_N) \\ \displaystyle\sum_{m=1}^{M} \mu_m B_m = B, \ (\mu_m, B_m) \text{ satisfy } (H_M) \end{cases} \qquad . \qquad (21)$$

Combining (20), (21) and Step 1 we get

$$\begin{cases} \varepsilon + \alpha R'f(A) + (1 - \alpha)R'f(B) \\ \qquad \geq \displaystyle\sum_{n=1}^{N} \alpha \lambda_n f(A_n) + \displaystyle\sum_{m=1}^{M} (1 - \alpha)\mu_m f(B_m) \\ \displaystyle\sum_{n=1}^{N} \alpha \lambda_n A_n + \displaystyle\sum_{m=1}^{M} (1 - \alpha)\mu_m B_m = \alpha A + (1 - \alpha)B \\ ((\alpha \lambda_n, A_n)_{1 \leq n \leq N}, ((1 - \alpha)\mu_m, B_m)_{1 \leq m \leq M}) \text{ satisfy } (H_{N+M}) \end{cases} \qquad . \qquad (22)$$

Using the definition of $R'f$, (22) and the arbitrariness of ε, we have indeed obtained that $R'f$ is rank one convex.

Step 3: Note first that if f is rank one convex, then, by Proposition 1.4 of Chapter 4, we have $R'f = f$. Combining Step 2 and this last observation we have $R'(R'f) = R'f$. Let $g \leq f$ be rank one convex then

$$g = R'g \leq R'(R'f) = R'f \leq f \ ,$$

thus $R'f \geq Rf$ and since $R'f$ is rank one convex we have indeed $R'f = Rf$.
Formula of Qf. We first recall that (cf. Chapter 2) for $\Omega \subset \mathbb{R}^n$ a bounded open set, we let

$$\text{Aff}_0(\Omega; \mathbb{R}^m) = \{\varphi \in W_0^{1,\infty}(\Omega; \mathbb{R}^m); \varphi \quad \text{piecewise affine}\} \ .$$

We first let, for $D \subset \mathbb{R}^n$ a bounded open set,

$$Q'f(A) \equiv \inf \left\{ \frac{1}{\text{meas}\,D} \int_D f(A + \nabla\varphi(x))\,dx \ : \ \varphi \in \text{Aff}_0(D; \mathbb{R}^m) \right\} \ . \quad (23)$$

We decompose the proof into five steps.

Step 1: We first show that the definition of $Q'f$ is independent of the choice of D.

Step 2: We then establish that

$$\int_D Q'f(A + \nabla\psi(x))\,dx \geq Q'f(A)\operatorname{meas} D$$

for every $A \in \mathbb{R}^{nm}$ and for every $\psi \in \operatorname{Aff}_0(D;\mathbb{R}^m)$.

Step 3: We next show that $Q'f$ is quasiconvex.

Step 4: We then deduce that $Q'f = Qf$.

Step 5: We finally establish (4).

Step 1: Let, for $D \subset \mathbb{R}^n$ a bounded open set,

$$Q'f_D(A) \equiv \inf\left\{\frac{1}{\operatorname{meas} D}\int_D f(A + \nabla\varphi(x))\,dx : \varphi \in \operatorname{Aff}_0(D;\mathbb{R}^m)\right\} . \quad (24)$$

We wish to show that given two such sets D_1,D_2 then $Q'f_{D_1} = Q'f_{D_2}$. We first define an equivalence relation between two sets D_1, D_2, by setting

$$D_1 \sim D_2$$

if there exists $x_0 \in \mathbb{R}^n$, $\lambda > 0$ such that

$$D_2 = x_0 + \lambda D_1 .$$

A change of variables gives immediately that

$$D_1 \sim D_2 \Rightarrow Q'f_{D_1} = Q'f_{D_2} . \quad (25)$$

If D_1 and D_2 are not equivalent in the above sense, one approximates D_2 by a disjoint union of D_1^i such that, for every $\varepsilon > 0$

$$\begin{cases} D_1^i \sim D_1,\ 1 \leq i \leq I^\varepsilon \\ \operatorname{meas}\left(D_2 - \bigcup_{i=1}^{I^\varepsilon} D_1^i\right) \leq \varepsilon \end{cases} .$$

Using (24) and (25) we get that there exist $\varphi_i^\varepsilon \in \operatorname{Aff}_0(D_1^i;\mathbb{R}^m)$ such that

$$\int_{D_1^i} f(A + \nabla\varphi_i^\varepsilon(x))\,dx \leq (\varepsilon + Q'f_{D_1}(A))\operatorname{meas} D_1^i .$$

Define next $\varphi^\varepsilon \in \mathrm{Aff}_0(D_2; \mathbb{R}^m)$ by

$$\varphi^\varepsilon(x) = \begin{cases} \varphi_i^\varepsilon(x) & \text{if } x \in D_1^i \\ 0 & \text{if } x \in D_2 - \bigcup_{i=1}^{I^\varepsilon} D_1^i \end{cases} \quad .$$

Using again (24) we have

$$Q'f_{D_2}(A) \operatorname{meas} D_2 \leq \int_{D_2} f(A + \nabla \varphi^\varepsilon(x))\, dx$$

$$\leq \sum_{i=1}^{I^\varepsilon} \int_{D_1^i} f(A + \nabla \varphi_i^\varepsilon(x))\, dx$$

$$+ f(A) \operatorname{meas}\left(D_2 - \bigcup_{i=1}^{I^\varepsilon} D_1^i \right)$$

$$\leq \left(\varepsilon + Q'f_{D_1}(A)\right) \operatorname{meas}\left(\bigcup_{i=1}^{I_\varepsilon} D_1^i \right) + \varepsilon f(A) \quad .$$

Since ε is arbitrary we have indeed shown that

$$Q'f_{D_2} \leq Q'f_{D_1} \quad .$$

A similar argument establishes the reverse inequality and therefore $Q'f_{D_2} = Q'f_{D_1}$.

Step 2: We now wish to show that

$$\int_D Q'f(A + \nabla \psi(x))\, dx \geq Q'f(A) \operatorname{meas} D \tag{26}$$

for every $A \in \mathbb{R}^{nm}$ and $\psi \in \mathrm{Aff}_0(D; \mathbb{R}^m)$. Note that (26) ensures, up to a density argument (cf. Step 3), that $Q'f$ is quasiconvex. Since $\psi \in \mathrm{Aff}_0(D; \mathbb{R}^m)$, there exist $D_i \subset D$ with $\bigcup_{i=1}^{I} D_i = D$ and $B_i \in \mathbb{R}^{nm}$, $1 \leq i \leq I$, such that

$$\int_D Q'f(A + \nabla \psi(x))\, dx = \sum_{i=1}^{I} Q'f(A + B_i) \operatorname{meas} D_i \quad . \tag{27}$$

Fixing $\varepsilon \geq 0$ and using (23) we have that there exists $\varphi_i^\varepsilon \in \mathrm{Aff}_0(D_i; \mathbb{R}^m)$ such that

$$Q'f(A + B_i) + \varepsilon \geq \frac{1}{\operatorname{meas} D_i} \int_{D_i} f(A + B_i + \nabla \varphi_i^\varepsilon(x))\, dx \quad . \tag{28}$$

Let $\xi \in \text{Aff}_0(D; \mathbb{R}^m)$ be defined by

$$\xi(x) = \psi(x) + \varphi_i^\varepsilon(x) \text{ if } x \in D_i, \ i = 1, \ldots, I \ .$$

We have therefore, using (27) and (28), that

$$\int_D Q'f(A + \nabla\psi(x))\, dx + \varepsilon \operatorname{meas} D \geq \int_D f(A + \nabla\xi(x))\, dx$$

$$\geq Q'f(A) \operatorname{meas} D \ ,$$

where we have used (23) in the last inequality. ε being arbitrary we have obtained (26).

Step 3: We next want to show that $Q'f$ is quasiconvex. It will be sufficient to show that (26) implies that $Q'f$ is continuous and therefore combining (26), the continuity of $Q'f$, the fact that $\text{Aff}_0(D; \mathbb{R}^m)$ is dense in $W_0^{1,\infty}(D; \mathbb{R}^m)$ (cf. Part 2 of Theorem 1.8, Chapter 2) and Lebesgue dominated convergence theorem, we shall have that $Q'f$ is quasiconvex. In order to show that $Q'f$ is continuous we prove that (26) implies that $Q'f$ is convex in each of its variables (in fact one can show that $Q'f$ is rank one convex), then the continuity will follow from standard properties of convex functions (cf. Theorem 2.3 of Chapter 2). Without loss of generality it is sufficient to prove that if $\lambda \in [0,1]$, $A = (A_{ij}), B = (B_{ij}) \in \mathbb{R}^{nm}$ are such that $A_{11} = a, B_{11} = b$ and $A_{ij} = B_{ij}$ otherwise, then

$$Q'f(\lambda A + (1 - \lambda)B) \leq \lambda Q'f(A) + (1 - \lambda)Q'f(B) \ , \tag{29}$$

i.e. $Q'f$ is convex in its first variable. Observe that if $\varepsilon > 0$, then there exists $\varphi^\varepsilon : D \equiv (0,1)^n \to \mathbb{R}$, $D_1^\varepsilon, D_2^\varepsilon \subset D$ with $D_1^\varepsilon \cap D_2^\varepsilon = \emptyset$ and $\varphi^\varepsilon \in \text{Aff}_0(D; \mathbb{R}^m)$ such that

$$\begin{cases} \operatorname{meas} D_1^\varepsilon \to \lambda, \ \operatorname{meas} D_2^\varepsilon \to (1 - \lambda) \text{ as } \varepsilon \to 0 \\ \|\operatorname{grad} \varphi^\varepsilon\|_{L^\infty} \leq K = K(a,b) \\ \operatorname{grad} \varphi^\varepsilon(x) = \begin{cases} (1 - \lambda)(a - b, 0, \ldots, 0) & \text{if } x \in D_1^\varepsilon \\ -\lambda(a - b, 0, \ldots, 0) & \text{if } x \in D_2^\varepsilon \end{cases} \end{cases} \tag{30}$$

Such a construction is standard in numerical analysis and we give below (Step 3′) a hint on this construction. We may now define $\psi^\varepsilon : D \to \mathbb{R}^m$ by $\psi^\varepsilon(x) = (\varphi^\varepsilon(x), 0, \ldots, 0)$ where φ^ε satisfies (30). We then use (26) to obtain

$$\int_D Q'f(\lambda A + (1 - \lambda)B + \nabla\psi^\varepsilon(x))\, dx \geq Q'f(\lambda A + (1 - \lambda)B) \ .$$

Letting $\varepsilon \to 0$, we have indeed obtained (29) and thus the continuity of $Q'f$ and consequently the quasiconvexity of $Q'f$.

Step 3': Construction of φ^ε satisfying (30). We give here a possible one in $\mathbb{R}^2$ and we then show how to extend it to higher dimensions (note the similarity with Step 1, Theorem 3.1 of Chapter 3). Let $\varepsilon = \frac{1}{N}$ and define

$$\begin{cases} I_N^K = \left(\dfrac{K}{N}, \dfrac{K+\lambda}{N}\right), \ 0 \le K \le N-1 \text{ and } I_N = \displaystyle\bigcup_{k=0}^{N-1} I_N^K \\[4mm] J_N^K = \left(\dfrac{K+\lambda}{N}, \dfrac{K+1}{N}\right), \ 0 \le K \le N-1 \text{ and } J_N = \displaystyle\bigcup_{k=0}^{N-1} J_N^K \end{cases},$$

then

$$\bar{I}_N \cup \bar{J}_N = [0,1] \text{ and } \operatorname{meas} I_N = \lambda, \ \operatorname{meas} J_N = (1-\lambda) \ .$$

Define (see Figure 5.1)

$$E_N^K = \left\{ (x,y) \in \mathbb{R}^2 : x \in I_N^K \text{ and } x - \frac{K}{N} \le y \le -\left(x - \frac{K}{N}\right) + 1\right\}$$

$$F_N^K = \left\{ (x,y) \in \mathbb{R}^2 : x \in J_N^K \right.$$

$$\left. \text{and } \frac{\lambda}{(1-\lambda)}\left(-x + \frac{K+1}{N}\right) \le y \le \frac{\lambda}{1-\lambda}\left(x - \frac{K+1}{N}\right) + 1\right\}$$

and for $\varepsilon = \frac{1}{N}$

$$D_1^\varepsilon = \bigcup_{K=0}^{N-1} E_N^K, \quad D_2^\varepsilon = \bigcup_{K=0}^{N-1} F_N^K.$$

We then define

$$\varphi^\varepsilon(x,y) = \begin{cases} (1-\lambda)(a-b)\left(x - \dfrac{K}{N}\right) & \text{if } x \in E_N^K \\[3mm] -\lambda(a-b)\left(x - \dfrac{K+1}{N}\right) & \text{if } x \in F_N^K \\[3mm] (1-\lambda)(a-b)y & \text{if } (x,y) \notin D_1^\varepsilon \cup D_2^\varepsilon \\ & \text{and } 0 < y < \dfrac{\lambda}{N} \\[3mm] (1-\lambda)(a-b)(-y+1) & \text{if } (x,y) \notin D_1^\varepsilon \cup D_2^\varepsilon \\ & \text{and } 1 - \dfrac{\lambda}{N} < y < 1 \end{cases}$$

We now give an idea of how to extend the following construction to higher dimensions and we give the example of $\mathbb{R}^3$. We want to construct $\varphi^\varepsilon(x,y,z)$

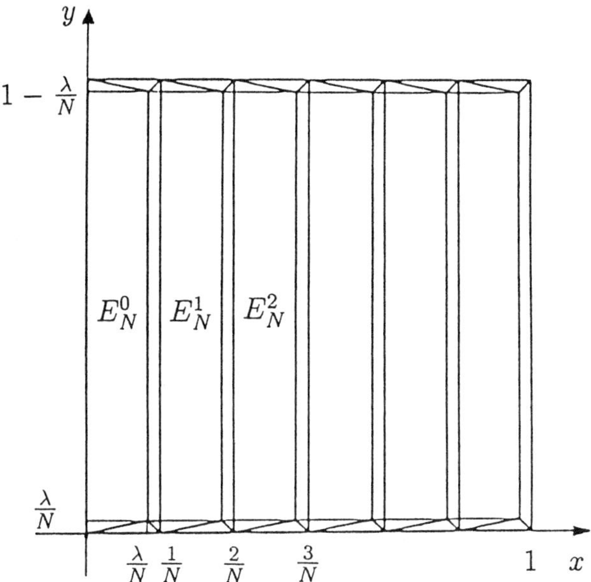

Figure 5.1

satisfying (30); by abuse of notations we denote by $\varphi^\varepsilon(x,y)$ the function constructed above in $\mathbb{R}^2$ and satisfying (30). We then define

$$\varphi^\varepsilon(x,y,z) = \begin{cases} \varphi^\varepsilon(x,y) & \text{if } (x,y,z) \in L \\ \varphi^\varepsilon(x,z) & \text{if } (x,y,z) \in M \end{cases},$$

where

$$L = \left\{ (x,y,z) \in [0,1]^3 : 0 \le x,y \le 1,\ \frac{\lambda}{N} \le z \le 1 - \frac{\lambda}{N} \right\}$$

$$\cup \left\{ (x,y,z) : 0 \le x \le 1,\ 0 \le y \le z \le \frac{\lambda}{N} \right\}$$

$$\cup \left\{ (x,y,z) : 0 \le x \le 1,\ 0 \le y \le \frac{\lambda}{N},\ 1 - \frac{\lambda}{N} \le z \le -y + 1 \right\}$$

$$\cup \left\{ (x,y,z) : 0 \le x \le 1,\ 1 - \frac{\lambda}{N} \le z \le y \le 1 \right\}$$

$$\cup \left\{ (x,y,z) : 0 \le x \le 1,\ 1 - \frac{\lambda}{N} \le y \le 1,\ -y + 1 \le z \le \frac{\lambda}{N} \right\}$$

and a symmetrical definition for M with the role of y and z interchanged.

The extension to $\mathbf{R}^n$ is then done in the same way. This completes Step $3'$ and thus Step 3.

Step 4: We now show that $Q'f = Qf$. Observe that if h is quasiconvex then, trivially, $Q'h = h$. Therefore let $h \leq f$ and quasiconvex then

$$h = Q'h \leq Q'(Q'f) = Q'f \leq f \ .$$

Hence $Q'f \geq Qf$ and since $Q'f$ itself is quasiconvex we have indeed established that $Q'f = Qf$.

Step 5: It remains to establish (4), i.e. if

$$\tilde{Q}f(A) = \inf\left\{ \frac{1}{\operatorname{meas} D} \int_D f(A + \nabla\varphi(x))\,dx \ : \ \varphi \in W_0^{1,\infty}(D;\mathbf{R}^m) \right\}$$

then $Qf = \tilde{Q}f$. From Step 4 we also have

$$Qf(A) = \inf\left\{ \frac{1}{\operatorname{meas} D} \int_D f(A + \nabla\varphi(x))\,dx \ : \ \varphi \in \operatorname{Aff}_0(D;\mathbf{R}^m) \right\} \ .$$

Since $\operatorname{Aff}_0(D;\mathbf{R}^m) \subset W_0^{1,\infty}(D;\mathbf{R}^m)$, we deduce that

$$Qf(A) \geq \tilde{Q}f(A) \ . \tag{31}$$

Since Qf is quasiconvex we immediately obtain that

$$\tilde{Q}(Qf) = Qf \ .$$

Therefore combining (31) and the above identity we have

$$Qf \geq \tilde{Q}f \geq \tilde{Q}(Qf) = Qf \ .$$

This indeed establishes the result.

Formula for f^{PP}. As pointed out in Section 5.1 we always have $f^{PP} \leq Pf$. We now wish to prove the reverse inequality. We divide the proof into two steps.

Step 1: We first show that if f is polyconvex and finite then

$$f^{PP} = f \ . \tag{32}$$

From Theorem 1.3 of Chapter 4, we have that there exists $g : \mathbf{R}^\tau \to \mathbf{R}$, $\tau = \tau(n,m)$, convex and finite such that

$$f(A) = g(T(A)) \ .$$

It is obvious from the definition that

$$g^*(Y) = \sup_{X \in \mathbf{R}^\tau} \{\langle X; Y \rangle - g(X)\}$$

$$\geq \sup_{A \in \mathbf{R}^{nm}} \{\langle T(A); Y \rangle - g(T(A))\} \equiv f^P(Y) \ ,$$

and therefore

$$g^{**}(X) \leq (f^P)^*(X) \ .$$

However since g is convex and finite we have

$$f(A) = g(T(A)) = g^{**}(T(A)) \leq (f^P)^*(T(A)) \equiv f^{PP}(A) \ .$$

Since the reverse inequality is trivial, we have indeed established (32).

Step 2: Applying Step 1 to Pf which is polyconvex and finite we get

$$Pf = (Pf)^{PP} \ .$$

We thus deduce

$$Pf = (Pf)^{PP} \leq f^{PP} \leq f$$

and the result follows. $\qquad\qquad\square$

5.1.2 Examples

We now turn our attention to some examples, where one can compute explicitly the different envelopes. Usually one is interested, cf. Section 5.2, in computing Qf (Cf in the scalar case), which is in general a difficult problem. One way of doing so is to compute Pf and Rf and then show that they are equal; that this is not always true will be shown at the end of this section.

5.1.2.1 Some more Properties of the Envelopes

We start with a result which allows to separate in some cases the computation of the different envelopes.

Theorem 1.2. *Let $u = u(x,t) : \mathbf{R}^n \times \mathbf{R}^N \to \mathbf{R}^m$, $\nabla u = (\nabla_x u, \nabla_t u) \in \mathbf{R}^{nm} \times \mathbf{R}^{Nm}$. Let $A = (B,C) \in \mathbf{R}^{nm} \times \mathbf{R}^{Nm} = \mathbf{R}^{(n+N)m}$, and*

$$f(A) = g(B) + h(C)$$

where g and h satisfy the hypotheses of Theorem 1.1. Then

$$Cf = Cg + Ch$$

$$Pf = Pg + Ph$$

$$Qf = Qg + Qh$$

$$Rf = Rg + Rh \; .$$

REMARKS.

i) This result is standard for Cf and has been established by Dacorogna [6] for Qf;

ii) the above theorem may be useful when one deals with integrands of the type $f(\nabla u) = \frac{1}{2}\left(\frac{\partial u}{\partial t}\right)^2 - h(\nabla_x u)$, cf. Dacorogna [6].

PROOF. The proof for Cf is standard and results for example from any of the three others. All the three formulas follow directly from Theorem 1.1 and we shall prove here only those for Qf and Rf (the formula for Pf is proved in exactly the same way).

Formula for Rf. It is clear that if $B \in \mathbf{R}^{nm}$ and $C \in \mathbf{R}^{Nm}$ then

$$Rg(B) + Rh(C) \le Rf(B,C) \; . \tag{1}$$

We wish to prove the reverse inequality. For this we prove first that

$$Rf(B,C) \le Rg(B) + h(C) \; . \tag{2}$$

Fix $\varepsilon \ge 0$, then by Theorem 1.1 there exist $(\lambda_i, B_i)_{1 \le i \le I}$ such that

$$\begin{cases} \lambda_i \ge 0 \text{ with } \displaystyle\sum_{i=1}^{I} \lambda_i = 1 \\ B_i \in \mathbf{R}^{nm} \text{ with } (\lambda_i, B_i) \text{ satisfying } (H_I) \\ \displaystyle\sum_{i=1}^{I} \lambda_i B_i = B \\ \varepsilon + Rg(B) \ge \displaystyle\sum_{i=1}^{I} \lambda_i g(B_i) \end{cases}$$

It is clear that if $C \in \mathbf{R}^{Nm}$ then $(\lambda_i, (B_i, C))_{1 \le i \le I}$ satisfy (H_I), therefore

$$\begin{cases} \sum_{i=1}^{I} \lambda_i (B_i, C) = (B, C), \ (\lambda_i, (B_i, C)) \text{ satisfy } (H_I) \\ \varepsilon + Rg(B) + h(C) \ge \sum_{i=1}^{I} \lambda_i (g(B_i) + h(C)) \end{cases} .$$

Using again Theorem 1.1 and the fact that ε is arbitrary we obtain (2). A similar argument shows that

$$Rf(B, C) \le g(B) + Rh(C) . \tag{3}$$

We now combine Theorem 1.1 and (2) – (3) to get

$$Rf(B, C) = R(Rf(B, C)) \le R(Rg(B) + h(C)) \le Rg(B) + Rh(C) ,$$

which, combined with (1), is the claimed result.

Formula of Qf. We establish it exactly in the same way. We first prove that $Qf \le Qg + h$, then that $Qf \le g + Qh$ and conclude as above. Therefore we shall only show that

$$Qf(B, C) \le Qg(B) + h(C) . \tag{4}$$

From Theorem 1.1, we have that if $D \subset \mathbf{R}^n$ and $\Omega \subset \mathbf{R}^N$ are unit hypercubes, then

$$Qf(B, C) = \inf\Big\{ \int_D \int_\Omega [g(B + \nabla_x \varphi(x, t)) $$
$$+ h(C + \nabla_t \varphi(x, t))] \, dx dt : \varphi \in W_0^{1,\infty}(D \times \Omega; \mathbf{R}^m) \Big\} . \tag{5}$$

Let $\varepsilon > 0$ be fixed, then Theorem 1.1 implies that there exists $\sigma \in W_0^{1,\infty}(D; \mathbf{R}^m)$ such that

$$\int_D g(B + \nabla_x \sigma(x)) \, dx \le \varepsilon + Qg(B) .$$

On extending σ by periodicity from D to $\mathbf{R}^n$, we trivially have that for $\nu \in \mathbf{N}$.

$$\int_D g(B + \nabla_x \sigma(\nu x)) \, dx \le \varepsilon + Qg(B) . \tag{6}$$

Let $\Omega_\nu \subset \Omega$ be a hypercube with the same centre as Ω and such that

$$\mathrm{dist}(\partial\Omega; \Omega_\nu) = \frac{1}{\nu} \ .$$

We then define $\psi \in W_0^{1,\infty}(\Omega)$, $0 \le \psi(t) \le 1$, such that

$$\psi(t) = \begin{cases} 1 & \text{if } t \in \Omega_\nu \subset \Omega \subset \mathbf{R}^N \\ 0 & \text{if } t \in \partial\Omega \end{cases}$$

and choose

$$\varphi(x,t) = \frac{1}{\nu}\sigma(\nu x)\,\psi(t) \ .$$

Observe that $\varphi \in W_0^{1,\infty}(D \times \Omega; \mathbf{R}^m)$. Using (5) we get

$$Qf(B,C) \le \int_D \int_\Omega g(B + \psi(t)\nabla_x\sigma(\nu x))\,dx\,dt$$

$$+ \int_D \int_\Omega h(C + \frac{1}{\nu}\sigma(\nu x) \otimes \mathrm{grad}\psi(t))\,dx\,dt \ , \qquad (7)$$

where $\sigma(\nu x) \otimes \mathrm{grad}\psi(t)$ denotes the tensorial product in $\mathbf{R}^{(n+N)m}$. We next use (6) to get, recalling that $\mathrm{meas}\,\Omega = \mathrm{meas}\,D = 1$,

$$\int_D \int_\Omega g(B + \psi(t)\nabla_x\sigma(\nu x))\,dx\,dt$$

$$= \int_D \int_{\Omega_\nu} g(B + \nabla_x\sigma(\nu x))\,dx\,dt$$

$$+ \int_D \int_{\Omega - \Omega_\nu} g(B + \psi(t)\nabla_x\sigma(\nu x))\,dx\,dt$$

$$\le \varepsilon + Qg(B) + \mathrm{meas}\,(\Omega - \Omega_\nu) \sup \{|g(B + \psi(t)\nabla_x\sigma(\nu x))|\} \ . \quad (8)$$

Similarly we have

$$\int_D \int_\Omega h(C + \frac{1}{\nu}\sigma(\nu x) \otimes \mathrm{grad}\psi(t))\,dx\,dt$$

$$\le h(C)\,\mathrm{meas}\,\Omega_\nu + \int_D \int_{\Omega - \Omega_\nu} h(C + \frac{1}{\nu}\sigma(\nu x) \otimes \mathrm{grad}\psi(t))\,dx\,dt$$

$$\le h(C) + \mathrm{meas}\,(\Omega - \Omega_\nu) \sup \left\{|h(C + \frac{1}{\nu}\sigma(\nu x) \otimes \mathrm{grad}\psi(t))|\right\} \ . \quad (9)$$

Combining (7), (8), (9), letting $\nu \to +\infty$ and using the arbitrariness of ε, we have indeed established (4) and thus the result. $\qquad\square$

5.1.2.2 Examples

We now give some examples which should be compared to those of Theorem 1.10 of Chapter 4.

Theorem 1.3. *Let $f : \mathbf{R}^{nm} \to \mathbf{R}$.*

i) Let $\Phi : \mathbf{R}^{nm} \to \mathbf{R}$ be quasiaffine not identically constant and $g : \mathbf{R} \to \mathbf{R}$ such that

$$f(A) = g(\Phi(A)) . \tag{1}$$

Then

$$Pf = Qf = Rf = Cg$$

and in general

$$Qf > Cf.$$

ii) Minimal surfaces: let $m = n + 1$. Let for $A \in \mathbf{R}^{n(n+1)}$

$$\mathrm{adj}_n A = (\det \hat{A}_1, - \det \hat{A}_2, \ldots, (-1)^{n+2} \det \hat{A}_{n+1})$$

where $\hat{A}_k$ is the $n \times n$ matrix obtained from A by suppressing the k-th line. Let $g : \mathbf{R}^{n+1} \to \mathbf{R}$ be such that

$$f(A) = g(\mathrm{adj}_n A) . \tag{2}$$

Then

$$Pf = Qf = Rf = Cg$$

and in general

$$Qf > Cf .$$

iii) For $A \in \mathbf{R}^{nm}$ let

$$|A| = \left(\sum_{i=1}^{m} \sum_{j=1}^{n} (A_j^i)^2 \right)^{1/2}$$

and let $g : \mathbf{R}_+ \to \mathbf{R}$ with $y(0) - \inf\{g(x) : x \geq 0\}$ be such that

$$f(A) = g(|A|) . \tag{3}$$

Then, in general,

$$Pf > Cf = Cg . \tag{4}$$

If, however, there exists $\alpha \geq 0$ such that

$$g(\alpha) = g(0) \ \text{and} \ Cg(x) = g(x) \ \text{for every} \ x \geq \alpha$$

then

$$Rf = Qf = Pf = Cf = Cg \ . \tag{5}$$

iv) Let $m = n = 2$, $A = (A^i_j)_{1 \leq i,j \leq 2}$. Let $g, h : \mathbf{R} \to \mathbf{R}$ with h convex such that

$$f(A) = g(A^1_1) + h(\det A) \tag{6}$$

then

$$Pf = Qf = Rf = Cg + h \ . \tag{7}$$

REMARKS.

i) The first, second and fourth examples have been established by Dacorogna [5,7,8], see also Acerbi-Fusco [1] for the second one. The first counterexample to (4) is due to Kohn-Strang [1,2] who consider the case where

$$f(A) = \begin{cases} 1 + |A|^2 & \text{if } A \neq 0 \\ 0 & \text{if } A = 0 \end{cases}$$

(cf. Lemma 2.7 in the Appendix). They show that in this case

$$Rf = Qf = Pf > Cf = Cg \ .$$

A simpler example, but showing only (4), was given in Dacorogna [7] (cf. also below in the proof of (4));

ii) the inequality (4) is at first sight surprising since in view of Theorem 1.10 of Chapter 4 if f and g satisfy (3) then the rank one convexity, quasiconvexity and polyconvexity of f are all equivalent to the convexity of f (or g);

iii) an example of functions f satisfying (3) and (5) is

$$f(A) = g(|A|) = (|A|^2 - 1)^2$$

then

$$Rf(A) = Qf(A) = Pf(A) = Cf(A)$$

$$= Cg(|A|) = \begin{cases} (|A|^2 - 1)^2 & \text{if } |A| \geq 1 \\ 0 & \text{if } |A| \leq 1 \end{cases} ;$$

iv) note that in the examples given in Theorem 1.3, since f is finite we always have $Cf = f^{**}$, $Cg = g^{**}$ and $Pf = f^{PP}$;

v) observe that one cannot expect that, in general, $Rf = Qf = Pf$. In Theorem 1.7 of Chapter 4 we have constructed, in the case $m = n = 3$, a function f, quadratic, which is rank one convex and quasiconvex, but not polyconvex, therefore for such a function one has $Rf = Qf = f > Pf$. In Proposition 1.13 and 1.14 of Chapter 4 we also gave an example, in the case $m = n = 2$, of a function f which is rank one convex but not polyconvex and therefore $Rf > Pf$ in this case.

Before proceeding with the proof we establish two preliminary lemmas.

Lemma 1.4. *Let $\Phi : \mathbb{R}^{nm} \to \mathbb{R}$ be quasiaffine and not identically constant. Let $A \in \mathbb{R}^{nm}$ be such that*

$$\nabla \Phi(A) \equiv \left(\frac{\partial \Phi}{\partial A_j^i} \right)_{1 \leq i \leq m,\, 1 \leq j \leq n} \neq 0 \,.$$

Let $\beta, \gamma \in \mathbb{R}$ and $\lambda \in [0, 1]$ be such that

$$\Phi(A) = \lambda \beta + (1 - \lambda)\gamma$$

then there exist $B, C \in \mathbb{R}^{nm}$ such that

$$\begin{cases} A = \lambda B + (1 - \lambda)C \\ \Phi(B) = \beta, \ \Phi(C) = \gamma \\ \operatorname{rank}\{B - C\} \leq 1 \end{cases} .$$

PROOF. Let $a \in \mathbb{R}^m$, $b \in \mathbb{R}^n$ be such that

$$\langle \nabla \Phi(A); a \otimes b \rangle = \gamma - \beta$$

where $\langle \bullet; \bullet \rangle$ denotes the scalar product in $\mathbb{R}^{nm}$ and $a \otimes b = (a^i b_j)_{1 \leq i \leq m,\, 1 \leq j \leq n}$. Since $\nabla \Phi(A) \neq 0$ one can always find such a and b. Define then

$$\begin{cases} B = A - (1 - \lambda)a \otimes b \\ C = A + \lambda a \otimes b \end{cases} .$$

In order to obtain the lemma it is therefore sufficient to show that $\Phi(B) = \beta$ and $\Phi(C) = \gamma$. Since Φ is quasiaffine we have (cf. Theorem 1.5 of Chapter 4)

$$\begin{cases} \Phi(B) = \Phi(A - (1 - \lambda)a \otimes b) = \Phi(A) - (1 - \lambda)\langle \nabla \Phi(A); a \otimes b \rangle = \beta \\ \Phi(C) = \Phi(A + \lambda a \otimes b) = \Phi(A) + \lambda \langle \nabla \Phi(A); a \otimes b \rangle = \gamma \end{cases} .$$

$\square$

Lemma 1.5. *Let $m = n + 1$, $A \in \mathbb{R}^{n(n+1)}$ and*

$$\operatorname{adj}_n A \neq 0 \,.$$

Let $I \in \mathbf{N}$, $\lambda_i \geq 0$ with $\sum_{i=1}^{I} \lambda_i = 1$, $\beta_i \in \mathbf{R}^{n+1}$ such that

$$\mathrm{adj}_n A = \sum_{i=1}^{I} \lambda_i \beta_i \ .$$

Then there exist $A_i \in \mathbf{R}^{n(n+1)}$ such that

$$\begin{cases} A = \sum_{i=1}^{I} \lambda_i A_i \\ \mathrm{adj}_n A_i = \beta_i \ 1 \leq i \leq I \\ (\lambda_i, A_i)_{1 \leq i \leq I} \ \textit{satisfy } (H_I) \end{cases} .$$

PROOF. We proceed by induction on I. The case $I = 2$ is precisely Lemma
1.11 of Chapter 4. Assume therefore that the lemma has been proved up to
the order $(I - 1)$ and we wish to prove it holds for I. We let

$$\gamma = \frac{1}{1 - \lambda_1} \sum_{i=2}^{I} \lambda_i \beta_i \ .$$

We may assume, upon a possible relabelling, that $\gamma \neq 0$. Observe also that
we have

$$\mathrm{adj}_n A = \lambda_1 \beta_1 + (1 - \lambda_1)\gamma \ .$$

We now apply Lemma 1.11 of Chapter 4 to β_1 and γ to get $A_1, C \in \mathbf{R}^{n(n+1)}$
such that

$$\begin{cases} A = \lambda_1 A_1 + (1 - \lambda_1)C \\ \mathrm{adj}_n A_1 = \beta_1, \ \mathrm{adj}_n C = \gamma \ . \\ \mathrm{rank}\{A_1 - C\} \leq 1 \end{cases} \qquad (1)$$

We may then use the hypothesis of induction to get that there exist $A_i \in
\mathbf{R}^{n(n+1)}$ such that

$$\begin{cases} C = \sum_{i=2}^{I} \frac{\lambda_i}{1 - \lambda_1} A_i \\ \mathrm{adj}_n A_i = \beta_i \ 2 \leq i \leq I \\ \left(\dfrac{\lambda_i}{1 - \lambda_1}, A_i \right)_{2 \leq i \leq I} \quad \text{satisfy } (H_{I-1}) \end{cases} \qquad (2)$$

Collecting (1) and (2) we have indeed obtained the lemma. $\qquad \square$

We may now proceed with the proof of the theorem.

PROOF of Theorem 1.3.

Part 1: It is easy to see that

$$Rf \geq Qf \geq Pf \geq Cg ,$$

it remains therefore to show that for every $A \in \mathbf{R}^{nm}$

$$Rf(A) \leq Cg(\Phi(A)) . \tag{8}$$

Case 1: $\nabla\Phi(A) \neq 0$. Fix $\varepsilon > 0$; from Theorem 1.1 we have that there exist $\alpha, \beta \in \mathbf{R}$, $\lambda \in [0,1]$ such that

$$\begin{cases} \lambda g(\alpha) + (1 - \lambda)g(\beta) \leq Cg(\Phi(A)) + \varepsilon \\ \lambda\alpha + (1 - \lambda)\beta = \Phi(A) \end{cases} .$$

Using Lemma 1.4 we have that there exist $B, C \in \mathbf{R}^{nm}$ satisfying the conclusions of the lemma. Using again Theorem 1.1 we have that

$$Rf(A) \leq \lambda f(B) + (1 - \lambda)f(C) = \lambda g(\alpha) + (1 - \lambda)g(\beta) \leq Cg(\Phi(A)) + \varepsilon .$$

Since ε is arbitrary we have indeed obtained (8).

Case 2: $\nabla\Phi(A) = 0$. Since Rf and Cg are continuous and Φ is not identically constant we have (cf. Theorem 1.5 of Chapter 4) that for every $\varepsilon > 0$ there exists $B \in \mathbf{R}^{nm}$ such that

$$\begin{cases} \nabla\Phi(B) \neq 0 \\ Cg(\Phi(B)) \leq Cg(\Phi(A)) + \varepsilon \\ Rf(A) \leq Rf(B) + \varepsilon \end{cases} .$$

Applying Case 1 to B we have that

$$Rf(A) \leq Rf(B) + \varepsilon = Cg(\Phi(B)) + \varepsilon \leq Cg(\Phi(A)) + 2\varepsilon ;$$

the arbitrariness of ε implies then (8).

It remains therefore to show that in general $Qf > Cf$. Choosing for example $m = n$ and

$$f(A) = (\det A)^2$$

we have immediately

$$Rf(A) = Qf(A) = Pf(A) = Cg(\det A) = f(A) > Cf(A) = 0 .$$

The identity $Cf(A) \equiv 0$ is a consequence of the fact that

$$0 \leq Cf(A) \leq \inf \left\{ \lambda(\det B)^2 + (1 - \lambda)(\det C)^2 : \lambda B + (1 - \lambda)C = A \right\}$$

and that the infimum in the right hand side is exactly zero.

Part 2: We proceed similarly as before. We trivially have that $Rf \geq Qf \geq Pf \geq Cg$ and we shall show that for every $A \in \mathbf{R}^{n(n+1)}$

$$Rf(A) \leq Cg(\mathrm{adj}_n A) \ . \tag{9}$$

Case 1: $\mathrm{adj}_n A \neq 0$. Fix $\varepsilon > 0$, from Theorem 1.1 we have that there exist $\beta_i \in \mathbf{R}^{n+1}$, $\lambda_i \geq 0$ with $\sum_{i=1}^{n+2} \lambda_i = 1$ such that

$$\begin{cases} \displaystyle\sum_{i=1}^{n+2} \lambda_i g(\beta_i) \leq Cg(\mathrm{adj}_n A) + \varepsilon \\ \displaystyle\sum_{i=1}^{n+2} \lambda_i \beta_i = \mathrm{adj}_n A \end{cases} \ .$$

We then use Lemma 1.5 to get A_i satisfying the conclusions of the lemma. Using again Theorem 1.1 we have that

$$Rf(A) \leq \sum_{i=1}^{n+2} \lambda_i f(A_i) = \sum_{i=1}^{n+2} \lambda_i g(\beta_i) \leq Cg(\mathrm{adj}_n A) + \varepsilon \ ;$$

the arbitrariness of ε implies (9).
Case 2: $\mathrm{adj}_n A = 0$. Using the continuity of Rf and Cg and a similar argument as in Part 1 gives immediately (9).

In order to show that in general $Qf > Cf$, choose

$$f(A) = |\mathrm{adj}_n A|^2 \ .$$

We then immediately get

$$Rf(A) = Qf(A) = Pf(A) = Cg(\mathrm{adj}_n A) = f(A) > Cf(A) \equiv 0 \ .$$

Part 3:
Step 1: We first show that $Cf = Cg$. Observe first that one always has $Cg \leq Cf$. We wish to show the reverse inequality. Let $\varepsilon > 0$ be fixed, then from Theorem 1.1 we get that there exist $\lambda \in [0,1]$, $b, c \in \mathbf{R}_+$ such that

$$\begin{cases} \varepsilon + Cg(|A|) \geq \lambda g(b) + (1 - \lambda)g(c) \\ |A| = \lambda b + (1 - \lambda)c \end{cases} \ .$$

Choose then

$$B = \frac{bA}{|A|}, \quad C = \frac{cA}{|A|} \ .$$

We therefore get

$$\begin{cases} \varepsilon + Cg(|A|) \geq \lambda f(B) + (1 - \lambda)f(C) \geq Cf(A) \\ A = \lambda B + (1 - \lambda)C \end{cases}.$$

Since ε is arbitrary, we have indeed obtained the claimed result.

Step 2: We now construct a function g such that (4) holds. We choose $m = n = 2$ and $g : \mathbb{R}_+ \to \mathbb{R}$ continuous and such that

$$\begin{cases} g(0) = \inf\{g(x) : x \geq 0\} \\ g(x) \geq a|x|^{\alpha}, \ a > 0 \text{ and } \alpha > 2 \\ Cg \text{ strictly increasing} \\ Cg \not\equiv g \end{cases},$$

and for $A \in \mathbb{R}^4$, we let

$$f(A) = g(|A|) . \tag{10}$$

One can choose for example

$$g(x) = \begin{cases} x & \text{if } x \in [0, 2] \\ -x + 4 & \text{if } x \in [2, 3] \\ x^3 - 26 & \text{if } x \geq 3 \end{cases}$$

then

$$Cg(x) = \begin{cases} \dfrac{1}{3}x & \text{if } x \in [0, 3] \\ x^3 - 26 & \text{if } x \geq 3 \end{cases}.$$

We shall now show that $Pf > Cg$. We decompose the proof into two steps.
Step 2.1: We first show that for every $A \in \mathbb{R}^4$ there exist $\lambda_i \in [0, 1]$, $\sum_{i=1}^{6} \lambda_i = 1$, $A_i \in \mathbb{R}^4$ such that

$$\begin{cases} Pf(A) = \displaystyle\sum_{i=1}^{6} \lambda_i f(A_i) = \sum_{i=1}^{6} \lambda_i g(|A_i|) \\ T(A) = (A, \det A) = \displaystyle\sum_{i=1}^{6} \lambda_i T(A_i) \end{cases}. \tag{11}$$

Let $\varepsilon > 0$, then, from Theorem 1.1, there exist $\lambda_i^{\varepsilon} \in [0, 1]$, $\sum_{i=1}^{6} \lambda_i^{\varepsilon} = 1$ and

$A_i^\varepsilon \in \mathbb{R}^4$ such that

$$\begin{cases} \displaystyle\sum_{i=1}^{6} \lambda_i^\varepsilon f(A_i^\varepsilon) - \varepsilon \leq Pf(A) \\ \displaystyle\sum_{i=1}^{6} \lambda_i^\varepsilon T(A_i^\varepsilon) = T(A) \end{cases} \tag{12}$$

Up to a subsequence, we infer from (12) that

$$\begin{cases} \lambda_i^\varepsilon \to \bar{\lambda}_i, \ i = 1,\ldots,6 \ \displaystyle\sum_{i=1}^{6} \bar{\lambda}_i = 1 \\ A_i^\varepsilon \to \bar{A}_i \ \ \text{if } i \in I \\ |A_i^\varepsilon| \to \infty \ \ \text{if } i \in J \\ I \cup J = \{1,2,\ldots,6\} \end{cases}$$

Using the coercivity of g we get

$$a \sum_{i=1}^{6} \lambda_i^\varepsilon |A_i^\varepsilon|^\alpha \leq \sum_{i=1}^{6} \lambda_i^\varepsilon f(A_i^\varepsilon) \leq f(A) = g(|A|) \ .$$

Combining (12) with the above inequality and the fact that $\alpha > 2$, we obtain

$$\lambda_i^\varepsilon |T(A_i^\varepsilon)| \to 0 \ \text{if } i \in J$$

and thus

$$\sum_{i\in I} \bar{\lambda}_i T(\bar{A}_i) = T(A) \ \text{and} \ \sum_{i\in I} \bar{\lambda}_i = 1 \ . \tag{13}$$

To obtain (11) we only need to show that

$$Pf(A) = \sum_{i\in I} \bar{\lambda}_i f(\bar{A}_i) \ . \tag{14}$$

From Theorem 1.1 and (13) we have immediately that

$$Pf(A) \leq \sum_{i\in I} \bar{\lambda}_i f(\bar{A}_i) \ .$$

Using (12) and the fact that $f \geq 0$ we get

$$\sum_{i\in I} \lambda_i^\varepsilon f(A_i^\varepsilon) - \varepsilon \leq \sum_{i=1}^{6} \lambda_i^\varepsilon f(A_i^\varepsilon) - \varepsilon \leq Pf(A) \leq \sum_{i\in I} \bar{\lambda}_i f(\bar{A}_i) \ .$$

Letting $\varepsilon \to 0$, using the continuity of f, we have indeed (14) and thus (11).

Step 2.2: We now show that $Pf > Cg$. We proceed by contradiction; assume that for every $A \in \mathbb{R}^4$

$$Pf(A) = Cg(|A|) \ .$$

(Note that trivially $Pf \geq Cg$). From Step 2.1, there exist $\lambda_i \in [0,1]$, $\sum_{i=1}^{6} \lambda_i = 1$ (we may assume without loss of generality that $\lambda_i > 0$) and $A_i \in \mathbb{R}^4$ such that

$$\begin{cases} Cg(|A|) = Pf(A) = \sum_{i=1}^{6} \lambda_i g(|A_i|) \geq \sum_{i=1}^{6} \lambda_i Cg(|A_i|) \\ T(A) = \sum_{i=1}^{6} \lambda_i T(A_i) \end{cases} \tag{15}$$

Using the fact that Cg is convex and increasing we have

$$Cg(|A|) = Pf(A) \geq \sum_{i=1}^{6} \lambda_i Cg(|A_i|) \geq Cg\left(\sum_{i=1}^{6} \lambda_i |A_i|\right)$$

$$\geq Cg\left(\left|\sum_{i=1}^{6} \lambda_i A_i\right|\right) = Cg(|A|) \ .$$

Since Cg is strictly increasing we deduce that

$$\sum_{i=1}^{6} \lambda_i |A_i| = \left|\sum_{i=1}^{6} \lambda_i A_i\right| = |A| \ .$$

However this may happen only if there exist $\alpha_i \geq 0$ such that

$$A_i = \alpha_i A \quad i = 1,\ldots,6 \ .$$

Since $T(A) = \sum \lambda_i T(A_i)$ and if $\det A \neq 0$ we obtain therefore that

$$\sum_{i=1}^{6} \lambda_i \alpha_i = \sum_{i=1}^{6} \lambda_i \alpha_i^2 = 1 \ .$$

Since the function $x \to x^2$ is strictly convex and $\lambda_i > 0$ we deduce that $\alpha_i = 1$, $1 \leq i \leq 6$ and hence from (15) that

$$Cg(|A|) = g(|A|)$$

for every $A \in \mathbf{R}^4$ with $\det A \neq 0$. This, however, contradicts the construction of g and hence $Pf > Cg$.

Step 3: It remains now to show (5), i.e. that if there exists $\alpha \geq 0$ such that

$$Cg(x) = \begin{cases} g(x) & \text{if } x \geq \alpha \\ g(0) = g(\alpha) & \text{if } x \leq \alpha \end{cases} , \tag{16}$$

then $Rf = Qf = Pf = Cf = Cg$ (note that the function g considered in Step 2 does not satisfy (16)). It is obvious that $Cg(|A|) \leq Rf(A)$, therefore we need only to show that for every $A \in \mathbf{R}^{nm}$

$$Rf(A) \leq Cg(|A|) . \tag{17}$$

Note also that if $|A| \geq \alpha$, then (17) is trivially satisfied. Therefore we only need to consider the case where $0 < |A| < \alpha$. From (16) we then obtain

$$Cg(|A|) = g(\alpha) .$$

Let $A = (A_j^i)_{1 \leq i \leq m, 1 \leq j \leq n}$ (we may assume without loss of generality that $A_1^1 \neq 0$) and

$$\lambda = \frac{1}{2} \left(1 + \frac{|A_1^1|}{(\alpha^2 - |A|^2 + (A_1^1)^2)^{1/2}} \right) ,$$

then $\frac{1}{2} < \lambda < 1$. Let $E = (E_j^i)_{1 \leq i \leq m, 1 \leq j \leq n}$ such that

$$E_1^1 = \frac{2A_1^1}{1 - 2\lambda}, \quad E_j^i = 0 \text{ otherwise} .$$

Finally let

$$\begin{cases} B = A - (1 - \lambda)E \\ C = A + \lambda E \end{cases} .$$

Then it is easy to see that

$$\begin{cases} A = \lambda B + (1 - \lambda)C \\ |B| = |C| = \alpha \\ \text{rank}\{B - C\} \leq 1 \end{cases} .$$

From Theorem 1.1 we have that

$$Rf(A) \leq \lambda f(B) + (1 - \lambda)f(C) = \lambda g(\alpha) + (1 - \lambda)g(\alpha) = Cg(|A|) ,$$

which is precisely (17).

Part 4: Let $A = \left(\begin{smallmatrix} \alpha & \beta \\ \gamma & \delta \end{smallmatrix}\right)$ and

$$f(A) = g(\alpha) + h(\det A) .$$

Since h is convex one has immediately that

$$Cg + h \leq Pf \leq Qf \leq Rf .$$

We next show the reverse inequality. From Theorem 1.1, for every $\varepsilon > 0$, there exist $\lambda \in [0,1]$, $\alpha_1, \alpha_2 \in \mathbb{R}$ such that

$$\begin{cases} \lambda g(\alpha_1) + (1 - \lambda)g(\alpha_2) \leq Cg(\alpha) + \varepsilon \\ \lambda\alpha_1 + (1 - \lambda)\alpha_2 = \alpha \end{cases} .$$

Case 1: $\beta \neq 0$, we then choose

$$A_1 = \begin{pmatrix} \alpha_1 & \beta \\ \gamma + \dfrac{\alpha_1 - \alpha}{\beta}\delta & \delta \end{pmatrix} , \quad A_2 = \begin{pmatrix} \alpha_2 & \beta \\ \gamma + \dfrac{\alpha_2 - \alpha}{\beta}\delta & \delta \end{pmatrix} .$$

We deduce that

$$\begin{cases} \lambda A_1 + (1 - \lambda)A_2 = A \\ \det A_1 = \det A_2 = \det A \\ \operatorname{rank}\{A_1 - A_2\} \leq 1 \end{cases} .$$

Therefore, using Theorem 1.1, we have

$$Rf(A) \leq \lambda f(A_1) + (1 - \lambda)f(A_2) = \lambda g(\alpha_1) + (1 - \lambda)g(\alpha_2) + h(\det A)$$

$$\leq Cg(\alpha) + h(\det A) + \varepsilon .$$

Since ε is arbitrary we have obtained the result.
Case 2: $\beta = 0$, $\gamma \neq 0$ is identical to Case 1.
Case 3: $\beta = \gamma = 0$ and therefore $A = \left(\begin{smallmatrix} \alpha & 0 \\ 0 & \delta \end{smallmatrix}\right)$. We choose

$$A_1 = \begin{pmatrix} \alpha_1 & 1 \\ \delta(\alpha_1 - \alpha) & \delta \end{pmatrix} , \quad A_2 = \begin{pmatrix} \alpha_2 & 1 \\ \delta(\alpha_2 - \alpha) & \delta \end{pmatrix}$$

$$A_3 = \begin{pmatrix} \alpha_1 & -1 \\ -\delta(\alpha_1 - \alpha) & \delta \end{pmatrix} , \quad A_4 = \begin{pmatrix} \alpha_2 & 1 \\ -\delta(\alpha_2 - \alpha) & \delta \end{pmatrix}$$

with

$$\lambda_1 = \lambda_3 = \frac{\lambda}{2} \quad \text{and} \quad \lambda_2 = \lambda_4 = \frac{1 - \lambda}{2} .$$

We then have

$$
\begin{cases}
\displaystyle\sum_{i=1}^{4}\lambda_i A_i = A \\[2mm]
\det A_i = \alpha\delta = \det A, \; i = 1,2,3,4 \\[2mm]
\det(A_1 - A_2) = \det(A_3 - A_4) \\[2mm]
\qquad\qquad = \det\left(\dfrac{\lambda_1 A_1 + \lambda_2 A_2}{\lambda_1 + \lambda_2} - \dfrac{\lambda_3 A_3 + \lambda_4 A_4}{\lambda_3 + \lambda_4}\right) = 0
\end{cases}
$$

The three last identities show that $(\lambda_i, A_i)_{1\le i\le 4}$ satisfy (H_4) and therefore, using Theorem 1.1, we have

$$
Rf(A) \le \sum_{i=1}^{4}\lambda_i f(A_i) = \lambda g(\alpha_1) + (1 - \lambda)g(\alpha_2) + h(\det A)
$$

$$
\le Cg(\alpha) + h(\det A) + \varepsilon \; .
$$

The arbitrariness of ε gives the result. $\square$

5.2 Relaxation Theorems

5.2.1 Relaxation Theorems

5.2.1.1 Relaxation of Variational Problems

We now proceed with the relaxation theorems. We first start with the case $f(x, u, \nabla u) = f(\nabla u)$.

Theorem 2.1. *Let $\Omega \subset \mathbf{R}^n$ be a bounded open set with Lipschitz boundary. Let $f : \mathbf{R}^{nm} \to \mathbf{R}$ be continuous and satisfying*

$$
(C) \qquad a + \sum_{i=1}^{I} b_i |\Phi_i(A)|^{\beta_i} \le f(A) \le c + \sum_{i=1}^{I} d_i |\Phi_i(A)|^{\beta_i}
$$

for every $A \in \mathbf{R}^{nm}$ and for some $a, c \in \mathbf{R}$, $I \ge 1$ (an integer), $\beta_i > 1$, $d_i \ge b_i > 0$ and where $\Phi_i : \mathbf{R}^{nm} \to \mathbf{R}$, $i = 1,\ldots,I$ are quasiaffine. Let $u \in W^{1,\infty}(\Omega; \mathbf{R}^m)$, then there exists $\{u^s\}_{s=1}^{\infty}$, $u^s \in W^{1,\infty}(\Omega; \mathbf{R}^m)$ such that

i) $u^s = u$ on $\partial\Omega$,
ii) $\Phi_i(\nabla u^s) \rightharpoonup \Phi_i(\nabla u)$ in $L^{\beta_i}(\Omega)$, $i = 1,\ldots,I$, as $s \to \infty$,
iii) $\int_\Omega f(\nabla u^s(x))\,dx \to \int_\Omega Qf(\nabla u(x))\,dx$ as $s \to \infty$,
where $Qf = \sup\{g \le f : g \text{ quasiconvex}\}$.

Before making some remarks we give two examples.

EXAMPLES.

i) The case where f satisfies a condition of the type

$$a + b|A|^\beta \le f(A) \le c + d|A|^\beta$$

is a particular case of (C). It suffices to choose $I = nm$, $\beta_i = \beta > 1$, $d_i = d \ge b_i = b > 0$, $i = 1, \ldots, I$ and for $A = (A_j^i)_{1\le i\le m, 1\le j\le n}$,

$$\begin{cases} \Phi_1(A) = A_1^1, \ldots, \Phi_n(A) = A_n^1 \\ \Phi_{n+1}(A) = A_1^2, \ldots, \Phi_{2n}(A) = A_n^2 \\ \qquad \vdots \\ \Phi_{(m-1)n+1}(A) = A_1^m, \ldots, \Phi_{mn}(A) = A_n^m \end{cases}$$

which are all quasiaffine. And in this case ii) of the theorem becomes

$$u^s \rightharpoonup u \text{ in } W^{1,\beta}(\Omega; \mathbf{R}^m), \text{ as } s \to \infty.$$

Note also that if $n = 1$ or $m = 1$ (i.e. in the scalar case), then $Qf = Cf = f^{**}$;

ii) if $m = n$ and

$$a + b|\det A|^\beta \le f(A) \le c + d|\det A|^\beta$$

then choose in (C), $I = 1$ and $\Phi_i(A) = \det A$ which is quasiaffine. And in this case ii) is read

$$\det \nabla u^s \rightharpoonup \det \nabla u \text{ in } L^\beta(\Omega), \text{ as } s \to \infty,$$

and if $f(A) = g(\det A)$ then iii) can be rewritten as

$$\int_\Omega g(\det \nabla u^s(x))\, dx \to \int_\Omega Cg(\det \nabla u(x))\, dx$$

since by Theorem 1.3, $Qf = Cg = g^{**}$.

REMARKS.

i) The history of Theorem 2.1 is two folds. First the scalar case ($m = 1$ or $n = 1$); recall that in this case $Qf = f^{**} = Cf$. As stated the result was established by L.C. Young [1-3] when $m = n = 1$ and then generalized by Ioffe-Tihomirov [1], Ekeland-Témam [1] in the case $m = 1$ (or $n = 1$); see also Berliochi-Lasry [1,2], Marcellini-Sbordone [1]. However this theorem is also reminiscent of control theory (see Gamkrelidze [1], Warga [1,2], Clarke [1], Mac Shane [5]) where the idea of relaxation is standard. The result for the vectorial case (i.e. $m, n > 1$, recall also that in this case $Qf > Cf = f^{**}$) was established by Dacorogna [1-3]. It was then proved following a different approach by Acerbi-Fusco [1]. See also for related results Acerbi-Buttazzo-Fusco [1], Buttazzo [1,2], Buttazzo-Dal Maso [1], De Giorgi [5];

ii) this approach of relaxing non convex (non quasiconvex, in the vectorial case) problems is not the only one. There is a closely related idea due to L.C. Young [1-3] (and in fact prior to the one presented here), see also Mac Shane [3,4], which instead of replacing f by Qf, enlarges the space of admissible functions from Sobolev spaces to spaces of parametrized measures (called, generalized curve, surface, by L.C. Young). This idea of L.C. Young has been very fruitful in the calculus of variations as well as in optimal control theory and recently in partial differential equations;

iii) Theorem 2.1 implies in particular that

$$\inf(P) = \inf(QP)$$

where

$$(P) \qquad \inf\left\{ I(u) = \int_\Omega f(\nabla u(x))\,dx \ : \ u \in u_0 + W_0^{1,p}(\Omega; \mathbb{R}^m) \right\}$$

$$(QP) \ \inf\left\{ \bar{I}(u) = \int_\Omega Qf(\nabla u(x))\,dx \ : \ u \in u_0 + W_0^{1,p}(\Omega; \mathbb{R}^m) \right\} \ ;$$

note also that if (QP) has a solution $\bar{u}$, then the theorem asserts that there exists a minimizing sequence $\{u^s\}$ for (P) with Properties i), ii) and iii) of the theorem. Reciprocally, since every minimizing sequence $\{u^s\}$ for (P) is also a minimizing sequence of (QP), then, up to the extraction of a subsequence, u^s converges weakly to a solution $\bar{u}$ of (QP) (provided it exists). It is in this sense that one should understand that solutions of (QP) are generalized solutions of (P), while in the approach of L.C. Young, mentioned above, it is the limit in the strong convergence, in the sense of parametrized measures, which is called generalized solution of (P);

iv) it is obvious that in the theorem u need not be in $W^{1,\infty}$ but only in $W^{1,p}$ where p is sufficiently large to ensure that $\Phi_i(\nabla u) \in L^{\beta_i}$, $1 \le i \le I$. For example if $a + b|A|^\beta \le f(A) \le c + d|A|^\beta$, then one can take $p = \beta$;

v) note also that in the scalar case (QP) and (P^{**}) are the same problem where

$$(P^{**}) \quad \inf\left\{ I^{**}(u) = \int_\Omega f^{**}(\nabla u(x))\,dx \ : \ u \in u_0 + W_0^{1,p}(\Omega; \mathbb{R}^m) \right\} \ .$$

However this is not true in the vectorial case, one has in general

$$\inf(P) = \inf(QP) > \inf(P^{**}) \; .$$

For example if $m = n$, $n \geq 2$ and $f(A) = (\det A)^2$, then

$$f(\nabla u) = Qf(\nabla u) = (\det \nabla u)^2 > f^{**}(\nabla u) \equiv 0$$

(cf. Theorem 1.3) and therefore if $\det \nabla u_0 > 0$, then, using Jensen inequality, we have

$$\inf(P) = \inf(QP) \geq \operatorname{meas} \Omega \left(\frac{1}{\operatorname{meas} \Omega} \int_\Omega \det \nabla u_0(x) \, dx \right)^2$$

$$> 0 = (\inf P^{**}) \; .$$

One can prove (cf. Theorem 2.4 in the Appendix) that for some u_0 and Ω one in fact has

$$\inf(P) = \left[\frac{1}{\operatorname{meas} \Omega} \int_\Omega \det \nabla u_0(x) \, dx \right]^2 \operatorname{meas} \Omega$$

and the infimum is attained;

vi) note that the above theorem does not apply to the minimal surface case since the growth condition (C) holds in this case with $\beta_i = 1$, however the minimal surface problem in parametric form can be handled in a similar way, cf. Dacorogna [2], Acerbi-Fusco [1].

We now proceed with the proof of the theorem.

PROOF. We decompose the proof into three steps. We first reduce the problem to consider piecewise affine functions u and then construct u^s in small cubes and finally piece up together the u^s to get the result.

Step 1: Let $\varepsilon > 0$ and observe that there is no loss of generality in supposing that u is piecewise affine (i.e. $u \in \operatorname{Aff}(\Omega; \mathbf{R}^m)$) otherwise we may find $\tilde{\Omega} \subset \Omega$ an open set and $v \in W^{1,\infty}(\Omega; \mathbf{R}^m)$ such that (cf. Theorem 1.8 of Chapter 2)

$$\begin{cases} \operatorname{meas}(\Omega - \tilde{\Omega}) \leq \varepsilon \\ v \in \operatorname{Aff}(\Omega; \mathbf{R}^m) \text{ and } v = u \text{ on } \partial\Omega \\ |v(x) - u(x)| \leq \varepsilon \text{ for all } x \in \Omega \\ \|\nabla v - \nabla u\|_{L^\alpha} \leq \varepsilon \text{ for every } \alpha \geq 1 \end{cases} \; .$$

Therefore if we can prove the theorem for v and $\tilde{\Omega}$, defining $v = u^\nu$ in $\Omega - \tilde{\Omega}$, we shall have proved the theorem for every $u \in W^{1,\infty}(\Omega; \mathbf{R}^m)$.

We may therefore assume that $u \in \mathrm{Aff}(\Omega; \mathbf{R}^m)$. We may decompose Ω into open sets Ω_j $1 \le j \le J$ so that ∇u is constant in Ω_j. We then decompose Ω_j into small cubes D_j^p $1 \le p \le P_j$ so that

$$\left| \int_{\Omega_j - \bigcup_{p=1}^{P_j} D_j^p} Qf(\nabla u(x))\, dx \right| \le \left| \int_{\Omega_j - \bigcup_{p=1}^{P_j} D_j^p} f(\nabla u(x))\, dx \right|$$

$$\le \frac{\varepsilon}{3J}, \quad 1 \le j \le J \ . \tag{1}$$

Step 2: We now construct the sequence u^s on each of the D_j^p (denoted by D in this step). From Theorem 1.1 we have that

$$Qf(\nabla u) = \inf \left\{ \frac{1}{\mathrm{meas}\, D} \int_D f(\nabla u + \nabla \varphi(x))\, dx \ : \ \varphi \in W_0^{1,\infty}(D; \mathbf{R}^m) \right\} \ .$$

Therefore there exists a sequence $\varphi^\nu \in W_0^{1,\infty}(D; \mathbf{R}^m)$ such that

$$Qf(\nabla u) + \frac{1}{\nu} \ge \frac{1}{\mathrm{meas}\, D} \int_D f(\nabla u + \nabla \varphi^\nu(x))\, dx \ge Qf(\nabla u) \ . \tag{2}$$

Note that from (2) and the hypothesis (C) of the theorem we have that there exists a constant K such that

$$\int_D |\Phi_i(\nabla u + \nabla \varphi^\nu(x))|^{\beta_i}\, dx \le K, \quad i = 1, \ldots, I \ . \tag{3}$$

We now extend φ^ν by periodicity from D to the whole of $\mathbf{R}^n$ (this is possible since $\varphi^\nu = 0$ on ∂D) and define

$$u^s(x) = u(x) + \frac{1}{s} \varphi^s(sx) \ . \tag{4}$$

Observe that $u^s = u$ on ∂D since φ^s is periodic of period equal to the edge length of D and $\varphi^s = 0$ on ∂D. We now wish to show that

$$\begin{cases} \Phi_i(\nabla u^s) \rightharpoonup \Phi_i(\nabla u) \text{ in } L^{\beta_i}(D), \ 1 \le i \le I \text{ as } s \to \infty \\[2mm] \displaystyle\int_D f(\nabla u^s(x))\, dx \\[2mm] \qquad \rightarrow \displaystyle\int_D Qf(\nabla u(x))\, dx = Qf(\nabla u)\, \mathrm{meas}\, D \text{ as } s \to \infty \ . \end{cases} \tag{5, 6}$$

It is easy to see (6) since

$$\int_D f(\nabla u^s(x))\, dx = \int_D f(\nabla u + \nabla \varphi^s(sx))\, dx$$

$$= \frac{1}{s^n} \int_{sD} f(\nabla u + \nabla \varphi^s(y))\, dy$$

$$= \int_D f(\nabla u + \nabla \varphi^s(y))\, dy \ , \tag{7}$$

the last equality being a consequence of the periodicity of φ^s.

Combining (2) and (7) we get immediately (6). It therefore remains to show (5). In view of Lemma 1.4 of Chapter 2 and of (3) it is sufficient to prove that if $E = x_0 + \delta D \subset D$ where $x_0 \in D$, $\delta > 0$ then

$$\int_E \Phi_i(\nabla u^s(x))\, dx \ \to \ \Phi_i(\nabla u)\, \mathrm{meas}\, E \ \text{ as } s \to \infty, \ 1 \le i \le I \ . \tag{8}$$

Note that

$$\int_E \Phi_i(\nabla u^s(x))\, dx = \int_{x_0+\delta D} \Phi_i(\nabla u^s(x))\, dx$$

$$= \int_{x_0+\delta D} \Phi_i(\nabla u + \nabla \varphi^s(sx))\, dx$$

$$= \frac{1}{s^n} \int_{s(x_0+\delta D)} \Phi_i(\nabla u + \nabla \varphi^s(y))\, dy$$

$$= \frac{1}{s^n} \int_{sx_0+[s\delta]D} \Phi_i(\nabla u + \nabla \varphi^s(y))\, dy$$

$$+ \frac{1}{s^n} \int_{sx_0+s\delta D-[s\delta]D} \Phi_i(\nabla u + \nabla \varphi^s(y))\, dy$$

$$= \left(\frac{[s\delta]}{s}\right)^n \int_D \Phi_i(\nabla u + \nabla \varphi^s(y))\, dy$$

$$+ \frac{1}{s^n} \int_{sx_0+s\delta D-[s\delta]D} \Phi_i(\nabla u + \nabla \varphi^s(y))\, dy \tag{9}$$

where $[s\delta]$ denotes the integer part of $s\delta$ and where we have used in (9) the periodicity of φ^s. Since $\varphi^s = 0$ on ∂D and Φ_i is quasiaffine (cf. definition) we have

$$\int_E \Phi_i(\nabla u^s(x))\, dx = \left(\frac{[s\delta]}{s}\right)^n \Phi_i(\nabla u)\, \mathrm{meas}\, D$$

$$+ \frac{1}{s^n} \int_{sx_0+s\delta D-[s\delta]D} \Phi_i(\nabla u + \nabla \varphi^s(y))\, dy \ .$$

Therefore using (3), the above identity and the periodicity of φ^s we deduce

$$\left| \int_E \Phi_i(\nabla u^s(x))\, dx - \Phi_i(\nabla u)\operatorname{meas} E \right|$$

$$\leq \left(\frac{[s\delta]^n}{s^n} - \delta^n \right) \Phi_i(\nabla u)\operatorname{meas} D + \frac{K}{s^n} \tag{10}$$

where K denotes a constant. Letting $s \to \infty$ we have indeed obtained (8) and thus Step 2.

Step 3: Summarizing the results we have constructed $u^s \in W^{1,\infty}(D_j^p; \mathbb{R}^m)$, $1 \leq p \leq P_j, 1 \leq j \leq J$ such that

$$\begin{cases} u^s = u \text{ on } \partial D_j^p \\[2mm] \Phi_i(\nabla u^s) \rightharpoonup \Phi_i(\nabla u) \text{ in } L^{\beta_i}(D_j^p), \text{ as } s \to \infty \\[2mm] \displaystyle\int_{D_j^p} f(\nabla u^s(x))\, dx \to Qf(\nabla u)\operatorname{meas} D_j^p, \text{ as } s \to \infty \end{cases} \tag{11}$$

We have therefore defined $u^s \in W^{1,\infty}\left(\bigcup_{p=1}^{P_j} D_j^p; \mathbb{R}^m \right)$. Defining $u^s = u$ on $\Omega_j - \bigcup_{p=1}^{P_j} D_j^p, 1 \leq j \leq J$, we have indeed constructed a sequence u^s satisfying i) and ii) of the theorem. It therefore remains to show iii). Consider

$$\left| \int_\Omega Qf(\nabla u(x))\, dx - \int_\Omega f(\nabla u^s(x))\, dx \right|$$

$$= \left| \sum_{j=1}^J \left\{ Qf(\nabla u)\operatorname{meas} \Omega_j - \int_{\Omega_j} f(\nabla u^s(x))\, dx \right\} \right|$$

$$\leq \sum_{j=1}^J |Qf(\nabla u)|\operatorname{meas}\left(\Omega_j - \bigcup_{p=1}^{P_j} D_j^p \right) + \sum_{j=1}^J \left| \int_{\Omega_j - \bigcup_{p=1}^{P_j} D_j^p} f(\nabla u(x))\, dx \right|$$

$$+ \left| \sum_{j=1}^J \sum_{p=1}^{P_j} \left\{ \int_{D_j^p} f(\nabla u^s(x))\, dx - Qf(\nabla u)\operatorname{meas} D_j^p \right\} \right| .$$

Combining (1) and (11) we have indeed obtained the result. $\qquad\square$

REMARK. Let

$$I(u) = \int_\Omega f(\nabla u(x))\, dx, \quad \bar{I}(u) = \int_\Omega Qf(\nabla u(x))\, dx .$$

With these notations one can prove that if f is continuous (and does not necessarily satisfy condition (C)) then

$$\bar{I}(u) = \inf_{\{u^s\}} \left\{ \liminf[I(u^s) : u^s \overset{*}{\rightharpoonup} u \text{ in } W_0^{1,\infty}(\Omega; \mathbb{R}^m)] \right\} .$$

The proof of this fact is almost identical to that of the theorem. Let φ^ν be as in (2) and extended by periodicity. Let

$$u^{s,\nu} = u(x) + \frac{1}{s}\varphi^\nu(sx)$$

for ν fixed. Since φ^ν is in $W_0^{1,\infty}$ we deduce that

$$u^{s,\nu} \overset{*}{\rightharpoonup} u \text{ in } W_0^{1,\infty} \text{ as } s \to \infty, \ \nu \text{ fixed } .$$

Furthermore we have as in (7)

$$\int_D f(\nabla u^{s,\nu}(x))\,dx = \int_D f(\nabla u + \nabla\varphi^\nu(x))\,dx .$$

Combining the above identity, (2) and the same argument as in Step 3 we obtain immediately the claimed result.

We now generalize the result of Theorem 2.1 to integrands which depend not only on ∇u but also on x and u.

Corollary 2.2. *Let $\Omega \subset \mathbb{R}^n$ be a bounded open set with Lipschitz boundary. Let $f : \Omega \times \mathbb{R}^{nm} \to \mathbb{R}$ be continuous and such that*

$$(C_1) \qquad a_1(x) + \sum_{i=1}^{I} b_i|\Phi_i(A)|^{\beta_i} \leq f(x, A) \leq a_2(x) + \sum_{i=1}^{I} d_i|\Phi_i(A)|^{\beta_i}$$

$$(L_1) \qquad |f(x, A) - f(y, A)| \leq \eta(|x - y|)\left(1 + \sum_{i=1}^{I} d_i|\Phi_i(A)|^{\beta_i}\right)$$

for every $x \in \Omega$, $A \in \mathbb{R}^{nm}$ and for some $a_1, a_2 \in L^1(\Omega)$, $I \geq 1$ (an integer), $\beta_i > 1$, $d_i \geq b_i > 0$, $\Phi_i : \mathbb{R}^{nm} \to \mathbb{R}$ quasiaffine and $\eta : \mathbb{R} \to \mathbb{R}$ continuous and increasing function such that $\eta(0) = 0$. Let $u \in W^{1,\infty}(\Omega, \mathbb{R}^m)$ then there exists $\{u^s\}_{s=1}^{\infty}$, $u^s \in W^{1,\infty}(\Omega; \mathbb{R}^m)$ such that

i) $u^s = u$ on $\partial\Omega$
ii) $\Phi_i(\nabla u^s) \rightharpoonup \Phi_i(\nabla u)$ in $L^{\beta_i}(\Omega)$, $i = 1, \ldots, I$ as $s \to \infty$
iii) $I(u^s) \equiv \int_\Omega f(x, \nabla u^s(x))\,dx \to \bar{I}(u) \equiv \int_\Omega Qf(x, \nabla u(x))\,dx$ as $s \to \infty$
where Qf is the quasiconvex envelope of f (with respect to the last variable).

REMARK. The same remarks as before apply to the corollary.

PROOF. Let Ω be approximated by a union of cubes Ω_p, $1 \leq p \leq P$ and let x_p denote the centre of the cube Ω_p. Fix $\varepsilon > 0$ and choose P sufficiently large so that

$$
\begin{cases}
\left| \left| \int_{\Omega - \bigcup_{p=1}^{P} \Omega_p} f(x, \nabla u(x)) \, dx \right| \leq \varepsilon \right. \\[2ex]
\left| \int_{\Omega - \bigcup_{p=1}^{P} \Omega_p} Qf(x, \nabla u(x)) \, dx \right| \leq \varepsilon \quad . \\[2ex]
\eta(|x - x_p|) \leq \dfrac{\varepsilon}{\operatorname{meas} \Omega} \ , x \in \bigcup_{p=1}^{P} \Omega_p
\end{cases}
\tag{1}
$$

Having fixed p in this way we may use the theorem on each of the Ω_p, to find $u_p^s \in W^{1,\infty}(\Omega_p; \mathbb{R}^m)$ such that for s sufficiently large

$$
\begin{cases}
u_p^s = u \ \text{on} \ \partial\Omega_p \\[1ex]
\Phi_i(\nabla u_p^s) \rightharpoonup \Phi_i(\nabla u) \ \text{in} \ L^{\beta_i}(\Omega_p), \ 1 \leq i \leq I \\[1ex]
\left| \int_{\Omega_p} [f(x_p, \nabla u_p^s(x)) - Qf(x_p, \nabla u(x))] \, dx \right| \leq \dfrac{\varepsilon}{P}
\end{cases}
\tag{2}
$$

We then define $u^s = u_p^s$ in Ω_p and $u^s = u$ in $\Omega - \bigcup \Omega_p$. We have therefore immediately i) and ii) of the corollary. It then remains to prove iii). Observe that from (1), (2) and (L$_1$) we have that

$$
|I(u^s) - \bar{I}(u)| \leq \left| \int_{\Omega - \bigcup \Omega_p} f(x, \nabla u(x)) \, dx \right| + \left| \int_{\Omega - \bigcup \Omega_p} Qf(x, \nabla u(x)) \, dx \right|
$$

$$
+ \sum_{p=1}^{P} \int_{\Omega_p} |f(x, \nabla u^s(x)) - f(x_p, \nabla u^s(x))| \, dx
$$

$$
+ \sum_{p=1}^{P} \int_{\Omega_p} |Qf(x, \nabla u(x)) - Qf(x_p, \nabla u(x))| \, dx
$$

$$
+ \sum_{p=1}^{P} \left| \int_{\Omega_p} [f(x_p, \nabla u^s(x)) - Qf(x_p, \nabla u(x))] \, dx \right|
$$

$$
\leq K\varepsilon \ ,
$$

where K is a constant independent of s. Since $\varepsilon > 0$ is arbitrary we have indeed obtained the result. $\qquad\square$

Corollary 2.3. *Let $\Omega \subset \mathbb{R}^n$ be a bounded open set with Lipschitz boundary. Let $f : \Omega \times \mathbb{R}^m \times \mathbb{R}^{nm} \to \mathbb{R}$ be continuous and such that*

$$(C_2) \qquad a_1(x) + b|A|^\beta \leq f(x,u,A) \leq a_2(x) + d(|u|^\beta + |A|^\beta)$$

$$(L_2) \qquad |f(x,u,A) - f(x,v,B)| \leq K(1 + |u|^{\beta-1} + |v|^{\beta-1} + |A|^{\beta-1}$$
$$+ |B|^{\beta-1})(|u-v| + |A-B|)$$

$$(L_2') \qquad |f(x,u,A) - f(y,u,A)| \leq \eta(|x-y|)(1 + |u|^\beta + |A|^\beta)$$

where $a_1, a_2 \in L^1(\Omega)$, $d \geq b > 0$, $\beta > 1$, $K > 0$, $\eta : \mathbb{R} \to \mathbb{R}$ is continuous, increasing and $\eta(0) = 0$. Let $u \in W^{1,\beta}(\Omega; \mathbb{R}^m)$ then there exist $\{u^s\}$, $u^s \in W^{1,\beta}(\Omega; \mathbb{R}^m)$ such that

i) $u^s = u$ *on* $\partial\Omega$
ii) $u^s \rightharpoonup u$ *in* $W^{1,\beta}(\Omega; \mathbb{R}^m)$
iii) $I(u^s) \equiv \int_\Omega f(x, u^s(x), \nabla u^s(x))\, dx \to \bar{I}(u) \equiv \int_\Omega Qf(x, u(x), \nabla u(x))\, dx$,
where Qf is the lower quasiconvex envelope of f (with respect to the last variable).

REMARK. It is obvious that one can replace in (C_2), (L_2) and (L_2') the exponent β of $|u|^\beta$ by γ (the Sobolev exponent) where
1) if $\beta < n$, then $1 \leq \gamma < \frac{n\beta}{n-\beta}$
2) if $\beta = n$, then $1 \leq \gamma < \infty$
3) if $\beta > n$, then replace $u_2(x) + d|u|^\beta$ by any continuous function $g(x,u)$.

PROOF. The proof is similar to the above corollary. We apply Corollary 2.2 to the functions $f(x, u(x), \bullet)$ and $Qf(x, u(x), \bullet)$ to get $u^s \in W^{1,\beta}(\Omega; \mathbb{R}^m)$ such that

$$\begin{cases} u^s = u \text{ on } \partial\Omega \\ u^s \rightharpoonup u \text{ in } W^{1,\beta}(\Omega; \mathbb{R}^m) \\ \int_\Omega f(x, u(x), \nabla u^s(x))\, dx \to \int_\Omega Qf(x, u(x), \nabla u(x))\, dx \end{cases} \qquad (1)$$

In order to get the corollary it therefore remains to show that

$$\lim_{s \to \infty} \int_\Omega |f(x, u(x), \nabla u^s(x)) - f(x, u^s(x), \nabla u^s(x))|\, dx = 0 .$$

This is ensured by (L_2), (1) and Hölder inequality. $\qquad\square$

REMARK. Note that the hypotheses of the corollary can be weakened by suppressing the hypotheses (L_2) and (L_2') and by considering Carathéodory functions; see for more details Acerbi-Fusco [1].

5.2.1.2   Relaxation for Euler Equations

We now apply the results of the preceding section to Euler equations associated to variational problems. We first start with an abstract result (see Ekeland-Témam [1]).

Theorem 2.4. *Let V be a Banach space, V^* its dual, $\langle \bullet; \bullet \rangle$ the bilinear canonical pairing over $V \times V^*$. Let $F : V \to \mathbb{R}$ be lower semicontinuous, Gâteaux-differentiable and such that $-\infty < \inf F < +\infty$. Let $\varepsilon > 0$ and $\bar{u} \in V$ such that*

$$F(\bar{u}) \leq \inf F + \varepsilon \ .$$

Then there exists $v \in V$ such that

$$\begin{cases} F(v) \leq F(\bar{u}) \\ \|\bar{u} - v\|_V \leq \sqrt{\varepsilon} \\ \|F'(v)\|_{V^*} \leq \sqrt{\varepsilon} \end{cases} .$$

We next use the above result for the following problem. Let

$$(P) \qquad \inf \left\{ I(u) = \int_\Omega f(x, \nabla u(x))\, dx \ : \ u \in u_0 + W_0^{1,p}(\Omega; \mathbb{R}^m) \right\}$$

$$(QP) \qquad \inf \left\{ \bar{I}(u) = \int_\Omega Qf(x, \nabla u(x))\, dx \ : \ u \in u_0 + W_0^{1,p}(\Omega; \mathbb{R}^m) \right\}$$

where

i) $\Omega \subset \mathbb{R}^n$ is a bounded open set with Lipschitz boundary,

ii) $f : \Omega \times \mathbb{R}^{nm} \to \mathbb{R}$ is continuous, C^1 in the last variable and satisfy

$$(H) \qquad \begin{cases} a_1(x) + b_1|A|^p \leq f(x, A) \leq a_2(x) + b_2|A|^p \\ |f(x, A) - f(y, A)| \leq \eta(|x - y|)(1 + |A|^p) \\ |\mathrm{grad}_A f(x, A)| \leq a_3(x) + b_3|A|^{p-1} \end{cases}$$

where $b_2 \geq b_1 > 0$, $b_3 \geq 0$, $a_1, a_2, a_3 \in L^1(\Omega)$, $p > 1$, $\eta : \mathbb{R} \to \mathbb{R}$ is continuous, increasing and $\eta(0) = 0$;

iii) $Qf(x, \bullet) = \sup\{g(x, \bullet) \leq f(x, \bullet), g(x, \bullet) \text{ quasiconvex } \}$ and C^1 in the last variable.

REMARKS.

i) Recall that if $m = 1$ (or $n = 1$), then $Qf = f^{**}$ and the hypothesis that $Qf \in C^1$ in the last variable is not necessary since then (cf. Chapter 2) f^{**} is necessarily C^1 if f is C^1;

ii) one can also consider the case where the integrand is of the form $f(x, u(x), \nabla u(x))$. Then a similar result holds, provided that one makes the natural growth conditions in order to ensure the differentiability of I (cf. Chapter 3, Theorem 4.4).

Theorem 2.5. *Under the above hypotheses and assuming that (QP) admits a solution $\bar{u}$ (which is ensured for example if $p > 1$) then there exists a minimizing sequence u_s of (P) such that*

$$\begin{cases} u_s \rightharpoonup \bar{u} \ in \ W^{1,p}(\Omega; \mathbb{R}^m) \\ I(u_s) \to \bar{I}(\bar{u}) = \inf(P) = \inf(QP) \\ I'(u_s) \to 0 = \bar{I}'(\bar{u}) \ in \ W^{-1,p'}(\Omega; \mathbb{R}^m) \end{cases} \tag{1}$$

where $\frac{1}{p} + \frac{1}{p'} = 1$ and

$$\begin{cases} I'(u) = \left(-\sum_{\alpha=1}^{n} \frac{\partial}{\partial x_\alpha} \left(\frac{\partial f}{\partial A_\alpha^i}(x, \nabla u(x)) \right) \right)_{1 \leq i \leq m} \\ \bar{I}'(u) = \left(-\sum_{\alpha=1}^{n} \frac{\partial}{\partial x_\alpha} \left(\frac{\partial Qf}{\partial A_\alpha^i}(x, \nabla u(x)) \right) \right)_{1 \leq i \leq m} \end{cases} \tag{2}$$

If furthermore $m = 1$ (or $n = 1$) then for every $\bar{u} \in u_0 + W_0^{1,p}(\Omega)$ there exist $u_s \in u_0 + W_0^{1,p}(\Omega)$ such that

$$\begin{cases} u_s \rightharpoonup \bar{u} & in \ W^{1,p}(\Omega) \\ I(u_s) \to \bar{I}(\bar{u}) \\ I'(u_s) \to \bar{I}'(\bar{u}) & in \ W^{-1,p'}(\Omega) \end{cases} \tag{3}$$

REMARKS.

i) Note that in (3) $\bar{u}$ need not be a minimum for (QP);

ii) the result (3) is not known to hold a) in the scalar case ($m = 1$ or $n = 1$) if f depends explicitly on u (i.e. $f - f(x, u, \nabla u)$) or b) in the vectorial case even if f does not depend explicitly on u and $\bar{u}$ is not a minimum of (QP), for reasons which will be obvious after the proof.

PROOF.

i) Note that (H_3) ensures, cf. Chapter 3 and 4, the differentiability of I and $\bar{I}$. By making a translation, if necessary, we can consider only the case where $u_0 = 0$. From Corollary 2.2 we have that there exists v_s such that

$$\begin{cases} v_s \rightharpoonup \bar{u} \text{ in } W_0^{1,p}(\Omega; \mathbb{R}^m) \\ I(v_s) \to \bar{I}(\bar{u}) = \inf(P) = \inf(QP) \end{cases}.$$

Since $\bar{u}$ is a minimum we have also $\bar{I}'(\bar{u}) = 0$ (cf. Theorem 4.4 of Chapter 3). We now apply Theorem 2.4 to the sequence v_s to get the result;

ii) to prove (3) we first note that since $m = 1$ (or $n = 1$) then $Qf(x, A) = f^{**}(x, A)$ and therefore $\bar{I}$ is convex. Define

$$\begin{cases} F(w) = I(w) - \bar{I}(\bar{u}) - \langle \bar{I}'(\bar{u}); w - \bar{u} \rangle \\ \bar{F}(w) = \bar{I}(w) - \bar{I}(\bar{u}) - \langle \bar{I}'(\bar{u}); w - \bar{u} \rangle \end{cases}.$$

We therefore have from the convexity of $\bar{I}$ that

$$F(w) \geq \bar{F}(w) \geq 0 = \bar{F}(\bar{u}) = \inf\{\bar{F}(u) : u \in V\}$$

where $V = W_0^{1,p}(\Omega)$. Using now Corollary 2.2 we have that there exists $v_s \in W_0^{1,p}(\Omega)$ such that

$$\begin{cases} v_s \rightharpoonup \bar{u} \text{ in } W_0^{1,p}(\Omega) \\ I(v_s) \to \bar{I}(\bar{u}) \end{cases}$$

i.e.

$$\begin{cases} v_s \rightharpoonup \bar{u} \text{ in } W_0^{1,p}(\Omega) \\ F(v_s) \to \bar{F}(\bar{u}) = 0 = \inf\{\bar{F}(u) : u \in V\} = \inf\{F(u) : u \in V\} \end{cases}.$$

Applying now Theorem 2.4 to the sequence v_s, to F and noting that $F'(v) = I'(v) - \bar{I}'(\bar{u})$, we have indeed established the result. $\qquad\square$

5.2.2 Existence and Non-Existence of Solutions

We conclude this chapter with some examples dealing with existence and non-existence of solutions of

$$(P) \qquad \inf\left\{ I(u) = \int_\Omega f(x, u(x), \nabla u(x)) \, dx \ : \ u \in u_0 + W_0^{1,p}(\Omega; \mathbb{R}^m) \right\}$$

when f fails to be convex in the last variable. We shall consider only the scalar case. The direct methods developed in this book cannot be applied and we shall give examples where the minimum is not attained. We shall also give examples where, even though f is not convex, the minimum is nevertheless attained.

There is no known unified way of dealing with non convex problems, except from relaxation (in the sense presented above or in the sense of parametrized measure). We intend to give here few examples and we refer for more details to the literature quoted below.

5.2.2.1 The Case of Single Integrals

We let $\Omega = (0,1)$, $u : (0,1) \rightarrow \mathbf{R}^m$, $m \geq 1$ and

$$(P) \qquad \inf\left\{ I(u) = \int_0^1 f(x, u(x), u'(x))\, dx \ : \ u(0) = 0, \right.$$
$$\left. u(1) = \alpha, \ u \in W^{1,p}(\Omega; \mathbf{R}^m) \right\} .$$

That this problem does not have a solution when f is not convex in the last variable, was observed long ago and we give here an example due to Bolza.

EXAMPLE. Let $m = 1$, $\alpha = 0$,

$$f(u, \xi) = u^2 + (\xi^2 - 1)^2$$

and

$$(P) \qquad \inf\left\{ I(u) = \int_0^1 f(u(x), u'(x))\, dx \ : \ u \in W_0^{1,4}(0,1) \right\} .$$

It is easy to see that

$$\inf(P) = 0 .$$

Since choosing for N an integer and $0 \leq k \leq N - 1$

$$u^N(x) = \begin{cases} x - \dfrac{k}{N} & \text{if } x \in \left[\frac{2k}{2N}, \frac{2k+1}{2N}\right] \\ -x + \dfrac{k+1}{N} & \text{if } x \in \left[\frac{2k+1}{2N}, \frac{2k+2}{2N}\right] \end{cases}$$

then $u^N \in W^{1,\infty}(0,1)$ and

$$\begin{cases} 0 \leq u^N(x) \leq \dfrac{1}{2N} & \text{for every } x \in (0,1) \\ u^N(0) = u^N(1) = 0 \\ |(u^N(x))'| = 1 & \text{a.e. in } (0,1) \end{cases}$$

Therefore

$$0 \leq \inf P \leq I(u^N) \leq \frac{1}{4N^2} \ .$$

Letting $N \to \infty$, we have indeed $\inf(P) = 0$. However no function $u \in W_0^{1,4}(0,1)$ can verify $I(u) = 0$, since then $u = 0$ a.e. and $|u'| = 1$ a.e., which is in contradiction with the fact that u is in $W^{1,4}(0,1)$. Therefore the problem (P) does not have a solution.

Note that the relaxed problem is

$$(QP) \qquad \inf \left\{ \bar{I}(u) = \int_0^1 Qf(u(x), u'(x)) \, dx \ : \ u \in W_0^{1,4}(0,1) \right\}$$

where

$$Qf(u, \xi) = \begin{cases} f(u, \xi) & \text{if } |\xi| \geq 1 \\ u^2 & \text{if } |\xi| \leq 1 \end{cases} \ .$$

(QP) has at least the solution $\bar{u} \equiv 0$ and we have that the sequence u^N constructed above satisfy

$$\begin{cases} u^N \rightharpoonup \bar{u} \equiv 0 \text{ in } W^{1,4}(0,1) \\ I(u^N) \to \bar{I}(0) = 0 = \inf(P) = \inf(QP) \end{cases} \ .$$

This simple example shows that one cannot expect to have a general existence theory for non convex problems. However a great deal of work has been done in this direction and we mention here only some of the recent results which all concern the scalar case and usually the case of single integrals: Aubert [1], Aubert-Tahraoui [1,2], Cesari [4], Ekeland [2], Klotzler [1], Marcellini [2,3], Mascolo-Schianchi [1,2], Olech [1], Raymond [1], Tahraoui [2,3].

We here single out a simple result which concerns the case of integrands of the form $f(u'(x))$. As seen in the above counterexample integrands of the form $f(x, u(x), u'(x))$ require a different treatment.

Theorem 2.6. *Let $f : \mathbb{R}^m \to \mathbb{R}$ be continuous and let*

$$(P) \qquad \inf \left\{ I(u) = \int_0^1 f(u'(x)) \, dx \ : \ u(0) = 0, \right.$$

$$\left. u(1) = \alpha, \ u \in (W^{1,\infty}(0,1))^m \right\} \ .$$

Then, (P) admits a solution if and only if there exist $\lambda_\nu \geq 0$ with $\sum_{\nu=1}^{m+1} \lambda_\nu =$

1, $\beta_\nu \in \mathbb{R}^m$, $1 \leq \nu \leq m+1$ *such that*

$$
\begin{cases}
f^{**}(\alpha) = \displaystyle\sum_{\nu=1}^{m+1} \lambda_\nu f(\beta_\nu) \\[2ex]
\alpha = \displaystyle\sum_{\nu=1}^{m+1} \lambda_\nu \beta_\nu
\end{cases}
\tag{1}
$$

where $f^{**} = \sup\{g \leq f : g \text{ convex }\}$. *Furthermore if (1) is satisfied then*

$$
\bar{u}(x) = \beta_p x + \sum_{\nu=1}^{p} \lambda_\nu(\beta_\nu - \beta_p),
$$

$$
x \in I_p \equiv \left[\sum_{\nu=1}^{p-1} \lambda_\nu, \sum_{\nu=1}^{p} \lambda_\nu \right], \quad 1 \leq p \leq m+1 ,
\tag{2}
$$

is a solution of (P).

REMARKS.

i) The sufficiency of (1) is implicitly or explicitly proved in the above mentioned papers. The necessity is less known but is also implicit in some of the above papers;

ii) recall that by Carathéodory theorem (cf. Theorem 1.1) we always have

$$
f^{**}(\alpha) = \inf \left\{ \sum_{\nu=1}^{m+1} \lambda_\nu f(\beta_\nu) : \sum_{\nu=1}^{m+1} \lambda_\nu \beta_\nu = \alpha \right\} .
\tag{3}
$$

Therefore (1) states that a necessary and sufficient condition for existence of solutions is that the infimum in (3) be attained. Note also that if f is *convex* or f *coercive* (in the sense that $f(\alpha) \geq a|\alpha|^p + b$ with $p > 1$, $a > 0$) then the infimum in (3) is always attained;

iii) therefore if $f(x, u, u') = f(u')$, counterexamples for existence must be non convex and non coercive; for example

$$
(P) \qquad \inf \left\{ I(u) = \int_0^1 e^{\,|u'(x)|}\, dx \; : \; u \in W_0^{1,\infty}(0,1) \right\}
$$

i.e. $f(\alpha) = e^{-|\alpha|}$, then $f^{**}(\alpha) \equiv 0$ and therefore by the relaxation theorem

$$
\inf(P) = \inf(QP) = 0 .
$$

However it is obvious that $I(u) \neq 0$ for every $u \in W_0^{1,\infty}(0,1)$ and hence the infimum of (P) is not attained;

iv) a similar proof to that of Theorem 2.6 (see for example Marcellini [2]) shows that a sufficient condition to ensure existence of minima to

$$(P) \qquad \inf\left\{ I(u) = \int_0^1 f(x, u'(x))\, dx \; : \; u(0) = 0, \right.$$

$$\left. u(1) = \alpha, \; u \in (W^{1,\infty}(0,1))^m \right\}$$

is (1) where λ_ν and β_ν are then measurable functions. Of course if f depends explicitly on u, the example of Bolza shows that the theorem is then false.

PROOF of Theorem 2.6.
Sufficiency: The sufficiency part is elementary. Let

$$(QP) \qquad \inf\left\{ \bar{I}(u) = \int_0^1 f^{**}(u'(x))\, dx \; : \; u(0) = 0, \right.$$

$$\left. u(1) = \alpha, \; u \in (W^{1,\infty}(0,1))^m \right\}$$

then $\tilde{u}(x) = \alpha x$ is trivially a solution of (QP) and therefore

$$\inf(QP) = f^{**}(\alpha) \ .$$

Let now $\bar{u}$ be as in (2). Observe first that $\bar{u} \in (W^{1,\infty}(0,1))^m$ and $\bar{u}(0) = 0$, $\bar{u}(1) = \alpha$. We now compute

$$I(\bar{u}) = \int_0^1 f(\bar{u}'(x))\, dx = \sum_{p=1}^{m+1} \int_{I_p} f(\bar{u}'(x))\, dx = \sum_{p=1}^{m+1} f(\beta_p)\, \mathrm{meas}\, I_p$$

$$= \sum_{p=1}^{m+1} \lambda_p f(\beta_p) = f^{**}(\alpha) = \inf(QP) \leq \inf(P) \ . \qquad (4)$$

Necessity: We decompose the proof into three steps.
Step 1: Let $\bar{v}$ be such that $\bar{v} \in (W_0^{1,\infty}(0,1))^m$ and, using (4),

$$I(\alpha x + \bar{v}) = \inf(P) = \inf(QP) = f^{**}(\alpha) \qquad (5)$$

i.e. $\alpha x + \bar{v}$ is a solution of (P). We now approximate $\bar{v}$ by piecewise affine functions. Let $\varepsilon > 0$, then there exists $v_\varepsilon \in (\mathrm{Aff}_0(0,1))^m$ such that

$$
\begin{cases}
v_\varepsilon \to \bar{v} \text{ in } (W^{1,p}(0,1))^m, \ 1 \le p < \infty \\
\|v_\varepsilon\|_{W^{1,\infty}} \le \|\bar{v}\|_{W^{1,\infty}} + 1 \\
\left| \int_0^1 [f(\alpha + v_\varepsilon'(x)) - f(\alpha + \bar{v}'(x))] \, dx \right| \le \varepsilon
\end{cases}
\tag{6}
$$

Since v_ε is piecewise affine, there exist I^ε an integer, $\lambda_i^\varepsilon \in [0,1]$, $a_i^\varepsilon \in \mathbf{R}^m$, $1 \le i \le I^\varepsilon$ such that

$$
\begin{cases}
\displaystyle \int_0^1 f(\alpha + v_\varepsilon'(x)) \, dx = \sum_{i=1}^{I^\varepsilon} \lambda_i^\varepsilon f(\alpha_i^\varepsilon) \\
\displaystyle \alpha = \sum_{i=1}^{I^\varepsilon} \lambda_i^\varepsilon \alpha_i^\varepsilon, \ \sum_{i=1}^{I^\varepsilon} \lambda_i^\varepsilon = 1
\end{cases}
\tag{7}
$$

Step 2: We next show that combining together some of the α_i^ε we can restrict ourselves to the case $I^\varepsilon \le m + 2$, this will be done in a very similar way as in Carathéodory theorem. Denote by

$$
M = \{1\} \times \mathrm{epi} f = \{1\} \times \{(x, \alpha) \in \mathbf{R}^m \times \mathbf{R} : f(x) \le \alpha\} \subset \mathbf{R}^{m+2} \ .
$$

We have therefore that

$$
\sum_{i=1}^{I^\varepsilon} \lambda_i^\varepsilon (1, \alpha_i^\varepsilon, f(\alpha_i^\varepsilon)) \in \mathrm{co} M = \{1\} \times \mathrm{co}(\mathrm{epi} f) \ ,
$$

where $\mathrm{co} M$ denotes the convex hull of M. We then proceed as in Carathéodory theorem (cf. Chapter 2) to reduce I^ε to $m + 2$.

Step 3: Combining (6), (7) and the fact that $I^\varepsilon \le m + 2$ we have that λ_i^ε and α_i^ε are bounded independently of ε and therefore up to the extraction of a subsequence

$$
\lambda_i^\varepsilon \to \bar{\lambda}_i, \ \alpha_i^\varepsilon \to \bar{\alpha}_i, \ 1 \le i \le m + 2
$$

and

$$
\begin{cases}
\displaystyle \int_0^1 f(\alpha + \bar{v}'(x)) \, dx = \sum_{i=1}^{m+2} \bar{\lambda}_i f(\bar{\alpha}_i) \\
\displaystyle \alpha = \sum_{i=1}^{m+2} \bar{\lambda}_i \bar{\alpha}_i, \ \sum_{i=1}^{m+2} \bar{\lambda}_i = 1
\end{cases}
\tag{10}
$$

Combining (5) and (10) we have

$$
\begin{cases}
f^{**}(\alpha) \;=\; \displaystyle\sum_{i=1}^{m+2} \bar{\lambda}_i f(\bar{\alpha}_i) \\[2ex]
\alpha \;=\; \displaystyle\sum_{i=1}^{m+2} \bar{\lambda}_i \bar{\alpha}_i, \quad \sum_{i=1}^{m+2} \bar{\lambda}_i = 1
\end{cases}
\tag{11}
$$

To obtain the theorem it therefore remains to show that one can take in (11) only $(m+1)$ elements. This can be done exactly as in Theorem 1.3 of Chapter 4. We show that there exist μ_i, $1 \leq i \leq m+1$, such that

$$
\begin{cases}
\mu_i \geq 0, \;\; \displaystyle\sum_{i=1}^{m+2} \mu_i = 1, \;\; \text{at least one of the } \mu_i = 1 \\[2ex]
\displaystyle\sum_{i=1}^{m+2} \mu_i f(\bar{\alpha}_i) \leq \sum_{i=1}^{m+2} \bar{\lambda}_i f(\bar{\alpha}_i), \;\; \alpha = \sum_{i=1}^{m+2} \mu_i \bar{\alpha}_i
\end{cases}
\tag{12}
$$

Combining (11) and (12) with the fact that $f^{**}(\alpha) \leq \sum \mu_i f(\bar{\alpha}_i)$ we shall have the result. Assume that $\bar{\lambda}_i > 0$, $1 \leq i \leq m+2$, otherwise (12) is trivial. To prove (12) observe that $\alpha \in \mathrm{co}\{\bar{\alpha}_1, \ldots, \bar{\alpha}_{m+2}\} \subset \mathbb{R}^m$. Thus it follows from Carathéodory theorem that there exist $\lambda_i \geq 0, 1 \leq i \leq m+1$ with $\sum \lambda_i = 1$ and at least one of the $\lambda_i = 0$, such that

$$
\alpha \;=\; \sum_{i=1}^{m+2} \lambda_i \bar{\alpha}_i \;.
$$

Assume, without loss of generality, that

$$
\sum_{i=1}^{m+2} \lambda_i f(\bar{\alpha}_i) \;>\; \sum_{i=1}^{m+2} \bar{\lambda}_i f(\bar{\alpha}_i) \;=\; f^{**}(\alpha) \;;
\tag{13}
$$

otherwise choosing $\mu_i = \lambda_i$ we would have immediately (12). Let

$$
J \;=\; \{i \in \{1, \ldots, m+2\} : \bar{\lambda}_i - \lambda_i < 0\} \;.
$$

Observe that $J \neq \emptyset$, since otherwise $\bar{\lambda}_i \geq \lambda_i \geq 0$ for every i and since at least one of the $\lambda_i = 0$, we would have a contradiction with $\sum \lambda_i = \sum \bar{\lambda}_i = 1$ and $\bar{\lambda}_i > 0$ for every i. We then define

$$
\gamma \;=\; \min_{i \in J}\left\{ \frac{\bar{\lambda}_i}{\lambda_i - \bar{\lambda}_i} \right\} \;.
$$

We clearly have that $\gamma > 0$. Finally let

$$\mu_i = \bar{\lambda}_i + \gamma(\bar{\lambda}_i - \lambda_i), \ 1 \le i \le m + 2 \ .$$

We immediately get that

$$\mu_i \ge 0, \ \sum_{i=1}^{m+2} \mu_i = 1, \ \text{at least one of the } \mu_i = 0 \ . \tag{14}$$

From (13) we obtain

$$\sum_{i=1}^{m+2} \mu_i f(\bar{\alpha}_i) = \sum_{i=1}^{m+2} \bar{\lambda}_i f(\bar{\alpha}_i) + \gamma \left(\sum_{i=1}^{m+2} \bar{\lambda}_i f(\bar{\alpha}_i) - \sum_{i=1}^{m+2} \lambda_i f(\bar{\alpha}_i) \right)$$

$$\le \sum_{i=1}^{m+2} \bar{\lambda}_i f(\bar{\alpha}_i) \ . \tag{15}$$

The combination of (14) and (15) give immediately (12) and thus the theorem.
$\square$

5.2.2.2 The Case of Multiple Integrals

We here only give two examples which show that in the case of multiple integrals the situation is intrinsically more complicated and one cannot expect general existence theorem without a convexity hypothesis even if $f(x, u(x), \text{grad } u(x)) = f(\text{grad } u(x))$. We shall not mention here any existence theorem and we refer to the above papers for such results.

EXAMPLE 1: Let $m = 1$, $n = 2$, $\Omega = (0,1)^2$, $u_0 \equiv 0$ and

$$f(\text{grad } u) = (u_x^2 - 1)^2 + u_y^4$$

where $u_x \equiv \frac{\partial u}{\partial x}$, $u_y \equiv \frac{\partial u}{\partial y}$. Let

$$(P) \qquad \inf \left\{ I(u) = \int_\Omega f(\text{grad } u(x,y)) \, dx dy \ : \ u \in W_0^{1,4}(\Omega) \right\} \ .$$

Let

$$f^{**}(\text{grad } u) = \begin{cases} f(u_x, u_y) & \text{if } |u_x| \ge 1 \\ u_y^4 & \text{if } |u_x| \le 1 \end{cases}$$

and

$$(QP) \qquad \inf \left\{ \bar{I}(u) = \int_\Omega f^{**}(\text{grad } u(x,y)) \, dx dy \ : \ u \in W_0^{1,4}(\Omega) \right\} \ .$$

We then deduce from the relaxation theorem that

$$\inf(P) = \inf(QP) = \bar{I}(0) = 0 \ .$$

However it is clear that no $u \in W_0^{1,4}(\Omega)$ can satisfy $I(u) = 0$, since we should then have

$$u_x^2 = 1 \ \text{a.e. in} \ \Omega, \ u_y = 0 \ \text{a.e. in} \ \Omega \ \text{and} \ u = 0 \ \text{on} \ \partial\Omega \ ,$$

which is impossible.

EXAMPLE 2: We conclude with an example of Marcellini [3] which shows that there is a relationship between unicity of solutions of the relaxed problem and non existence of solutions. Let $m = 1$, $n = 2$, $\Omega = (0,1)^2$ and

$$f(\operatorname{grad} u) = g(|\operatorname{grad} u|)$$

where $|\operatorname{grad} u|^2 = u_x^2 + u_y^2$ and where g satisfies

$$\begin{cases} g(0) = \inf\{g(x) : x \geq 0\} \\ g^{**} \ \text{is strictly increasing} \\ g^{**}(1) < g(1) \end{cases} \ .$$

We let also $u_0(x,y) = x$ and consider

$$(P) \quad \inf\left\{ I(u) = \int_\Omega g(|\operatorname{grad} u(x,y)|)\, dx\, dy \ : \ u \in u_0 + W_0^{1,\infty}(\Omega) \right\}$$

$$(QP) \quad \inf\left\{ \bar{I}(u) = \int_\Omega g^{**}(|\operatorname{grad} u(x,y)|)\, dx\, dy \ : \ u \in u_0 + W_0^{1,\infty}(\Omega) \right\}$$

where we have used, for (QP), Theorem 1.3. We want to show that (P) does not have a solution. To this end we first show that (QP) admits only one solution which is $u = u_0$ in Ω. It is obvious that for every $v \in W^{1,\infty}(\Omega)$ with $v = u_0 = x$ on $\partial\Omega$ we have

$$\int_\Omega |\operatorname{grad} u_0(x,y)|\, dx\, dy = 1 = \int_0^1 \int_0^1 v_x(x,y)\, dx\, dy$$

$$\leq \int_0^1 \int_0^1 |\operatorname{grad} v(x,y)|\, dx\, dy \ . \tag{1}$$

Therefore to have equality in (1) we should have $v_x = |\operatorname{grad} v|$ a.e. in Ω and $v(x,y) = x$ on $\partial\Omega$, which is possible only if $v = x = u_0$ in Ω. Hence for every $v \in u_0 + W_0^{1,\infty}(\Omega)$ we have since g^{**} is strictly increasing

$$\bar{I}(v) = \int_\Omega g^{**}(|\operatorname{grad} v(x,y)|)\, dx\, dy \geq g^{**}\left(\int_\Omega |\operatorname{grad} v(x,y)|\, dx\, dy\right)$$

$$\geq g^{**}(1) = \bar{I}(u_0) \tag{2}$$

and with strict inequality if $v \neq u_0$. Returning to the problem (P) we have by the relaxation theorem that

$$\inf(P) = \inf(QP) = \bar{I}(u_0) = g^{**}(1) \ . \tag{3}$$

Assume now, for contradiction, that (P) has a solution $\bar{u}$ then from (2) and (3) we have

$$\bar{I}(\bar{u}) = \int_\Omega g^{**}(|\operatorname{grad} \bar{u}|)\, dx\, dy \geq g^{**}(1) = I(\bar{u}) = \int_\Omega g(|\operatorname{grad} \bar{u}|)\, dx\, dy$$

$$\geq \int_\Omega g^{**}(|\operatorname{grad} \bar{u}|)\, dx\, dy = \bar{I}(\bar{u}) \ .$$

Since u_0 is the unique minimum of $\bar{I}$ we deduce that $\bar{u} = u_0$. However we trivially have

$$I(u_0) = g(1) > g^{**}(1) = I(u_0) = \bar{I}(u_0) \ ,$$

which is absurd and therefore (P) does not have a solution.

Applications

A.0 Introduction

As mentioned earlier the theory developed in this book has many applications. We have decided to present some which are related to nonlinear elasticity and optimal design. We do not intend to make any exhaustive presentation, there are excellent books on these subjects; we rather present some typical examples.

The appendix is divided into two parts. The first one begins with some basic background from nonlinear elasticity. We then present some existence theorems of Ball on a fundamental problem of elasticity: the existence of minimizers for the energy of hyperelastic materials. Finally we give a uniqueness result of classical solutions for the equilibrium equations, due to Knops and Stuart. Both results use the notions and theorems introduced in Chapter 4.

The second part of the appendix is devoted to the applications of the relaxation theorems of Chapter 5 to elasticity and optimal design. We give several examples where the minimization problem under consideration involves energy integrals which are not weakly lower semicontinuous; this is usually the case when phase transitions occur. Therefore, in general, no solution is to be expected. However one can use the relaxation techniques to have a more precise information on the minimizing sequences of the original problem as well as the behaviour with respect to weak convergence of the original energy integral.

A.1 Existence and Uniqueness Theorems in Nonlinear Elasticity

A.1.1 Setting of the Problem

A.1.1.1 General Notations

We start by presenting some of the basic notions from elasticity. We refer for more details to Ciarlet [1,2] and Gurtin [1]. We here consider only equilibrium problems.

We let $\Omega \subset \mathbf{R}^n$ $(n = 1, 2, 3)$ be a bounded domain, considered as the *reference configuration* of a given elastic material. We let $u : \Omega \to \mathbf{R}^n$ be a deformation of the body, $u(\Omega)$ being the actual configuration, and we denote by F the deformation gradient

$$F = \nabla u = \begin{pmatrix} \operatorname{grad} u_1 \\ \vdots \\ \operatorname{grad} u_n \end{pmatrix} = \left(\frac{\partial u_i}{\partial x_\alpha} \right)_{1 \le i, \, \alpha \le n} .$$

We assume that the deformation preserves the orientation and is locally invertible, i.e.,

$$\det \nabla u(x) > 0, \ x \in \Omega .$$

Usually, one would require that u be globally invertible, but we shall here only consider local invertibility. Moreover, if the material under consideration is incompressible, we impose the pointwise constraint

$$\det \nabla u(x) \equiv 1, \ x \in \Omega .$$

The internal forces due to such a deformation are usually described in two ways. If one considers the reference configuration they are characterized at a point $x \in \Omega$ by the *Piola-Kirchhoff stress tensor* $S(x)$. In the deformed configuration $u(\Omega)$, they are given at a point $y \in u(\Omega)$ by the *Cauchy stress tensor* $T(y)$. They are related by the formula

$$S(x)(\nabla u(x))^t = \det \nabla u(x) \cdot T(u(x)), \ x \in \Omega ,$$

where F^t denotes the transpose of F.

In absence of applied forces acting on the interior of the body, the equilibrium equations are

$$\begin{cases} \operatorname{div} S = 0 \Leftrightarrow \displaystyle\sum_{j=1}^{n} \frac{\partial}{\partial x_j}\{S_{ij}(x)\} = 0, \ 1 \leq i \leq n, \ x \in \Omega \\ T(y) = (T(y))^t, \ y \in u(\Omega) \Leftrightarrow S(x)(\nabla u(x))^t = (\nabla u(x))(S(x))^t, \ x \in \Omega \end{cases}$$

We shall restrict ourselves to the case of *hyperelastic materials*, which are characterized by a *stored energy function* $W : \Omega \times \mathbb{R}_+^{n \times n} \to \mathbb{R}$ satisfying

$$S_{i\alpha} = \frac{\partial W}{\partial F_{i\alpha}}$$

where $\mathbb{R}^{n \times n}$ denotes the set of $n \times n$ matrices and $\mathbb{R}_+^{n \times n} = \{F \in \mathbb{R}^{n \times n} : \det F > 0\}$.

It is also assumed that the function W is *objective* i.e.

$$W(x, QF) = W(x, F)$$

for every proper orthogonal matrix Q and for every $F \in \mathbb{R}_+^{n \times n}$. The above condition just means that $W(x, \bullet)$ is independent of the choice of coordinates.

In addition to the objectivity condition, W is often required to be *isotropic*, that is

$$W(x; QFQ^t) = W(x; F)$$

for every orthogonal matrix Q and for every $F \in \mathbb{R}_+^{n \times n}$. This allows the function W to be rewritten in the following form (for a proof see the above mentioned books).

Proposition 1.1. *If* $W : \Omega \times \mathbb{R}_+^{n \times n} \to \mathbb{R}$ *is objective and isotropic, then there exists* $\Phi : \Omega \times \mathbb{R}_+ \times \ldots \times \mathbb{R}_+ \to \mathbb{R}$ *satisfying*

$$W(x; F) = \Phi(x; v_1; \ldots; v_n)$$

with Φ *symmetric in the* v_i *and where* v_i *denote the eigenvalues of* $V = (FF^t)^{1/2}$.

REMARK. In elasticity, the eigenvalues v_i are sometimes called the *principal stretches*.

EXAMPLES.

i) Observe that if V is as in the proposition then
  Case 1: $n = 2$

$$\|F\|^2 = \sum_{i,j=1}^{2} F_{ij}^2 = v_1^2 + v_2^2$$

$$\det F = F_{11}F_{22} - F_{12}F_{21} = v_1 v_2$$

Case 2: $n = 3$

$$\|F\|^2 = v_1^2 + v_2^2 + v_3^2$$

$$\|\operatorname{adj} F\|^2 = (v_1 v_2)^2 + (v_1 v_3)^2 + (v_2 v_3)^2$$

$$\det F = v_1 v_2 v_3 \ .$$

ii) The eigenvalues v_i can be computed explicitly when one considers *radial deformations*, that is

$$\begin{cases} u : \Omega = \{x \in \mathbf{R}^n : |x| < 1\} \to \mathbf{R}^n \\ u(x) = \gamma(|x|)\dfrac{x}{|x|}, \ x \in \Omega \end{cases} ,$$

where $\gamma(t), \gamma'(t) > 0$ for $t > 0$. An easy computation gives that the eigenvalues of $V = (\nabla u (\nabla u)^t)^{1/2}$ are

$$\left(\gamma'(|x|), \frac{\gamma(|x|)}{|x|}, \ldots, \frac{\gamma(|x|)}{|x|}\right) \ .$$

Since isotropic materials are often considered in elasticity and since we shall usually require that the function W be polyconvex or rank one convex, it is interesting to relate these different conditions on W with corresponding ones on Φ.

Proposition 1.2. *Let*

$$W(F) = \Phi(v_1, \ldots, v_n)$$

where Φ is symmetric and v_i are the eigenvalues of $V = (FF^t)^{1/2}$.

i) *W is convex on $\mathbf{R}^{n \times n}$ if and only if Φ is convex and nondecreasing in each variable v_i;*
ii) *if W is rank one convex and $\Phi \in C^2(\mathbf{R}^n)$ then*

$$\frac{\partial^2 \Phi}{\partial v_i^2}(v) \equiv \Phi_{v_i v_i}(v) \geq 0 \tag{1}$$

for every $v \in \mathbb{R}^n_+ = (0,\infty)^n$ and every $i = 1,\dots,n$;

$$\frac{v_i \Phi_{v_i}(v) - v_j \Phi_{v_j}(v)}{v_i - v_j} \geq 0 \tag{2}$$

for every $v \in \mathbb{R}^n_+$ with $v_i \neq v_j$ and for every $1 \leq i \neq j \leq n$;

iii) if $n = 2$, then W is rank one convex if and only if in addition to (1) and (2) the following hold

$$[\Phi_{v_1 v_1}(v) \Phi_{v_2 v_2}(v)]^{1/2} + \Phi_{v_1 v_2}(v) + \frac{\Phi_{v_1}(v) - \Phi_{v_2}(v)}{v_1 - v_2} \geq 0 \tag{3}$$

for every $v = (v_1, v_2) \in \mathbb{R}^2_+$ with $v_1 \neq v_2$,

$$[\Phi_{v_1 v_1}(v) \Phi_{v_2 v_2}(v)]^{1/2} - \Phi_{v_1 v_2}(v) + \frac{\Phi_{v_1}(v) + \Phi_{v_2}(v)}{v_1 + v_2} \geq 0 \tag{4}$$

for every $v \in \mathbb{R}^2_+$ and

$$\Phi_{v_1 v_1}(v_1, v_1) - \Phi_{v_1 v_2}(v_1, v_1) + \frac{\Phi_{v_1}(v_1, v_1)}{v_1} \geq 0 \tag{5}$$

for every $v_1 \in \mathbb{R}_+$.

iv) Let $n = 2$ and

$$W(F) = \Phi(v_1, v_2) = \Psi(v_1, v_2, v_1 v_2) \tag{6}$$

with $\Psi : \mathbb{R}^2_+ \times \mathbb{R}_+ \to \mathbb{R}$ convex and satisfying

$$\Psi(x_1, x_2, \delta) = \Psi(x_2, x_1, \delta) \tag{7}$$

$$\Psi(x_1, x_2, \delta) \text{ is non decreasing in } x_1, x_2 \tag{8}$$

then W is polyconvex.

v) Let $n = 3$ and

$$W(F) = \Phi(v_1, v_2, v_3) = \Psi(v_1, v_2, v_3, v_2 v_3, v_3 v_1, v_1 v_2, v_1 v_2 v_3) \tag{9}$$

with $\Psi : \mathbb{R}^6_+ \times \mathbb{R}_+ \to \mathbb{R}$ convex and satisfying

$$\Psi(P_1 x; P_2 y; \delta) = \Psi(x; y; \delta) \tag{10}$$

for every $x, y \in \mathbb{R}^3_+$, $\delta \in \mathbb{R}_+$ and every $P_1, P_2 \in \mathcal{P}_3$ (the set of all permutations of 3 vectors) and

$$\Psi(x_1; x_2; x_3; y_1; y_2; y_3; \delta) \text{ is non decreasing in } x_i, y_i \, , \tag{11}$$

then W is polyconvex.

REMARKS.

i) The first result is due to Thompson-Freede [1] and to Ball [2], see also Hill [1];
ii) for a proof of the second and third statements we refer to Aubert [1], Aubert-Tahraoui [3], Knowles-Sternberg [1];
iii) the fourth and fifth results can be found in Ball [2] and can be easily generalized to the case $n > 3$.

A.1.1.2 Examples of Stored Energy Functions

We now briefly discuss some realistic stored energy functions that are considered in elasticity. This is a highly debated problem and answers are sometimes divergent. This difficulty comes from the fact that experimental devices to determine the function W are hard to implement. We shall discuss here two classes of materials, the first one involving rank one convex (or even polyconvex) functions W and the second one W which are not rank one convex, this one will be considered in connection with the relaxation theorems in the next section.

Usually it is easier to determine W for isotropic functions. A large class of materials are the so-called *Ogden materials* (cf. Ogden [1]) which are expressed as

$$W(x, F) = a(x) + \sum_{i=1}^{M} b_i(x)(v_1^{\alpha i} + v_2^{\alpha i} + v_3^{\alpha i} - 3)$$

$$+ \sum_{j=1}^{N} c_j(x)((v_1 v_2)^{\beta j} + (v_1 v_3)^{\beta j} + (v_2 v_3)^{\beta j} - 3)$$

$$+ d(\det F)$$

where $d : (0, \infty) \to \mathbb{R}$ is convex and satisfies

$$\lim_{\delta \to 0} d(\delta) = \lim_{\delta \to \infty} d(\delta) = +\infty .$$

(The conditions on d mean that it takes an infinite energy to compress a volume to zero or to expand it to infinity). Usually Ogden materials are defined only for incompressible materials and therefore $d(\det F) = 0$ and one requires that $\det F = 1$.

As a subclass of Ogden materials there are the so-called *Mooney-Rivlin materials* where

$$W(x, F) = a(x) + b(x)(v_1^2 + v_2^2 + v_3^2 - 3)$$

$$+ c(x)((v_1 v_2)^2 + (v_1 v_3)^2 + (v_2 v_3)^2 - 3)$$

and the *Neo-Hookean material* where $c(x) \equiv 0$.

In view of Proposition 1.2 all these cases give rise to polyconvex functions W and the Neo-Hookean materials to convex W.

REMARK. Usually functions W are best determined if the material is assumed to be incompressible. If it is slightly compressible one sometimes uses the following device (cf. Charrier-Dacorogna-Hanouzet-Laborde [1] for more details). Since changes of volumes are hard to determine experimentally, first one makes some experiments (such as simple traction, biaxial traction, simple shear, ...) for which one can assume that volume changes are negligible.

In this way one finds

$$W^*(F) \;=\; W(F)\Big|_{\det F-1} \;.$$

Then one postulates that the pressure is related to change of volumes in the following way

$$P(\nabla u) \;=\; g'(\det \nabla u)$$

where $g : \mathbb{R}_+ \to \mathbb{R}$ is convex, C^1 and $g(1) = g'(1) = 0$. It is then easy (cf. the article quoted above) to show that

$$W(F) \;=\; W^*\left(\frac{F}{(\det F)^{1/3}}\right) + g(\det F)$$

i.e., W is the sum of an energy due to distortion and another due to compression (the polyconvexity of such W is studied in Theorem 1.10 of Chapter 4).

As mentioned before there are other realistic W which are not rank one convex; these are used in models involving phase transitions. We shall see some examples in the second section of the appendix. Another such material is the *St. Venant-Kirchhoff material* (cf. Ciarlet [2] for a treatment of such material), where W is given by

$$W(F) \;=\; -\frac{3\lambda + \mu}{4}(v_1^2 + v_2^2 + v_3^2) + \frac{\lambda + 2\mu}{8}(v_1^4 + v_2^4 + v_3^4)$$

$$+ \frac{\lambda}{4}\left[(v_1 v_2)^2 + (v_1 v_3)^2 + (v_2 v_3)^2\right] + \frac{6\mu + 9\lambda}{8} \;.$$

λ and μ are called the *Lamé constants* and are positive. It is clear that such a W is not polyconvex.

A.1.2 Existence Theorems

We present here the results of Ball [2,3]. We consider a mixed displacement, zero traction boundary value problem. This means that we have $\Omega \subset \mathbf{R}^n$ with a strongly Lipschitz boundary $\partial\Omega$ which is decomposed into two disjoint parts

$$\partial\Omega = (\partial\Omega_1) \cup (\partial\Omega_2), \ \text{meas}\,(\partial\Omega_1) > 0$$

and

$$u(x) = u_0(x) \ \text{if} \ x \in \partial\Omega_1$$

and traction free on $\partial\Omega_2$.

We consider the functional

$$I(u) = \int_\Omega [W(x, \nabla u(x)) + \psi(x, u(x))]\, dx$$

where W is the stored energy function defined above and $\psi : \Omega \times \mathbf{R}^n \to \mathbf{R}$ is the body force per unit volume.

The problem is therefore to seek for a minimum of (P) where

$$(P) \qquad\qquad\qquad \inf\{I(u) : u \in \mathcal{W}\}$$

with

$$\mathcal{W} = \{u \in W^{1,p}(\Omega; \mathbf{R}^n) : u(x) = u_0(x) \ \text{on} \ \partial\Omega_1 \ \text{and} \ \det \nabla u > 0 \ \text{a.e.}\}\ .$$

It is clear that if a solution $\bar{u}$ of (P) is sufficiently regular and if W and ψ satisfy the conditions of Proposition 4.3 of Chapter 3, then $\bar{u}$ satisfies the equilibrium equations

$$(E) \quad \text{div}\left(\frac{\partial W}{\partial F_i}\right) - \frac{\partial}{\partial u}\psi \equiv \left(\sum_{\alpha=1}^{n} \frac{\partial}{\partial x_\alpha}\frac{\partial W}{\partial F_{i\alpha}}(x, \nabla\bar{u}) - \frac{\partial}{\partial u_i}\psi(x, \bar{u})\right)_{1\leq i\leq n} = 0.$$

Note that if we restrict ourselves to incompressible materials, the problem under consideration will then be

$$(P_0) \qquad\qquad\qquad \inf\{I(u) : u \in \mathcal{W}_0\}$$

where

$$\mathcal{W}_0 = \{u \in W^{1,p}(\Omega; \mathbf{R}^n) : u = u_0 \text{ on } \partial\Omega \text{ and } \det \nabla u = 1 \text{ a.e.}\} \ .$$

We now make the following hypotheses on W and ψ

$(H1)$ W is polyconvex, i.e. there exists $g : \Omega \times \mathbf{R}^{3\times 3} \times \mathbf{R}^{3\times 3} \times \mathbf{R}_+ \to \mathbf{R}$ continuous, $g(x, \bullet, \bullet, \bullet)$ convex for every x such that

$$W(x, F) = g(x, F, \operatorname{adj} F, \det F) \ ;$$

$(H2)$ there exist $K > 0$, $C \in \mathbf{R}$, $p \geq 2$, $q \geq \frac{p}{p-1}$, $r > 1$ such that

$$g(x, F, G, \delta) \geq C + K(|F|^p + |G|^q + \delta^r)$$

for every $x \in \Omega$, $(F, G, \delta) \in \mathbf{R}^{3\times 3} \times \mathbf{R}^{3\times 3} \times \mathbf{R}_+$. Furthermore

$$\lim_{\delta \to 0} g(x, F, G, \delta) = +\infty \ ;$$

$(H3)$ $\psi : \Omega \times \mathbf{R}^3 \to \mathbf{R}_+$ is continuous.

REMARKS.

i) Note that Ogden materials defined in the preceding section satisfy the above properties provided $\alpha_i \geq 2$ and $\beta_i \geq \alpha_i/\alpha_i - 1$;

ii) Ball-Murat [1] have weakened hypothesis $(H2)$ in the sense that they only need
$$W(x, F) \geq C + K(|F|^p + |\operatorname{adj} F|^q)$$
with $p \geq 2$ and $q \geq p/p - 1$;

iii) note also that the vector $(F, \operatorname{adj} F, \det F)$ with $F = \nabla u$ measures, for its first argument, changes of length, through its second argument changes of surface and the last one measures changes of volume.

We now state the main theorem for compressible materials.

Theorem 1.3. *Assume that W and ψ satisfy $(H1)$, $(H2)$, $(H3)$ and that there exists $\tilde{u} \in \mathcal{W}$ so that $I(\tilde{u}) < \infty$, then (P) admits at least one solution $\bar{u} \in \mathcal{W}$ with $\operatorname{adj} \nabla \bar{u} \in L^q$ and $\det \nabla \bar{u} \in L^r$.*

PROOF. This is a direct consequence of Theorem 2.10 of Chapter 4. $\square$

For incompressible materials a similar result holds. Let

$(H1')$ there exists $g : \Omega \times \mathbf{R}^{3\times 3} \times \mathbf{R}^{3\times 3} \to \mathbf{R}$ continuous, $g(x, \bullet, \bullet)$ convex for every $x \in \Omega$ such that

$$W(x, F) = \begin{cases} g(x, F, \operatorname{adj} F) & \text{if } \det F = 1 \\ +\infty & \text{if } \det F \neq 1 \end{cases} \ ;$$

$(H2')$ there exist $K > 0$, $C \in \mathbf{R}$, $p \geq 2$, $q \geq p/p - 1$ such that

$$g(x, F, G) \geq C + K(|F|^p + |G|^q) \ ;$$

$(H3)$ $\psi : \Omega \times \mathbf{R}^3 \to \mathbf{R}_+$ is continuous.

Theorem 1.4. *Under the above hypotheses and if there exists $\tilde{u} \in \mathcal{W}_0$ such that $I(\tilde{u}) < \infty$, then (P_0) admits at least one solution $\bar{u} \in \mathcal{W}_0$ with $\operatorname{adj} \nabla \bar{u} \in L^q$.*

PROOF. This is also a consequence of Theorem 2.10. $\qquad\qquad\square$

REMARK. No regularity result is known for the solutions $\bar{u}$ found in Theorem 1.3 or 1.4. Hence it is not known if the minimum $\bar{u}$ satisfies the equilibrium equations, even in a weak sense.

A.1.3 Unicity of Classical Solutions of Equilibrium Equations

We now show that for some special boundary conditions and in absence of external forces the equilibrium equations (E) of the preceding section, have a unique classical solution. The result is due to Knops-Stuart [1] and we follow their proof.

Consider the energy

$$I(u) = \int_\Omega W(\nabla u(x)) \, dx$$

and the associated equation

$$\operatorname{div}\left(\nabla_F W(\nabla u(x))\right) = \left(\sum_{\alpha=1}^{n} \frac{\partial}{\partial x_\alpha}\left(\frac{\partial W}{\partial F_{i\alpha}}(\nabla u(x))\right)\right)_{1 \leq i \leq n} = 0, \ x \in \Omega \ .$$

Assume that $W : \mathbf{R}^{n \times n} \to \mathbf{R}$ is C^2 and *strictly quasiconvex*, i.e.

$$\int_\Omega W(\xi + \nabla\varphi(x)) \, dx \geq W(\xi) \operatorname{meas} \Omega \ ,$$

for every $\xi \in \mathbf{R}^{n \times n}$, $\varphi \in W_0^{1,\infty}(\Omega; \mathbf{R}^n)$ and with equality holding only if $\varphi \equiv 0$.

It is clear that if $\xi \in \mathbf{R}^{n \times n}$, $a \in \mathbf{R}^n$ are given, then

$$(P) \qquad \inf\{I(u) : u(x) = \xi x + a \text{ on } \partial\Omega, \ u \in C^1(\bar{\Omega}) \cap C^2(\Omega)\}$$

$$= I(\xi x + a)$$

and $\bar{u}(x) \equiv \xi x + a$ in $\bar{\Omega}$ is the unique minimum of (P).

The result we shall prove asserts that, under some geometrical restrictions on Ω, $\bar{u}$ is also the unique regular $(C^2(\Omega; \mathbf{R}^n) \cap C^1(\bar{\Omega}; \mathbf{R}^n))$ solution of (E) with boundary condition $u = \bar{u}$ on $\partial\Omega$.

Theorem 1.5. *Let Ω be star-shaped with respect to 0 and with piecewise C^1 boundary. Let $\bar{u}(x) = \xi x + a$, for $x \in \bar{\Omega}$ where $\xi \in \mathbf{R}^{n \times n}$, $a \in \mathbf{R}^n$ are given. Let $u \in C^2(\Omega; \mathbf{R}^n) \cap C^1(\bar{\Omega}; \mathbf{R}^n)$ be a solution of (E) satisfying $u = \bar{u}$ on $\partial\Omega$.*

i) If W is rank one convex, then

$$I(u) \leq I(\bar{u}) \ .$$

ii) If W is strictly quasiconvex, then $u \equiv \bar{u}$ in $\bar{\Omega}$.

REMARKS.

i) Ω is said to be star-shaped with respect to $x_0 \in \Omega$ if for every $y \in \mathbf{R}^n \setminus \{0\}$, the half line $\{x_0 + \lambda y : \lambda \geq 0\}$ intersects the boundary $\partial\Omega$ at one point only. If ν is the outward normal to $\partial\Omega$, we find that the star shapedness of Ω implies that $\langle \nu(x); x - x_0 \rangle \geq 0$ for every $x \in \partial\Omega$ where $\nu(x)$ is well defined;

ii) it is clear that the result of the theorem is still valid if Ω is star shaped with respect to x_0 and not only for $x_0 = 0$. A translation will give the theorem;

iii) a well known example of non uniqueness is given by F. John [1] when Ω is an annulus in $\mathbf{R}^2$. Therefore some restrictions on the geometry of Ω is needed;

iv) note that in elasticity deformations such as $\bar{u}$ $(\bar{u}(x) = \xi x + a)$ are called *homogeneous deformations*;

v) no unicity result is known in the class of weak solutions of (E).

Before proving the theorem we need two lemmas.

Lemma 1.6. *Let $\partial\Omega$ be piecewise C^1, ν be the outward unit normal to $\partial\Omega$, $W \in C^2(\mathbf{R}^{n \times n})$. Let $u \in C^2(\Omega; \mathbf{R}^n) \cap C^1(\bar{\Omega}; \mathbf{R}^n)$ be a solution of (E). Then*

$$I(u) = \frac{1}{n} \int_{\partial\Omega} \{W(\nabla u(x))\langle x; \nu(x)\rangle$$

$$+ \langle \nabla_F W(\nabla u(x)); (u(x) - \langle x; \nabla u(x)\rangle) \otimes \nu(x)\rangle\} \, d\sigma \ .$$

PROOF. This lemma has been established by Green [1]. Multiply (E) by $u(x) - \langle x; \nabla u(x)\rangle$ and integrate by parts. We have

$$0 = \int_{\partial\Omega} \langle \nabla_F W(\nabla u); (u(x) - \langle x; \nabla u(x)\rangle) \otimes \nu(x)\rangle \, d\sigma$$

$$- \int_{\Omega} \langle \nabla_F W(\nabla u); \nabla(u(x) - \langle x; \nabla u(x)\rangle)\rangle \, dx \ .$$

Observe now that

$$-\langle \nabla_x [W(\nabla u(x))]; x\rangle = \langle \nabla_F W(\nabla u(x)); \nabla(u(x) - \langle x; \nabla u(x)\rangle)\rangle \ .$$

Therefore a new integration by parts leads to

$$0 = \int_{\partial\Omega} \{W(\nabla u)\langle x; \nu\rangle + \langle \nabla_F W(\nabla u); (u(x) - \langle x; \nabla u(x)\rangle) \otimes \nu(x)\rangle\} \, d\sigma$$

$$- n \int_{\Omega} W(\nabla u(x)) \, dx \ .$$

Thus the result. $\qquad\qquad\qquad\qquad\qquad\qquad\qquad\qquad\qquad\qquad\qquad\qquad\square$

Lemma 1.7. *Let Ω and ν be as in the above lemma. Let $u, w \in C^1(\bar{\Omega})$ with $u = w$ on $\partial\Omega$, then*

i) $\nabla u(x) - \nabla w(x) = \frac{\partial}{\partial\nu}(u(x) - w(x)) \otimes \nu(x)$ *a.e. on $\partial\Omega$*

ii) $\langle x; \nabla u(x) - \nabla w(x)\rangle = \langle x; \nu(x)\rangle \frac{\partial}{\partial\nu}(u(x) - w(x))$ *a.e. on $\partial\Omega$.*

PROOF. The proof is elementary. $\qquad\qquad\qquad\qquad\qquad\qquad\qquad\qquad\qquad\square$

We are now in a position to prove the theorem.

PROOF. We decompose the proof into two steps.

Step 1: We first prove that if W is rank one convex and $u, \bar{u} \in C^2(\Omega; \mathbf{R}^n) \cap C^1(\bar{\Omega}; \mathbf{R}^n)$ are solutions of (E) with $u = \bar{u}$ on $\partial\Omega$, then

$$I(u) - I(\bar{u}) \leq \frac{1}{n} \int_{\partial\Omega} \langle \nabla_F W(\nabla u(x)) - \nabla_F W(\nabla \bar{u}(x)); \tag{1}$$

$$(\bar{u}(x) - \langle x; \nabla \bar{u}(x)\rangle) \otimes \nu\rangle \, d\sigma \ .$$

We use Lemma 1.6 to obtain

$$n[I(u) - I(\bar{u})] = \int_{\partial\Omega} \{(W(\nabla u) - W(\nabla \bar{u}))\langle x; \nu(x)\rangle\} \, d\sigma$$

$$+ \int_{\partial\Omega} \{\langle \nabla_F W(\nabla u); (u - \langle x; \nabla u(x)\rangle) \otimes \nu\rangle$$

$$- \langle \nabla_F W(\nabla \bar{u}); (\bar{u} - \langle x; \nabla \bar{u}(x)\rangle) \otimes \nu\rangle\} \, d\sigma \ . \tag{2}$$

Using Lemma 1.7 we deduce that $\nabla u - \nabla \bar{u}$ is of rank one on $\partial\Omega$. Combining this with the fact that W is C^2 and rank one convex we deduce that

$$W(\nabla \bar{u}) - W(\nabla u) \geq \langle \nabla_F W(\nabla u); \nabla \bar{u} - \nabla u \rangle \text{ a.e. on } \partial\Omega \ . \tag{3}$$

Combining (2), (3) and the fact that Ω is star shaped with respect to 0 (i.e., $\langle x; \nu(x) \rangle \geq 0$ on $\partial\Omega$) we obtain

$$n[I(u) - I(\bar{u})] \leq \int_{\partial\Omega} \{ \langle \nabla_F W(\nabla u);$$

$$(u - \langle x; \nabla u \rangle) \otimes \nu + (\nabla u - \nabla \bar{u})\langle x; \nu \rangle \rangle \} \, d\sigma$$

$$- \int_{\partial\Omega} \{ \langle \nabla_F W(\nabla u); (u - \langle x; \nabla \bar{u} \rangle) \otimes \nu \rangle \} \, d\sigma \ .$$

Using once more Lemma 1.7 and the fact that $u = \bar{u}$ on $\partial\Omega$ we obtain

$$n[I(u) - I(\bar{u})] \leq \int_{\partial\Omega} \langle \nabla_F W(\nabla u);$$

$$(\bar{u} - \langle x; \nabla u \rangle) \otimes \nu + \langle x; \nu \rangle \frac{\partial}{\partial \nu}(u - \bar{u}) \otimes \nu \rangle \, d\sigma$$

$$- \int_{\partial\Omega} \langle \nabla_F W(\nabla \bar{u}); (\bar{u} - \langle x; \nabla \bar{u} \rangle) \otimes \nu \rangle \, d\sigma$$

$$\leq \int_{\partial\Omega} \langle \nabla_F W(\nabla u);$$

$$[\bar{u} - \langle x; \nabla u \rangle + \langle x; \nu \rangle \frac{\partial}{\partial \nu}(u - \bar{u})] \otimes \nu \rangle \, d\sigma$$

$$- \int_{\partial\Omega} \langle \nabla_F W(\nabla \bar{u}); (\bar{u} - \langle x; \nabla \bar{u} \rangle) \otimes \nu \rangle \, d\sigma \ .$$

Applying Lemma 1.7 to the first integral, we have indeed obtained (1).

Step 2: The result now follows. Observe first that there is no loss of generality if we take $a = 0$ and hence $\bar{u}(x) = \xi x$.

i) Since $\bar{u}(x) = \xi x$ we deduce that $\bar{u} - \langle x; \nabla \bar{u} \rangle = 0$ for every $x \in \bar{\Omega}$; using then (1) we obtain

$$I(u) \leq I(\bar{u}) \ .$$

ii) Since W is quasiconvex and $\bar{u}(x) = \xi x$ we must have $I(\bar{u}) \leq I(u)$ and by i) $I(u) = I(\bar{u})$. The strict quasiconvexity of W implies then that $u \equiv \bar{u}$ in $\bar{\Omega}$. $\qquad\square$

A.2 Relaxation Theorems in Elasticity and Optimal Design

We give four examples; the first three are related to elasticity and the last one to optimal design.

A.2.1 Antiplane Shear Problem in Elasticity

We start with some results of Gurtin-Témam [1] and Bauman-Phillips [1] and we refer for more details to these articles.

Let $\Omega \subset \mathbb{R}^2$ be a sufficiently regular bounded open set and let

$$B = \Omega \times (0, L) = \{(x_1, x_2, x_3) \in \mathbb{R}^3 : (x_1, x_2) \in \Omega, \, x_3 \in (0, L)\} \ .$$

Assume that the elastic body occupying the reference configuration B is hyperelastic, incompressible, isotropic and homogeneous. We consider *antiplane shear deformations*, i.e.

$$u : B \to \mathbb{R}^3$$

$$u = u(x_1, x_2, x_3) = (x_1, x_2, x_3 + \varphi(x_1, x_2)) \ .$$

The deformation gradient is then

$$F = \nabla u = \begin{pmatrix} 1 & 0 & 0 \\ 0 & 1 & 0 \\ \dfrac{\partial \varphi}{\partial x_1} & \dfrac{\partial \varphi}{\partial x_2} & 1 \end{pmatrix}$$

and therefore the incompressibility condition, $\det F = 1$, is trivially satisfied.

We furthermore assume that the stored enery function W depends only on the first invariant of $F F^t$ and not on the second (note that the third is 1), i.e.

$$W(F) = \tilde{w}(|F|^2) = \tilde{w}\left(3 + \left(\frac{\partial \varphi}{\partial x_1}\right)^2 + \left(\frac{\partial \varphi}{\partial x_2}\right)^2\right)$$

$$= w\left(\sqrt{\left(\frac{\partial \varphi}{\partial x_1}\right)^2 + \left(\frac{\partial \varphi}{\partial x_2}\right)^2}\right) \ .$$

We assume that $w : \mathbb{R}_+ \to \mathbb{R}$ is C^2 and satisfies $w(0) = 0$ and

$$\alpha_1 |\xi|^p - \beta \leq w(\xi) \leq \alpha_2 |\xi|^p + \beta$$

for every $\xi \in \mathbb{R}_+$ and for some $p > 1$, $\alpha_2 \geq \alpha_1 > 0$ and $\beta \in \mathbb{R}$.

If there is no vertical force applied on B the problem to be studied is then

$$(P) \quad \inf \left\{ I(u) = \int_\Omega w(|\nabla\varphi(x)|) \, dx \ : \ \varphi = \varphi_0 \text{ on } \partial\Omega, \ \varphi \in W^{1,p}(\Omega) \right\}$$

where $\varphi : \Omega \subset \mathbb{R}^2 \to \mathbb{R}$ (φ_0 being given).

If w is convex, then the direct methods of Chapter 3 apply. However a more reasonable hypothesis seems to be $w'(\xi) \geq 0$. Often w is as in Figure A.1 below.

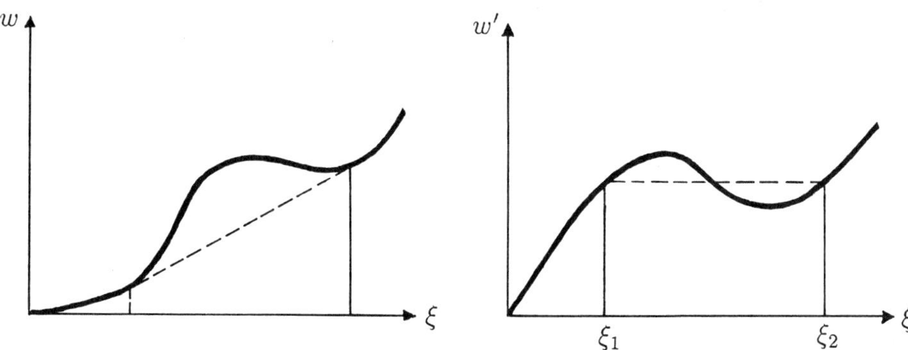

Figure A.1

Due to the lack of convexity, the problem (P) has in general no solution as we shall see below. One therefore needs to consider the relaxed problem

$$(P^{**}) \qquad \inf \left\{ I^{**}(u) = \int_\Omega w^{**}(|\nabla\varphi(x)|) \, dx \ : \right.$$

$$\left. \varphi = \varphi_0 \text{ on } \partial\Omega, \ \varphi \in W^{1,p}(\Omega) \right\}$$

where

$$w^{**} = \sup\{g \leq w \ : \ g \text{ convex}\} \ .$$

It is easy to see that if w is as in Figure 1, then

$$(w^{**})'(\xi) = \begin{cases} w'(\xi) & \text{if } \xi \in [0, \xi_1] \cup [\xi_2, +\infty) \\ w'(\xi_1) & \text{if } \xi \in [\xi_1, \xi_2] \end{cases} \ ,$$

where ξ_1 and ξ_2 are given by the equal area rule, i.e.

$$w(\xi_2) - w(\xi_1) = \int_{\xi_1}^{\xi_2} w'(\xi)d\xi = w'(\xi_1)(\xi_2 - \xi_1) = w'(\xi_2)(\xi_2 - \xi_1) \ .$$

The line $[(\xi_1, w'(\xi_1)); (\xi_2, w'(\xi_2))]$ is sometimes called the *Maxwell line*.

We then have the following theorem.

Theorem 2.1. *With the above hypotheses, then*

i) $\inf(P) = \inf(P^{**})$;

ii) (P^{**}) *admits at least one solution;*

iii) *every minimizing sequence* $\{\varphi_\nu\}$ *of* (P) *possesses a weakly convergent (in* $W^{1,p}$*) subsequence with limit* $\bar\varphi$*, which is a solution of* (P^{**})*;*

iv) *conversely for every solution* $\bar\varphi$ *of* (P^{**})*, there exists a minimizing sequence* $\{\varphi_\nu\}$ *of* (P) *such that* $\varphi_\nu \rightharpoonup \bar\varphi$ *in* $W^{1,p}$*.*

REMARKS.

i) The above theorem is just a restatement of Theorem 2.1 of Chapter 5;

ii) of course the theorem does not give any result on the existence or non existence of solution of (P). In general (P) does not attain its minimum and we give such an example below. When this is not the case the theorem gives the structure of the minimizing sequences as well as the "average" (in the sense of weak convergence) behaviour of these sequences, which are solutions of the relaxed problem. In physical terms these minimizing sequences are finer and finer mixtures of different phases.

In a recent article Bauman-Phillips [1] have refined the above theorem and we give their results without proofs. If we assume that $\Omega \subset \mathbb{R}^2$ is a sort of annulus, i.e.

$$\Omega = D_2 - \bar{D}_1$$

where $D_1 \subset D_2$ are two bounded convex domains of $\mathbb{R}^2$ with $C^{2,\alpha}$ boundaries. We set $\Gamma_1 = \partial D_1$, $\Gamma_2 = \partial D_2$ and the boundary condition φ_0 to be

$$\varphi_0(x) = \begin{cases} 0 & \text{if } x \in \Gamma_2 \\ h & \text{if } x \in \Gamma_1 \end{cases}$$

where $h > 0$ is a constant. In geometrical terms this means that the outer surface of the cylinder is kept fixed while the inner surface is displaced in the axial direction by a uniform distance h.

The result is then as follows (w being as in Figure 1).

Theorem 2.2.

i) (P^{**}) *has a unique solution.*

ii) *If* Γ_1 *is a circle, then there are constant* $h_2 > h_1 > 0$ *such that:*

    1) *if* $h \in [0, h_1] \cup [h_2, +\infty)$ *then* (P) *has a solution* $\bar\varphi$ *of class* $C^{2,\alpha}$ *with* $|\nabla\bar\varphi| \leq \xi_1$ *if* $h \leq h_1$ *and* $|\nabla\bar\varphi| \geq \xi_2$ *if* $h \geq h_2$;

    2) *if* $h_1 < h < h_2$ *and* Ω *is an annulus (i.e.* Γ_2 *is also a circle), then* (P) *has a radial solution;*

    3) *if* $h_1 < h < h_2$ *and* Ω *is not an annulus then* (P) *has no solution.*

REMARK. Note that the unicity of solutions for (P^{**}) is not obvious since w^{**} is not strictly convex. Observe that as in Chapter 5 non existence of solutions for (P) is related to uniqueness of solutions for (P^{**}).

A.2.2 A Problem of Equilibrium of Gases

The following application has been given in Dacorogna [1]. We adopt the notations of Section A.1.1. We let a gas occupy a reference volume $\Omega \subset \mathbb{R}^3$ and we consider the gas as an elastic material. A deformation $u : \Omega \subset \mathbb{R}^3 \to \mathbb{R}^3$ relates the position $x \in \Omega$ of a particle in the reference configuration to its actual position $u(x) \in \mathbb{R}^3$. We further suppose that the deformation is specified on $\partial\Omega$, i.e. $u = u_0$ on $\partial\Omega$. In absence of external forces, the energy of the deformation is given by

$$I(u) = \int_\Omega \left(\int^{\det \nabla u(x)} P(V)\,dV \right) dx$$

where P is the pressure and V the specific volume. The relationship $P = P(V)$ at constant temperature is given by an equation of state, which we assume to be the *Van der Waal's equation of state*. At constant temperature, the equation has the following form

$$P(V) = \frac{nRT}{V - nb} - \frac{an^2}{V^2}$$

where T denotes the temperature, n the number moles, a, b and R being constants. For some range of the temperature, P has the following shape $(b = 0)$

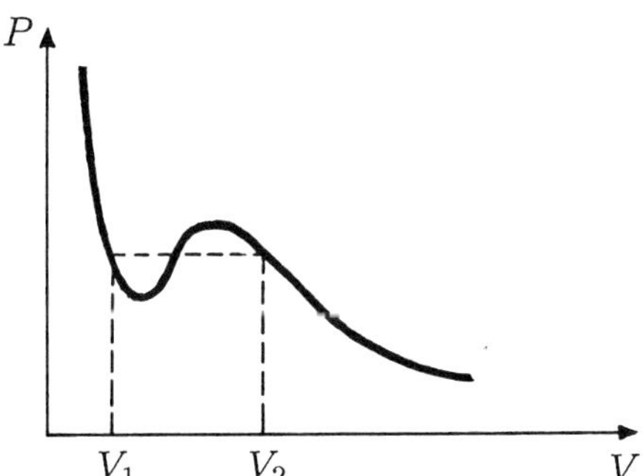

Figure A.2

We let

$$g(z) = - \int_{V_1}^{z} P(V) \, dV \; .$$

Observe that, for some range of the temperature T, the function P is not monotone and therefore g is not convex. We then define g^{**} to be the convex envelope of g. If P is as in Figure 2 we have

$$(g^{**})'(z) = \begin{cases} g'(z) = -P(z) & \text{if } z \in (0, V_1] \cup [V_2, +\infty) \\ g'(V_1) = -P(V_1) & \text{if } z \in [V_1, V_2] \end{cases}$$

where V_1 and V_2 are given by the *Maxwell line*, i.e.

$$\int_{V_1}^{V_2} P(V) \, dV = g(V_1) - g(V_2) = P(V_1)(V_2 - V_1) \; .$$

REMARK. The same type of problem occurs if instead of considering fixed temperature, we consider that the entropy is fixed.

The problem under consideration is then the following

$$(P) \ \inf \left\{ I(u) = \int_{\Omega} g(\det \nabla u(x)) \, dx \; : \; u = u_0 \text{ on } \partial\Omega, \; u \in C^{\infty}(\bar{\Omega}; \mathbf{R}^3) \right\}$$

and the associated relaxed problem is

(QP)
$$\inf \left\{ \bar{I}(u) = \int_{\Omega} g^{**}(\det \nabla u(x)) \, dx \; : \; u = u_0 \text{ on } \partial\Omega, \; u \in C^{\infty}(\bar{\Omega}; \mathbf{R}^3) \right\} \; .$$

Observe first that the existence of solutions for (QP) cannot be obtained by direct methods, since, although $\bar{I}$ is weakly lower semicontinuous, $\bar{I}$ is not coercive in any $W^{1,p}$ spaces. (A coercivity of g^{**} does not lead to a coercivity on $|\nabla u|$ but only on $|\det \nabla u|$.) To establish the existence of solutions for (QP) we shall need the following lemma.

Lemma 2.3. *Let $\Omega \subset \mathbf{R}^n$ be a bounded open set, $B \subset \mathbf{R}^n$ be the unit ball and $\varphi : \Omega \to B$ be a C^{∞} diffeomorphism $(\det \nabla \varphi > 0$ in $\bar{\Omega})$. Let $f \in C^{\infty}(\bar{\Omega})$ be strictly positive in $\bar{\Omega}$ and such that*

$$\int_{\Omega} f(x) \, dx = \text{meas}\, \Omega \; . \tag{1}$$

Then, there exists $u \in C^{\infty}(\bar{\Omega}; \mathbf{R}^n)$ satisfying

$$\begin{cases} \det \nabla u(x) = f(x), & x \in \Omega \\ u(x) = x, & x \in \partial\Omega \end{cases} \; . \tag{2}$$

REMARKS.

i) The above theorem has been proved by Moser [1] (see also Zehnder [1]) for manifolds without boundary. It was then extended by Banyaga [1] for manifolds with boundary and independently by Tartar and Dacorogna [1] for the case where Ω is the unit ball of $\mathbf{R}^2$ and $\mathbf{R}^3$;

ii) recently Dacorogna-Moser [1] have shown how to weaken the regularity assumptions on f and Ω. In particular it is proved that if $f \in C^{k,\alpha}(\bar{\Omega})$ with $k \geq 0$, an integer, and $0 < \alpha < 1$ ($C^{k,\alpha}$ denoting the usual space of Hölder continuous functions) then there exists a solution u in $C^{k+1,\alpha}(\bar{\Omega};\mathbf{R}^n)$. The proof given below produces only a $C^{k,\alpha}$ solution;

iii) no unicity is to be expected. For example, let $n = 2$ and $f \equiv 1$. Let in polar coordinates $u(x) = u(x_1, x_2) = (r\cos(\theta + 2k\pi r^2), r\sin(\theta + 2k\pi r^2))$, with $r \in [0,1]$, $k \in \mathbf{Z}$ (the set of relative integers); then u satisfies (2) for every k.

Before proceeding with the proof we introduce some notations.

Notations. Let $\Omega \subset \mathbf{R}^n$, $w \in C^1(\Omega; \mathbf{R}^{n(n-1)/2})$. We define $\mathrm{curl}^* w \in \mathbf{R}^n$ as follows:

$n = 2$: we let

$$\mathrm{curl}^* w = \left(\frac{\partial w}{\partial x_2}; -\frac{\partial w}{\partial x_1} \right) \; ;$$

$n = 3$: we let

$$\mathrm{curl}^* w = \mathrm{curl}\, w = \left(\frac{\partial w_3}{\partial x_2} - \frac{\partial w_2}{\partial x_3}; \frac{\partial w_1}{\partial x_3} - \frac{\partial w_3}{\partial x_1}; \frac{\partial w_2}{\partial x_1} - \frac{\partial w_1}{\partial x_2} \right) \; ;$$

$n > 3$: we let $w \in \mathbf{R}^{n(n-1)/2}$ with $w = (w_{ij})_{1 \leq i < j \leq n}$. We define $\mathrm{curl}^* w = ((\mathrm{curl}^* w)_\nu)_{1 \leq \nu \leq n}$ where

$$(\mathrm{curl}^* w)_\nu = (-1)^{\nu+1} \left\{ \sum_{1 \leq i < \nu} (-1)^{i+1} \frac{\partial w_{i\nu}}{\partial x_i} + \sum_{\nu < i \leq n} (-1)^i \frac{\partial w_{\nu i}}{\partial x_i} \right\} \; .$$

An easy computation shows that

$$\mathrm{div}\, \mathrm{curl}^* w = \sum_{\nu=1}^{n} \frac{\partial}{\partial r_\nu}(\mathrm{curl}^* w)_\nu = 0$$

for every $w \in C^2(\Omega; \mathbf{R}^{n(n-1)/2})$.

We may now proceed with the proof of the lemma.

PROOF. The proof is divided into three parts, the two first are established by Moser [1] (see also Banyaga [1], Zehnder [1]).

Part 1: We show that there is no loss of generality if we take $\Omega = B$, i.e. the unit ball.

Part 2: We then solve (2) for $\Omega = B$, provided a linearization of (2) can be solved.

Part 3: We show that the linearized problem (of (2))

$$\begin{cases} \operatorname{div} v(x) = f(x) - 1, & x \in B \\ v(x) = 0, & x \in \partial B \end{cases}$$

has at least one solution.

Part 1: To avoid a recurrent constant we assume that $\operatorname{meas} \Omega = \operatorname{meas} B$. Consider the differential equations

$$\begin{cases} \det \nabla \alpha(y) = \det \nabla \varphi^{-1}(y), & y \in B \\ \alpha(y) = y, & y \in \partial B \end{cases} \tag{3}$$

$$\begin{cases} \det \nabla \beta(y) = f(\varphi^{-1}(y)) \det \nabla \varphi^{-1}(y), & y \in B \\ \beta(y) = y, & y \in \partial B \end{cases} . \tag{4}$$

Observe that the compatibility conditions

$$\int_B \det \nabla \varphi^{-1}(y) \, dy = \int_B f(\varphi^{-1}(y)) \det \nabla \varphi^{-1}(y) \, dy = \operatorname{meas} B = \operatorname{meas} \Omega$$

are satisfied. Therefore, anticipating the result of Part 2, we can find $\alpha, \beta \in C^\infty(\bar{\Omega}; \mathbb{R}^n)$ satisfying (3) and (4). We then claim that

$$u(x) = \varphi^{-1} \circ \alpha^{-1} \circ \beta \circ \varphi(x), \quad x \in \Omega$$

satisfies (2). The boundary condition is obviously satisfied. We also have, using (3) and (4)

$$\begin{aligned} \det \nabla u(x) &= \det \nabla \varphi^{-1}(\alpha^{-1} \circ \beta \circ \varphi(x)) \det \nabla \alpha^{-1}(\beta \circ \varphi(x)) \\ &\quad \times \det \nabla \beta(\varphi(x)) \det \nabla \varphi(x) \\ &= \det \nabla \varphi^{-1}(\alpha^{-1} \circ \beta \circ \varphi(x)) \frac{\det \nabla \beta(\varphi(x)) \det \nabla \varphi(x)}{\det \nabla \alpha(\alpha^{-1} \circ \beta \circ \varphi(x))} \\ &= f(x) . \end{aligned}$$

This achieves Part 1.

Part 2: We may now take $\Omega = B$. Assume (cf. Part 3) that we can find $v \in C^{\infty}(\bar{B}; \mathbf{R}^n)$ such that

$$\begin{cases} \operatorname{div} v(x) = f(x) - 1 & \text{if } x \in B \\ v(x) = 0 & \text{if } x \in \partial B \end{cases} .$$

Note that (1) ensures that the above problem is at least well posed.

We divide the proof of Part 2 into two steps.

Step 1: We let for $t \in [0, 1]$, $z \in B$, and for v as above

$$v_t(z) = \frac{v(z)}{t + (1 - t)f(z)} . \tag{5}$$

We define $\Phi = \Phi_t(x) : [0, 1] \times B \to \mathbf{R}^n$ to be the solution of

$$\begin{cases} \dfrac{d}{dt}\Phi_t(x) = v_t(\Phi_t(x)) \equiv \dfrac{v(\Phi_t(x))}{t + (1 - t)f(\Phi_t(x))}, & t > 0 \\ \Phi_0(x) = x \end{cases} . \tag{6}$$

Observe that (6) has a unique global solution. Note also that

$$\Phi_t(x) \equiv x, \quad x \in \partial\Omega, \ t \in [0, 1] .$$

This follows from the observation that if $x \in \partial\Omega$, then x is a solution of (6) (since $v(x) = 0$ if $x \in \partial\Omega$) and hence by uniqueness $\Phi_t(x) \equiv x$ if $x \in \partial\Omega$.

We now show that $u(x) = \Phi_1(x)$ is a solution of (2). The boundary condition is already satisfied, we therefore only need to show that $\det \nabla\Phi_1(x) = f(x)$. To prove this let

$$h(t, x) \equiv \det \nabla\Phi_t(x)(t + (1 - t)f(\Phi_t(x))) . \tag{7}$$

If we show (cf. below) that

$$\frac{\partial}{\partial t}h(t, x) \equiv 0 \tag{8}$$

we shall have immediately the result since then

$$h(1, x) = \det \nabla\Phi_1(x) = h(0, x) = \det \nabla\Phi_0(x)f(\Phi_0(x)) = f(x) .$$

Thus the result.

Step 2: It therefore remains to show (8). Observe first that from (6) we have

$$\begin{cases} \dfrac{d}{dt}\nabla\Phi_t(x) = \nabla v_t(\Phi_t(x)) \cdot \nabla\Phi_t(x) \\ \nabla\Phi_0(x) = I \end{cases}$$

where I is the identity matrix. Note that by Abel's formula (see Coddington-Levinson [1], Section 7 of Chapter 1) we must then have

$$\det \nabla \Phi_t(x) = \exp \left\{ \int_0^t \text{trace}\left(\nabla v_s(\Phi_s(x))\right) ds \right\}$$

$$= \exp \left\{ \int_0^t \text{div}\, v_s(\Phi_s(x))\, ds \right\} \ .$$

We then deduce that

$$\frac{\partial}{\partial t}(\det \nabla \Phi_t(x)) = \det \nabla \Phi_t(x) \cdot \text{div}\, v_t(\Phi_t(x)) \ . \tag{9}$$

We now differentiate (7) with respect to t

$$\frac{\partial}{\partial t}h(t,x) = \frac{\partial}{\partial t}[\det \nabla \Phi_t(x)](t + (1-t)f(\Phi_t(x)))$$

$$+ \det \nabla \Phi_t(x)[1 - f(\Phi_t(x)) + (1-t)\langle \nabla f(\Phi_t(x)); \frac{d}{dt}\Phi_t(x)\rangle]$$

where $\langle \bullet; \bullet \rangle$ denotes the scalar product in $\mathbf{R}^n$. Using then (6) and (9) we get

$$\frac{\partial}{\partial t}h(t,x) = [\det \nabla \Phi_t(x)]\{[t + (1-t)f(\Phi_t(x))]\text{div}\, v_t(\Phi_t(x))$$

$$+ (1-t)\langle \nabla f(\Phi_t(x)); v_t(\Phi_t(x))\rangle + (1 - f(\Phi_t(x)))\} \ . \tag{10}$$

Using the definition of v_t (cf. (5)) we obtain

$$\text{div}\, v(y) = \text{div}\, \{[t + (1-t)f(y)]v_t(y)\}$$

$$= [t + (1-t)f(y)]\text{div}\, v_t(y) + (1-t)\langle \nabla f(y); v_t(y)\rangle \ .$$

Thus (10) becomes

$$\frac{\partial}{\partial t}h(t,x) = \det \nabla \Phi_t(x)[\text{div}\, v(\Phi_t(x)) + (1 - f(\Phi_t(x)))] \ .$$

From the choice of v the right hand side of the above equality is 0 and thus we have (8) and Part 2.

Part 3: It therefore remains to find $v \in C^\infty(\bar{B}; \mathbf{R}^n)$ satisfying

$$\begin{cases} \text{div}\, v(x) = f(x) - 1 & x \in B \\ \quad\ v(x) = 0 & x \in \partial B \end{cases} \tag{11}$$

provided $f \in C^\infty(\bar{B})$ and

$$\int_B (f(x) - 1)\, dx = 0 \ . \tag{12}$$

The above problem can be solved in many ways. One possibility is to define $a \in C^\infty(\bar{B})$ to be the solution of

$$\begin{cases} \Delta a = f - 1 & \text{in } B \\ \dfrac{\partial a}{\partial \nu} = 0 & \text{on } \partial B \end{cases} \tag{13}$$

where ν is the outward unit normal (note that (12) ensures the solvability of (13)).

We then let $b(x) \equiv -\mathrm{grad}\, a(x)$ (observe that $\langle b; \nu \rangle = 0$ on ∂B). We then show that we can find $w \in C^\infty(\bar{B}; \mathbb{R}^{n(n-1)/2})$ such that

$$\begin{cases} \mathrm{curl}^* w(x) = b(x), & x \in \partial B \\ \langle b(x); \nu(x) \rangle = 0, & x \in \partial B \end{cases} \ . \tag{14}$$

Note that if we can solve (14) we shall have (11) by setting

$$v(x) = \mathrm{grad}\, a(x) + \mathrm{curl}^* w(x) \ .$$

To find w satisfying (14) is easy (we do it in $\mathbb{R}^n$, the cases $n = 2$ and 3 being treated in a remark below). Let first $\xi : [0,1] \to [0,\infty)$ be C^∞ with $\xi \equiv 0$ near 0, $\xi(1) = 0$ and $\xi'(1) = 1$. Let $\beta : \partial B(\subset \mathbb{R}^n) \to \mathbb{R}^{n(n-1)/2}$, $\beta = (\beta_{i\nu})_{1 \le i < \nu \le n}$ where

$$\beta_{i\nu}(x) = (-1)^{i+\nu}[x_i b_\nu(x) - x_\nu b_i(x)], \ x \in \partial B \tag{15}$$

and where $b = (b_i)_{1 \le i \le n}$ satisfies (14) i.e.

$$\langle b; \nu \rangle = \sum_{i=1}^{n} x_i b_i(x) = 0, \ x \in \partial B \ . \tag{16}$$

Finally let, for $x \in \mathbb{R}^n$, w be written in polar coordinates, i.e.

$$w(x) = \xi(|x|)\beta\left(\frac{x}{|x|}\right) \ . \tag{17}$$

We now show that $\mathrm{curl}^* w = b$ on ∂B. Observe that

$$\left. \frac{\partial w_{i\nu}}{\partial x_i} \right|_{|x|=1} = \beta_{i\nu}(x)x_i \ .$$

Using then the definition of curl^* and (15) we have for $x \in \partial B$

$$
\begin{aligned}
(\mathrm{curl}^* w)_\nu &= (-1)^{\nu+1} \left\{ \sum_{1 \le i < \nu} (-1)^{i+1} \beta_{i\nu} x_i + \sum_{\nu < i \le n} (-1)^i \beta_{\nu i} x_i \right\} \\
&= \sum_{1 \le i < \nu} (x_i b_\nu - x_\nu b_i) x_i - \sum_{\nu < i \le n} (x_\nu b_i - x_i b_\nu) x_i \\
&= b_\nu \sum_{i \ne \nu} x_i^2 - x_\nu \sum_{i \ne \nu} b_i x_i \ .
\end{aligned}
$$

Using (16) and the fact that $|x| = 1$, we have indeed obtained $(\mathrm{curl}^* w)_\nu = b_\nu$ and this concludes Part 3 and the lemma. $\qquad\square$

REMARK. Part 3 can be read as follows when
$n = 2$: w written in polar coordinates is chosen so that

$$
w(x_1, x_2) = \xi(r)(-b_2(\cos\theta, \sin\theta)\cos\theta + b_1(\cos\theta, \sin\theta)\sin\theta) \ . \tag{17_2}
$$

The compatibility condition (16) is

$$
\langle b; \nu \rangle = b_1 \cos\theta + b_2 \sin\theta = 0 \ . \tag{16_2}
$$

We therefore have

$$
\mathrm{curl}^* w = \left(\frac{\partial w}{\partial x_2}, -\frac{\partial w}{\partial x_1} \right) = \left(\frac{\partial w}{\partial r}\frac{\partial r}{\partial x_2} + \frac{\partial w}{\partial \theta}\frac{\partial \theta}{\partial x_2}, -\frac{\partial w}{\partial r}\frac{\partial r}{\partial x_1} - \frac{\partial w}{\partial \theta}\frac{\partial \theta}{\partial x_1} \right) \ .
$$

On ∂B (i.e. $r = 1$) we have

$$
\mathrm{curl}^* w = ((-b_2 \cos\theta + b_1 \sin\theta)\sin\theta; -(-b_2 \cos\theta + b_1 \sin\theta)\cos\theta) \ .
$$

Using (16_2) we have indeed obtained $\mathrm{curl}^* w = (b_1, b_2)$.
$n = 3$: We let

$$
\begin{aligned}
w(x_1, x_2, x_3) = \xi(r)(&-b_3 \sin\theta \sin\varphi + b_2 \cos\varphi; -b_1 \cos\varphi \\
&+ b_3 \cos\theta \sin\varphi; b_1 \sin\theta \sin\varphi - b_2 \cos\theta \sin\varphi) \ .
\end{aligned} \tag{17_3}
$$

The compatibility condition (16) is

$$
\langle b; \nu \rangle = b_1 \cos\theta \sin\varphi + b_2 \sin\theta \sin\varphi + b_3 \cos\varphi = 0 \ . \tag{16_3}
$$

We also have

$$\operatorname{curl}^* w = \operatorname{curl} w = \left(\frac{\partial w_3}{\partial x_2} - \frac{\partial w_2}{\partial x_3}; \frac{\partial w_1}{\partial x_3} - \frac{\partial w_3}{\partial x_1}; \frac{\partial w_2}{\partial x_1} - \frac{\partial w_1}{\partial x_2} \right) .$$

On ∂B we have

$$\operatorname{curl}^* w = \operatorname{curl} w = \left(\frac{\partial w_3}{\partial r} x_2 - \frac{\partial w_2}{\partial r} x_3; \frac{\partial w_1}{\partial r} x_3 - \frac{\partial w_3}{\partial r} x_1; \frac{\partial w_2}{\partial r} x_1 - \frac{\partial w_1}{\partial r} x_2 \right) .$$

A direct computation, using (16_3), gives $\operatorname{curl}^* w = (b_1, b_2, b_3)$.

We can now state the main theorem.

Theorem 2.1. *Let $\Omega \subset \mathbb{R}^3$ be a bounded open set, $u_0 \in C^\infty(\bar{\Omega}; \mathbb{R}^3)$ be a diffeomorphism such that $u_0^{-1}(\Omega)$ is C^∞ diffeomorphic to the unit ball. Then*

i) (QP) admits at least one solution $\bar{u} \in C^\infty(\bar{\Omega}; \mathbb{R}^3)$;

ii) for every minimizing sequence $\{u_\nu\}$ of (P), there exist a subsequence (still denoted u_ν) and a solution $\bar{u} \in C^\infty(\bar{\Omega}; \mathbb{R}^3)$ of (QP) so that

$$\begin{cases} \det \nabla u_\nu \rightharpoonup \det \nabla \bar{u} \ \text{in} \ L^p \\ u_\nu = \bar{u} = u_0 \ \text{on} \ \partial\Omega \\ I(u_\nu) \to \bar{I}(\bar{u}) = \inf(P) = \inf(QP) \end{cases} \tag{1}$$

for some $p > 1$;

iii) conversely for every solution $\bar{u}$ of (QP) there exists a minimizing sequence $\{u_\nu\}$ of (P) satistying (1).

REMARKS.

i) The theorem can be interpreted as follows. If there is no equilibrium solution, i.e. there is no minimizer for (P), then the solutions of (QP) characterize in "average" (since weak convergence measures some kind of averages) the minimizing sequences of (P). They are then a finer and finer mixture of two different phases which in the limit give rise to equilibrium solutions of the relaxed problem;

ii) related results can be found in Mascolo-Schianchi [1].

PROOF.

i) Observe that under the above hypotheses on Ω und u_0 and with the help of Lemma 2.3 the problem

$$\begin{cases} \det \nabla \bar{u}(x) = \dfrac{1}{\operatorname{Vol} \Omega} \displaystyle\int_\Omega \det \nabla u_0(y)\, dy, & x \in \Omega \\ \bar{u}(x) = u_0(x), & x \in \partial\Omega \end{cases} \tag{2}$$

has a C^∞ solution. We then have from Jensen inequality that for every v with $v = u_0$ on $\partial\Omega$

$$
\begin{aligned}
g^{**}(\det \nabla \bar{u}(x)) &= g^{**} \left(\frac{1}{\operatorname{Vol}\Omega} \int_\Omega \det \nabla u_0(y)\,dy \right) \\
&= g^{**} \left(\frac{1}{\operatorname{Vol}\Omega} \int_\Omega \det \nabla v(y)\,dy \right) \qquad (3) \\
&\leq \frac{1}{\operatorname{Vol}\Omega} \int_\Omega g^{**}(\det \nabla v(y))\,dy \ .
\end{aligned}
$$

The combination of (2) and (3) establishes that $\bar{u}$ is a minimizer for (QP).

ii) The second part of Theorem 2.4 is a consequence of the relaxation theorem of Chapter 5 with the added difficulty that the function $g = g(z)$ is not defined for $z < 0$. We refer for more details to Dacorogna [1]. $\qquad\square$

A.2.3 Equilibrium of Elastic Crystals

We now present some results of Chipot-Kinderlehrer [1] and Fonseca [1,2] on equilibrium of crystals. We refer for a general bibliography on this subject to these articles as well as to Ericksen [2-4] and Kinderlehrer [1].

We start with a description of an elastic crystal. A crystal is considered as a countable set of identical atoms arranged periodically. A way of describing this periodic structure is to choose the origin at one of the atoms; then the position of any other atom is found using a choice of linearly independent lattice vectors $\{a_1, a_2, a_3\}$. We then let $A = (a_1, a_2, a_3)$ be the 3×3 matrix with column (a_i).

Let Φ be the *Helmholtz free energy*. We assume that Φ is frame indifferent and invariant under a change of lattice basis, i.e.

$$
\Phi = \Phi(A^t A) = \Phi(M^t A^t A M)
$$

for every $M \in GL(\mathbb{Z}^3) \equiv \{M \in \mathbb{R}^{3 \times 3} : \det M = \pm 1 \text{ and } m_{ij} \in \mathbb{Z}, i,j = 1,2,3\}$ and where $\mathbb{Z}$ denotes the set of relative integers.

Having fixed a basis of lattice vectors $\{a_1, a_2, a_3\}$ and thus a matrix A, the energy density, per unit reference volume, is given by

$$
W(F) = \Phi(A^t F^t F A)
$$

for every $F \in \mathbb{R}_+^{3 \times 3} \equiv \{F \in \mathbb{R}^{3 \times 3} : \det F > 0\}$.

We must therefore have

$$
(H1) \qquad\qquad\qquad W(F) = W(QFH)
$$

for every $F \in \mathbb{R}^{3\times3}_+$, $Q \in \mathcal{R}^+ = \{Q \in \mathbb{R}^{3\times3} : Q^t Q = I$ and $\det Q = 1\}$ (I denoting the identity matrix), $H \in A\,GL(\mathbb{Z}^3)A^{-1}$ which is a conjugate group of $GL(\mathbb{Z}^3)$.

As usual (cf. Section A.1.1) we also impose

$$(H2) \qquad \begin{cases} W(F) \geq 0 \quad W(I) = 0 \\ \lim_{\det F \to 0} W(F) = \infty \end{cases} .$$

An elementary computation shows that, because of the different invariances in the problem, the function W is neither coercive, nor rank one convex. Therefore the miminization of the energy subject, for example, to a pure displacement boundary value problem may not have a solution. One is therefore lead to introduce the relaxed problem. The above authors have proved the following results.

Theorem 2.5. *Let W satisfy $(H1)$ and $(H2)$. Let*

$$h(t) \equiv \inf\{W(F) : \det F = t\} .$$

If PW and RW denote the polyconvex and rank one convex envelope of W, then

i) for every $F \in \mathbb{R}^{3\times3}_+$

$$PW(F) = RW(F) = h^{**}(\det F),$$

ii) for every $F \in \mathbb{R}^{3\times3}_+$, the following holds

$$h^{**}(\det F) = \inf\left\{\frac{1}{\operatorname{meas}\Omega}\int_\Omega W(F + \nabla\varphi(x))\,dx : \varphi \in W_0^{1,\infty}(\Omega;\mathbb{R}^3)\right\}$$

where Ω is a bounded domain of $\mathbb{R}^3$.

REMARKS.

i) The function h is known as the subenergy density of W;
ii) the above result shows the following. Let Ω be the reference configuration and let $u_0(x) = Fx + b$ where $F \in \mathbb{R}^{3\times3}_+$, $b \in \mathbb{R}^3$ are fixed (i.e. u_0 is a homogeneous deformation). Therefore we are minimizing the energy subject to a prescribed homogeneous deformation on the boundary. Due to the lack of rank one convexity (as well as coercivity) of W the problem may not have any solution. However the minimal value of the energy is given by $h^{**}(\det F) \cdot \operatorname{meas}\Omega$, which also is the minimal value of the relaxed problem. As usual, this value as well as the solution of the relaxed problem (which is here, trivially, u_0) help characterizing the minimizing

sequences of the original problem. As before they are finer and finer phase mixture, which in "average" look like $u_0(x) = Fx + b$ the homogeneous deformation;

iii) Chipot-Kinderlehrer [1] and Fonseca [1,2] have also considered more general problems; in particular those with more general boundary conditions than the homogeneous ones;

iv) one should also remark that the second part of the theorem is not a direct consequence of the first part and of Theorem 1.1 of Chapter 5. This would be the case if W was finite on the whole of $\mathbf{R}^{3\times 3}$ and then $PW(F) = QW(F) = RW(F) = h^{**}(\det F)$; however we have here the restriction that $W(F) = +\infty$ if $\det F \leq 0$.

We shall now sketch (for more details, see the above articles) the proof of the first part of the theorem. We start with a preliminary lemma.

Lemma 2.6. *Let $A \in \mathbf{R}^{n\times n}$ with $\det A \neq 0$ and $f : \mathbf{R}_+^{n\times n} \to \mathbf{R}$ satisfy*

(S1) f is rank one convex,
(S2) for every $F \in \mathbf{R}_+^{n\times n}$ and for every $a, b \in \mathbf{Z}^n$ with $a \cdot b = \sum_{i=1}^n a_i b_i = 0$,

$$f(F) = f(FA(I + a \otimes b)A^{-1}),$$

where I is the identity matrix in $\mathbf{R}^{n\times n}$.

Then there exists $g : (0, \infty) \to \mathbf{R}$ convex, such that

$$f(F) = g(\det F)$$

for every $F \in \mathbf{R}_+^{n\times n}$.

REMARK. Note that since $a, b \in \mathbf{Z}^n$ with $a \cdot b = 0$ then $I + a \otimes b \in GL(\mathbf{Z}^n)$ and thus $A(I + a \otimes b)A^{-1} \in A\,GL(\mathbf{Z}^n)A^{-1}$. Hence the hypothesis $(H1)$ implies $(S2)$.

PROOF. We follow the proof of Fonseca [2]. We divide the proof into four steps.
Step 1: We first show that

$$f(F) = f(FA(I + ta \otimes b)A^{-1}) \tag{1}$$

for every $F \in \mathbf{R}_+^{n\times n}$, $t \in \mathbf{R}$, $a, b \in \mathbf{Z}^n$ with $a \cdot b = 0$.

Observe first that $\det(FA(I + ta \otimes b)A^{-1}) = \det F > 0$ for every $t \in \mathbf{R}$. Therefore, for fixed F, a and b, the function

$$\varphi(t) \equiv f(FA(I + ta \otimes b)A^{-1})$$

is well defined. Since f is rank one convex, we deduce that φ is convex. From

$(S2)$ we get that

$$\varphi(0) = \varphi(m), \text{ for every } m \in \mathbf{Z} . \tag{2}$$

The convexity of φ, coupled with (2), implies that φ is constant and thus (1) is proved.

Step 2: We now prove that, in fact,

$$f(F) = f(F(I + a \otimes b)) \tag{3}$$

for every $a, b \in \mathbf{R}^n$ with $a \cdot b = 0$. For every $a, b \in \mathbf{R}^n$ with $a \cdot b = 0$, there exist $a_\nu, b_\nu \in \mathbf{Q}^n$ ($\mathbf{Q}$ being the set of rational numbers) so that

$$\begin{cases} a_\nu \to a, \ b_\nu \to b \text{ as } \nu \to \infty \\ a_\nu \cdot b_\nu = 0, \text{ for every } \nu \end{cases} . \tag{4}$$

Since f is continuous on $\mathbf{R}_+^{n \times n}$ (f, being rank one convex and finite on $\mathbf{R}_+^{n \times n}$, is automatically continuous, cf. Theorem 2.3 of Chapter 2), we have that for every $\varepsilon > 0$, there exists ν such that

$$|f(FA(I + a_\nu \otimes b_\nu)A^{-1}) - f(FA(I + a \otimes b)A^{-1})| < \varepsilon . \tag{5}$$

Now observe that since $a_\nu, b_\nu \in \mathbf{Q}^n$, there exists $t \in \mathbf{R}$ (in fact $t \in \mathbf{Z}$) such that $ta_\nu, tb_\nu \in \mathbf{Z}^n$. Combining then (1) and (5) with the arbitrariness of ε we have obtained that

$$f(F) = f(FA(I + a \otimes b)A^{-1}) \tag{6}$$

for every $a, b \in \mathbf{R}^n$ with $a \cdot b = 0$. The identity (3) then follows, if we replace a by $A^{-1}a$ and b by $A^t b$ in (6).

Iterating (3), we can conclude that

$$f(F) = f\left(F \prod_{m=1}^{N} (I + a_m \otimes b_m) \right) \tag{7}$$

for every $F \in \mathbf{R}_+^{n \times n}$, $a_m, b_m \in \mathbf{R}^n$ with $a_m \cdot b_m = 0, 1 \leq m \leq N$.

Step 3: We then use an algebraic result (cf. Fonseca [2] for a proof). This result states that if $F \subset \mathbf{R}_+^{n \times n}$, there exist $a_m, b_m \in \mathbf{R}^n$ with $a_m \cdot b_m = 0$, $m = 1, 2, \ldots, 2n - 1$, such that

$$F = (\det F)^{1/n} \prod_{m=1}^{2n-1} (I + a_m \otimes b_m) .$$

Therefore applying this formula to F^{-1} we get

$$F^{-1} = (\det F)^{-1/n} \prod_{m=1}^{2n-1} (I + a_m \otimes b_m) \ .$$

From (7) we may then deduce that

$$f(F) = f\left(F \prod_{m=1}^{2n-1} (I + a_m \otimes b_m) \right) = f((\det F)^{1/n} I) \ , \qquad (8)$$

for every $F \in \mathbb{R}^{n \times n}$ with $\det F > 0$.

Step 4: Finally let for $x > 0$

$$g(x) = f(x^{1/n} I) \ . \qquad (9)$$

We then immediately have from (8) and (9) that

$$g(\det F) = f(F) \ . \qquad (10)$$

It therefore just remains to show that g is convex, i.e.

$$g(\lambda x + (1 - \lambda)y) \le \lambda g(x) + (1 - \lambda)g(y) \qquad (11)$$

for every $\lambda \in (0, 1)$, $x, y > 0$.

Let $F = (\lambda x + (1 - \lambda)y)^{1/n} I$. Using Lemma 1.4 of Chapter 5 (with $\Phi(F) = \det F$) we can find $G, H \in \mathbb{R}^{n \times n}$ so that

$$\begin{cases} F = \lambda G + (1 - \lambda)H \\ \det G = x, \ \det H = y \ . \\ \text{rank}\{G - H\} \le 1 \end{cases} \qquad (12)$$

Therefore using (9), (12) and the rank one convexity of f we get

$$g(\lambda x + (1 - \lambda)y) = f(F) = f(\lambda G + (1 - \lambda)H)$$

$$\le \lambda f(G) + (1 - \lambda)f(H) = \lambda g(x) + (1 - \lambda)g(y) \ . \quad \square$$

We can now proceed with the proof of the theorem.

PROOF.

i) We decompose the proof into two steps.

Step 1: Observe that if W satisfies $(H1)$, then the rank one convex envelope RW also satisfies $(H1)$; this follows, at once, from the representation formula for RW (cf. Theorem 1.1 of Chapter 5). Therefore RW satisfies $(S1)$ and $(S2)$ of the above lemma. Applying Lemma 2.6 to RW, we find that there exists $g : (0, \infty) \to \mathbb{R}$ convex such that

$$RW(F) = g(\det F) \, , \tag{1}$$

for every $F \in \mathbb{R}^{n \times n}_+$.

Step 2: Note that since $h(\det F) \leq W(F)$, then

$$h^{**}(\det F) \leq PW(F) \leq RW(F) = g(\det F) \tag{2}$$

for every $F \in \mathbb{R}^{n \times n}_+$.
However

$$g(\det F) = RW(F) \leq W(F) \tag{3}$$

for every $F \in \mathbb{R}^{n \times n}_+$, hence $g \leq h$. The convexity of g then implies that $g \leq h^{**}$, which, combined with (2), give the claimed result.

ii) The second part of the theorem is in the same spirit as Theorem 1.1 of Chapter 5 and we refer to Chipot-Kinderlehrer [1], Fonseca [2] for a complete proof. $\qquad\qquad\square$

A.2.4 Relaxation and Optimal Design

Recently Kohn and Strang [1,2] have shown how to apply the relaxation theorems to optimal design problems. We here present their results and we refer to their articles for more details and historical comments.

Optimal design consists in optimizing some functions subject to some geometrical constraints. One such case is the following. Consider the equations

$$\begin{cases} \operatorname{div} \sigma(x) = 0 & \text{in } \Omega \\ \sigma(x) = a(x)\nabla w(x) & \text{in } \Omega \\ \sigma \cdot n = f & \text{on } \partial\Omega \end{cases} \tag{1}$$

where $\Omega \subset \mathbb{R}^n$ is a bounded simply connected domain with outward unit normal n and f is given with $\int_{\partial\Omega} f \, ds = 0$.

Equations of the type (1) can be useful in various contexts and the above authors have considered different problems where the basic equations are as in (1). One simple example can be found in the context of electricity. In this case σ represents the current, w the potential and a the conductivity. The first equation is then Kirchoff's law while the second is Ohm's law. The rate at which energy is dissipated into heat is

$$J(a, \sigma) = \int_\Omega \frac{|\sigma|^2}{a(x)}\, dx \ .$$

One can then reformulate (1) as a variational problem

$$\inf\left\{ J(a, \tau) = \int_\Omega \frac{|\tau|^2}{a(x)}\, dx \ : \ \mathrm{div}\,\tau = 0, \ \tau \cdot n = f \ \text{on} \ \partial\Omega \right\} = J(a, \sigma) \ .$$

$$(2)$$

The design problem under consideration is to choose the conductivity $a(x)$ so that it minimizes a certain "cost". More precisely suppose that there are two available materials, the first one being a good conductor with conductivity $a(x) = 1$, the second one having a lower conductivity $a(x) = \delta, 0 < \delta < 1$ (or even a perfect insulator with $a = 0$) but being less expensive. Then an optimal design problem could be to find a design which uses as little expensive material as possible so as to minimize the cost $\int a\, dx$ but with the additional constraint that $J(a, \sigma)$ (the rate at which energy is dissipated) should be less than a constant. The last condition is a kind of efficiency condition required on the design. Therefore the problem is

$$(P) \qquad\qquad \inf\left\{ \int_\Omega a(x)\, dx \ : \ J(a) \leq c \right\} \ .$$

One can rewrite (P) in more geometrical terms, Suppose that $\delta = 0$, i.e. the inexpensive material being a perfect insulator, and let $S \subset \Omega$ be a closed set where $a \equiv 1$, i.e. S is the set where there is a good conductor, then

$$a(x) = \chi_S(x) = \begin{cases} 1 & \text{if } x \in S \\ 0 & \text{if } x \in \Omega - S \end{cases} \ .$$

Writing $J(S)$ instead of $J(a, \sigma)$ we can reformulate (2) as

$$J(S) = \int_\Omega |\sigma|^2\, dx$$

$$= \inf\left\{ \int_\Omega |\tau|^2 \ : \ \mathrm{div}\,\tau = 0, \ \tau \cdot n = f \ \text{on} \ \partial\Omega \ \text{and} \ \mathrm{supp}\,\tau \subset S \right\} \ .$$

We can then rewrite (P) as

$$(P_1) \qquad\qquad \inf\{\mathrm{meas}\,(S) \ : \ J(S) \leq c\} \ .$$

The problem can be generalized to a "many loads" problem, i.e. one has the equations

$$\begin{cases} \operatorname{div}\sigma_i = 0 & \text{in } \Omega & 1 \le i \le n \\ \sigma_i = a\nabla w_i & \text{in } \Omega & 1 \le i \le n \\ \sigma_i \cdot n = f_i & \text{on } \partial\Omega & 1 \le i \le n \end{cases} \qquad (3)$$

(P) becomes then

$$(P_n) \qquad\qquad \inf\{\operatorname{meas}(S) \, : \, J_i(S) \le c_i, 1 \le i \le n\} \, .$$

In more physical terms, (P_n) corresponds to minimization of the amount of good conductor used subject to constraints on the rate at which energy is dissipated under some given loads.

REMARK. As mentioned before one encounters problems of the type (P_n) not only when dealing with electrical current but in many other instances, in particular in elasticity, and we refer to the articles of Kohn-Strang for further developments.

One should note that in general (P_n), $n \ge 1$, has no solution, due to the lack of weak lower semicontinuity of a certain associated functional. However using the relaxation theorems one will be able to construct an almost optimal design.

Before giving the main theorem, we state the key lemma.

Lemma 2.7. *Let* $\xi \in \mathbf{R}^{n\times 2}$ *and let*

$$g(\xi) = \begin{cases} 1 + |\xi|^2 & \text{if } \xi \ne 0 \\ 0 & \text{if } \xi = 0 \end{cases}$$

where $|\xi|^2 = \sum_{j=1}^{2}\sum_{i=1}^{n}\xi_{ij}^2$.
Case 1: If $n = 1$, *then*

$$Pg(\xi) = Qg(\xi) = Rg(\xi) = g^{**}(\xi) = \begin{cases} 1 + |\xi|^2 & \text{if } |\xi| \ge 1 \\ 2|\xi| & \text{if } |\xi| \le 1 \end{cases} \, .$$

Case 2: If $n > 1$, *then*

$$Pg(\xi) = Qg(\xi) = Rg(\xi) = h(\xi)$$

where

$$h(\xi) = \begin{cases} 1 + |\xi|^2 & \text{if } |\xi|^2 + 2|\mathrm{adj}_2\xi| \ge 1 \\ 2(|\xi|^2 + 2|\mathrm{adj}_2\xi|)^{1/2} - 2|\mathrm{adj}_2\xi| & \text{if } |\xi|^2 + 2|\mathrm{adj}_2\xi| \le 1 \end{cases}$$

and where

$$|\mathrm{adj}_2\xi| = \left[\sum_{1\leq i<j\leq n} (\xi_{i_1}\xi_{j_2} - \xi_{j_1}\xi_{i_2})^2 \right]^{1/2} .$$

Furthermore

$$Qg(\xi) = h(\xi) > g^{**}(\xi) .$$

PROOF. We follow the proof of Kohn and Strang [1,2].
Case 1: The case $n = 1$ is easy and is proved exactly as Step 3 below.
Case 2: We only sketch the proof in the case $n = 2$ (the general case is proved similarly). Observe that the inequality $h(\xi) > g^{**}(\xi)$ is obvious. Note also that if $n = 2$, then $\mathrm{adj}_2\xi = \det\xi$ and therefore

$$h(\xi) = \begin{cases} 1 + |\xi|^2 & \text{if } |\xi|^2 + 2|\det\xi| \geq 1 \\ 2(|\xi|^2 + 2|\det\xi|)^{1/2} - 2|\det\xi| & \text{if } |\xi|^2 + 2|\det\xi| \leq 1 \end{cases} . \quad (1)$$

We divide the proof into four steps.
Step 1: It is not difficult to see (cf. Kohn-Strang) that h is polyconvex. Since $h \leq g$ we then have

$$h(\xi) \leq Pg(\xi) \leq Qg(\xi) \leq Rg(\xi) .$$

We therefore only need to show that

$$Rg(\xi) \leq h(\xi) . \quad (2)$$

Step 2: We now give some algebraic relations, whose proofs are straightforward. Let $a_i, b_i \in \mathbb{R}^2$, $i = 1,2$ be such that

$$\langle a_i; a_j \rangle = \langle b_i; b_j \rangle = \delta_{ij} = \begin{cases} 1 & \text{if } i = j \\ 0 & \text{if } i \neq j \end{cases} \quad (3)$$

where $\langle \bullet; \bullet \rangle$ denotes the scalar product in $\mathbb{R}^2$. Let for $\xi \in \mathbb{R}^{2\times 2}$ with $\xi = (\xi_{ij})_{1\leq i,j\leq 2}$

$$\tilde{\xi} = \begin{pmatrix} \xi_{22} & -\xi_{21} \\ -\xi_{12} & \xi_{11} \end{pmatrix}$$

(in the notations of the appendix to Chapter 4, $\tilde{\xi} = \mathrm{adj}_1\xi$).

We then have

$$(a_i \otimes b_i)(a_j \otimes b_j)^t = \begin{cases} a_i \otimes a_i & \text{if } i = j \\ 0 & \text{if } i \neq j \end{cases}$$

$$\langle \widetilde{a_1 \otimes a_1}; a_2 \otimes a_2 \rangle = 1$$

$$\langle a_i \otimes b_i; a_j \otimes b_j \rangle = \delta_{ij} = \begin{cases} 1 & \text{if } i = j \\ 0 & \text{if } i \neq j \end{cases}$$

where $a \otimes b = (\alpha_i \beta_j)_{1 \leq i,j \leq 2} \in \mathbb{R}^{2 \times 2}$ if $a = (\alpha_1, \alpha_2) \in \mathbb{R}^2, b = (\beta_1, \beta_2) \in \mathbb{R}^2$ and ξ^t denotes the transpose of the matrix $\xi \in \mathbb{R}^{2 \times 2}$.

From the above identities we immediately deduce that if $\gamma, \delta \geq 0$ and

$$\xi = \gamma a_1 \otimes b_1 + \delta a_2 \otimes b_2 \tag{4}$$

then

$$\xi \xi^t = \gamma^2 a_1 \otimes a_1 + \delta^2 a_2 \otimes a_2 \tag{5}$$

$$(\det \xi)^2 = \det(\xi \xi^t) = \gamma^2 \delta^2 \langle \widetilde{a_1 \otimes a_1}; a_2 \otimes a_2 \rangle = \gamma^2 \delta^2 \tag{6}$$

$$|\xi|^2 = \gamma^2 + \delta^2 \tag{7}$$

$$|\xi|^2 + 2|\det \xi| = (\gamma + \delta)^2 \ . \tag{8}$$

Step 3: Let $\theta \in (0,1), \gamma, \delta \geq 0, a, b, c, d \in \mathbb{R}^2$ with $|a|^2 = |b|^2 = 1, \langle a; b \rangle = 0$ and

$$\begin{cases} \xi_1 = \gamma a \otimes b - (1-\theta)c \otimes d \\ \xi_2 = \gamma a \otimes b + \theta c \otimes d \end{cases} .$$

Then

$$\inf_{\substack{\theta \in (0,1) \\ c,d \in \mathbb{R}^2}} \{\theta g(\xi_1) + (1-\theta)g(\xi_2)\} = \begin{cases} 2\gamma & \text{if } \gamma \leq 1 \\ 1 + \gamma^2 & \text{if } \gamma \geq 1 \end{cases} . \tag{9}$$

The above minimization follows from the following observations.
Case 1: If $c \otimes d = 0, \gamma \neq 0$, then

$$\theta g(\xi_1) + (1-\theta)g(\xi_2) = \theta(1 + \gamma^2) + (1-\theta)(1 + \gamma^2) = 1 + \gamma^2 \ .$$

Case 2: If $c \otimes d \neq 0, \gamma a \otimes b \neq -\theta c \otimes d$ and $\gamma a \otimes b \neq (1-\theta)c \otimes d$, then

$$\theta g(\xi_1) + (1-\theta)g(\xi_2) = \theta(1 + |\gamma a \otimes b - (1-\theta)c \otimes d|^2)$$

$$+ (1-\theta)(1 + |\gamma a \otimes b + \theta c \otimes d|^2) \geq 1 + \gamma^2 \ ;$$

the last inequality following from the convexity of the norm.

Case 3: If $c \otimes d \neq 0$, $\gamma a \otimes b = -\theta c \otimes d$ (similarly for the case $\gamma a \otimes b = (1 - \theta)c \otimes d$), then

$$\theta g(\xi_1) + (1 - \theta)g(\xi_2) = \theta(1 + |c \otimes d|^2) = \theta + \frac{\gamma^2}{\theta} \equiv \varphi(\theta) \ .$$

Case 3a: If $\gamma \geq 1$, then $\varphi'(\theta) \neq 0$ for every $\theta \in (0,1)$ and hence

$$\theta g(\xi_1) + (1 - \theta)g(\xi_2) \geq 1 + \gamma^2 = \varphi(1) \ .$$

Case 3b: If $\gamma < 1$, then $\varphi'(\gamma) = 0$ and thus

$$\theta g(\xi_1) + (1 - \theta)g(\xi_2) \geq \varphi(\gamma) = 2\gamma \ .$$

Thus (9) follows.

Step 4: We now show (2). Recall that from Theorem 1.1 of Chapter 5 we have

$$Rg(\xi) = \inf\left\{ \sum_{i=1}^{I} \lambda_i g(\xi_i) \ : \ \sum_{i=1}^{I} \lambda_i = 1; \ \sum_{i=1}^{I} \lambda_i \xi_i = \xi \right.$$

$$\left. \text{and } (\lambda_i, \xi_i)_{1 \leq i \leq I} \text{ satisfy } (H_I) \right\} \ . \tag{10}$$

It is clear that for every $\xi \in \mathbf{R}^{2 \times 2}$, there exist $\gamma, \delta \geq 0$, $a_i, b_i \in \mathbf{R}^2$, $i = 1, 2$ satisfying (3) and (4), i.e.

$$\begin{cases} \xi = \gamma a_1 \otimes b_1 + \delta a_2 \otimes b_2 \\ \langle a_i; a_j \rangle = \langle b_i; b_j \rangle = \delta_{ij} \end{cases} \ .$$

Therefore letting $\theta, \tilde{\theta} \in (0,1)$, $c, d \in \mathbf{R}^2$ and

$$\begin{cases} \xi_1 = \gamma a_1 \otimes b_1 - (1 - \tilde{\theta})c \otimes d & \lambda_1 = \theta\tilde{\theta} \\ \xi_2 = \gamma a_1 \otimes b_1 + \tilde{\theta} c \otimes d & \lambda_2 = \theta(1 - \tilde{\theta}) \\ \xi_3 = \gamma a_1 \otimes b_1 + \dfrac{\delta}{1 - \theta} a_2 \otimes b_2 & \lambda_3 = (1 - \theta) \end{cases}$$

we have

$$\begin{cases} \sum\limits_{i=1}^{3} \lambda_i = 1, \quad \sum\limits_{i=1}^{3} \lambda_i \xi_i = \xi \\ \det(\xi_1 - \xi_2) = \det\left(\xi_3 - \dfrac{\lambda_1 \xi_1 + \lambda_2 \xi_2}{\lambda_1 + \lambda_2} \right) = 0 \end{cases} \ . \tag{11}$$

Combining (10) and (11) we obtain

$$Rg(\xi) \leq \inf_{\substack{\theta,\tilde{\theta}\in(0,1) \\ c,d\in\mathbf{R}^2}} \{\theta\tilde{\theta}g(\xi_1) + \theta(1 - \tilde{\theta})g(\xi_2) + (1 - \theta)g(\xi_3)\} \ . \tag{12}$$

If we now prove that the right hand side of (12) is less than $h(\xi)$, we would obtain (2) and thus the lemma. To prove this observe the following.
Case 1: If $\gamma \geq 1$, then from (9) of Step 3 we deduce that

$$Rg(\xi) \leq \inf_{\theta\in(0,1)} \{\theta(1 + \gamma^2) + (1 - \theta)g(\xi_3)\} \ .$$

Since $g(\xi_3) \leq 1 + |\xi_3|^2$ we deduce from (7) and the above inequality that

$$Rg(\xi) \leq \inf_{\theta\in(0,1)} \left\{ \theta(1 + \gamma^2) + (1 - \theta)\left(1 + \gamma^2 + \frac{\delta^2}{(1 - \theta)^2}\right)\right\}$$

$$= 1 + \gamma^2 + \delta^2$$

$$= 1 + |\xi|^2 = h(\xi) \ ;$$

the last identity coming from the fact that $|\xi|^2 + 2|\det \xi| \geq \gamma^2 \geq 1$. Therefore (2) follows.
Case 2: If $\gamma \leq 1$, then from (9), (12) and the fact that $g(\xi_3) \leq 1 + |\xi_3|^2$ we deduce that

$$Rg(\xi) \leq \inf_{\theta\in(0,1)} \left\{ 2\gamma\theta + (1 - \theta)\left(1 + \gamma^2 + \frac{\delta^2}{(1 - \theta)^2}\right)\right\}$$

$$= \inf_{\theta\in(0,1)} \left\{ 1 + \gamma^2 - \theta(1 - \gamma)^2 + \frac{\delta^2}{(1 - \theta)}\right\} \ .$$

We therefore get

$$Rg(\xi) \leq \begin{cases} 1 + \gamma^2 + \delta^2 & \text{if } \gamma + \delta \geq 1 \\ 2[\gamma + \delta - \gamma\delta] & \text{if } \gamma + \delta \leq 1 \end{cases} \ .$$

Using (6), (7), (8) and the definition of h (cf. (1)) in the above inequality, we obtain (2), i.e. $Rg(\xi) \leq h(\xi)$.

This concludes the proof of the lemma. $\qquad\square$

We can now state the main theorem.

Theorem 2.8.

Case 1: One design constraint. Assume that $J(\Omega) < c$ and let

$$A = \inf(P_1) = \inf\{\mathrm{meas}\,(S) : J(S) \le c\}$$

$$\tilde{A} = \sup_{\lambda \ge 0} \inf_S \{\mathrm{meas}\,(S) + \lambda(J(S) - c)\} \ .$$

Let $F = \int_{\partial\Omega} f\,ds$ and g be defined as in the lemma. Then the least area of any optimal design satisfies

$$A = \tilde{A} = \sup_{\lambda \ge 0} \ \inf_{u = F \, on \, \partial\Omega} \left\{ \int_\Omega Qg\left(\sqrt{\lambda}\nabla u(x)\right) dx - \lambda c \right\} \ . \tag{1}$$

Furthermore the right hand side of the above identity has at least one solution $(\bar\lambda, \bar u)$ satisfying

$$A = \tilde{A} = \int_\Omega Qg\left(\sqrt{\bar\lambda}\nabla\bar u(x)\right) dx - \bar\lambda c \ . \tag{2}$$

Note that in this case $Qg = g^{**}$.

Case 2: Many design constraints. Let for $\xi \in \mathbb{R}^{n\times 2}$, $\xi = (\xi_1,\dots,\xi_n)$ with $\xi_i \in \mathbb{R}^2$ and for $\lambda = (\lambda_1,\dots,\lambda_n)$ with $\lambda_i \ge 0$

$$g_\lambda(\xi) = g\left(\sqrt{\lambda_1}\xi_1,\dots,\sqrt{\lambda_n}\xi_n\right) = \begin{cases} 1 + \displaystyle\sum_{i=1}^n \lambda_i|\xi_i|^2 & \text{if } \sum_{i=1}^n \lambda_i|\xi_i|^2 \ne 0 \\[2mm] 0 & \text{if } \sum_{i=1}^n \lambda_i|\xi_i|^2 = 0 \end{cases}$$

and let

$$Qg_\lambda(\xi) = Qg\left(\sqrt{\lambda_1}\xi_1,\dots,\sqrt{\lambda_n}\xi_n\right) \ .$$

Let $F_i = \int_{\partial\Omega} f_i\,ds$, $1 \le i \le n$. Then there exists $\bar u^\lambda = (\bar u_1^\lambda,\dots,\bar u_n^\lambda)$ such that

$$\int_\Omega Qg_\lambda(\nabla\bar u^\lambda(x))\,dx$$

$$= \inf\left\{ \int_\Omega Qg_\lambda(\nabla u(x))\,dx : u = F = (F_1,\dots,F_n) \text{ on } \partial\Omega \right\} \ . \tag{3}$$

Assume that for every $\lambda \in \mathbb{R}^n$ with $\lambda_i > 0$ fixed, the solution $\bar u^\lambda$ satisfies

$$\begin{cases} \text{for every } 1 \le i \le n,\ \nabla\bar u_i^\lambda \not\equiv 0 \text{ for } x \in R^\lambda \\ R^\lambda = \{x \in \Omega : Qg_\lambda(\nabla\bar u^\lambda(x)) \ne g_\lambda(\nabla\bar u^\lambda(x))\} \end{cases} \ . \tag{4}$$

Then for every $\lambda \in \mathbb{R}^n$ with $\lambda_i > 0$, there exists $c = (c_1,\dots,c_n)$ such that *the minimum area of any optimal design is given by*

$$A = \inf(P_n) = \inf\{\operatorname{meas}(S) : J_i(S) \le c_i,\ 1 \le i \le n\}$$

$$= \int_\Omega Qg_\lambda(\nabla \bar{u}^\lambda(x))\,dx - \sum_{i=1}^{n} \lambda_i c_i \ . \tag{5}$$

REMARKS.

i) In general one does not expect that (P_n) attains its minimum even if $n = 1$ (cf. Kohn-Strang). However the identities (2) for the "one design constraint" or (5) for the "many design constraints" problems give a way of constructing almost optimal designs (i.e. minimizing sequences) which in "average" look like $\bar{u}$ for (P_1) or $\bar{u}^\lambda$ for (P_n);

ii) note that in the "many design constraints" problem the existence of a minimizer in (3) is clearly obtained by the direct methods since Qg is quasiconvex and coercive;

iii) observe also that for (P_n) the result is weaker than for (P_1) in two aspects. First the hypothesis (4) on minimizers $\bar{u}^\lambda$ has to be added, this is a technical assumption. The second point is that the theorem can be proved for (P_n) only for a particular choice of the constants c_i, while for (P_1) the result is true for every c provided $J(\Omega) < c$.

PROOF. We do not prove the theorem and we refer to the articles of Kohn-Strang for details. We here only show where the relaxation theorem is applied. *Case 1*: Observe first that we trivially have

$$A = \inf\{\operatorname{meas}(S) : J(S) \le c\}$$

$$= \inf_{S} \sup_{\lambda \ge 0}\{\operatorname{meas}(S) + \lambda(J(S) - c)\} \ . \tag{6}$$

The aim is to show that one can reverse inf and sup in (6), i.e. to show that $A = \tilde{A}$. We can rewrite $\tilde{A}$, using the definition of $J(S)$, as

$$\tilde{A} = \sup_{\lambda \ge 0} \inf_{S}\{\operatorname{meas}(S) + \lambda(J(S) - c)\}$$

$$= \sup_{\lambda \ge 0}\left\{\inf_{S}\left\{\inf_{\tau} \int_\Omega [\chi_S(x) + \lambda|\tau|^2]\,dx : \operatorname{div}\tau = 0,\right.\right.$$

$$\left.\left. \tau \cdot n = f \text{ on } \partial\Omega,\ \operatorname{supp}\tau \subset S\right\} - \lambda c\right\}$$

$$= \sup_{\lambda \ge 0}\left\{\inf_{\tau}\left[\int_\Omega (\chi_{\operatorname{supp}\tau}(x) + \lambda|\tau|^2)\,dx : \operatorname{div}\tau = 0,\right.\right.$$

$$\left.\left. \tau \cdot n = f \text{ on } \partial\Omega\right] - \lambda c\right\} \ .$$

Since Ω is simply connected, we can find u such that

$$\begin{cases} \tau = \left(-\dfrac{\partial u}{\partial x_2}, \dfrac{\partial u}{\partial x_1} \right) \\[2mm] u = F = \displaystyle\int_{\partial\Omega} f\, ds \ \text{ on } \ \partial\Omega \end{cases}.$$

We therefore obtain

$$\tilde{A} = \sup_{\lambda \geq 0} \left\{ \inf_u \left[\int_\Omega (\chi_{\mathrm{supp}\nabla u}(x) + \lambda|\nabla u|^2)\, dx \ : \ u = F \text{ on } \partial\Omega \right] - \lambda c \right\}$$

$$= \sup_{\lambda \geq 0} \left\{ \inf_u \left[\int_\Omega g\left(\sqrt{\lambda}\nabla u(x) \right) dx \ : \ u = F \text{ on } \partial\Omega \right] - \lambda c \right\} .$$

The relaxation theorem of Chapter 5 and Lemma 2.7 give immediately that

$$\tilde{A} = \sup_{\lambda \geq 0} \left\{ \inf_u \left[\int_\Omega Qg\left(\sqrt{\lambda}\nabla u(x) \right) dx \ : \ u = F \text{ on } \partial\Omega \right] - \lambda c \right\} .$$

Note that since u is a scalar function $Qg = g^{**}$.

To show that $A = \tilde{A}$ and the existence of $(\bar{\lambda}, \bar{u})$ satisfying (2), one can use minimax theorems of convex analysis and we refer to the above articles for more details.

Case 2: The proof is very similar in this case. As above we have

$$A = \inf\{\mathrm{meas}\,(S) \ : \ J_i(S) \leq c_i, 1 \leq i \leq n\}$$

$$= \inf_S \sup_{\lambda_i \geq 0} \left\{ \mathrm{meas}\,(S) + \sum_{i=1}^n \lambda_i(J_i(S) - c_i) \right\} .$$

Similarly as before, letting $F_i = \int_{\partial\Omega} f_i\, ds$, $F = (F_1, \ldots, F_n)$, we have

$$\tilde{A} = \sup_{\lambda_i \geq 0} \inf_S \left\{ \mathrm{meas}\,(S) + \sum_{i=1}^n \lambda_i(J_i(S) - c_i) \right\}$$

$$= \sup_{\lambda_i \geq 0} \left\{ \inf_u \left[\int_\Omega g_\lambda(\nabla u(x))\, dx \ : \ u = F \text{ on } \partial\Omega \right] - \sum_{i=1}^n \lambda_i c_i \right\} .$$

Using the relaxation theorem and Lemma 2.7 we obtain

$$\tilde{A} = \sup_{\lambda_i \geq 0} \left\{ \inf_u \left[\int_\Omega Qg_\lambda(\nabla u(x))\, dx \ : \ u = F \text{ on } \partial\Omega \right] - \sum_{i=1}^n \lambda_i c_i \right\} .$$

To show that $A = \tilde{A}$ and the conclusions of the theorem is harder in the vectorial case since no equivalent to minimax theorems are known for quasiconvex functions. Therefore some ad hoc methods need to be used. It is for this reason that Kohn-Strang have added (4) as well as the restriction on the choice of c_i in the hypotheses of the theorem. $\qquad\square$

Bibliography

Acerbi, E., and Buttazzo, G.

[1] Semicontinuous envelopes of polyconvex integrals. *Proc. R. Soc. Edinb.* **96A** (1984) 51–54.

Acerbi, E., Buttazzo, G., and Fusco, N.

[1] Semicontinuity and relaxation for integrals depending on vector valued functions. *J. Math. Pures Appl.* **62** (1983) 371–387.

Acerbi, E., and Fusco, N.

[1] Semicontinuity problems in the calculus of variations. *Arch. Ration. Mech. Anal.* **86** (1984) 125–145.

[2] A regularity theorem for minimizers of quasiconvex functions. (To appear).

Adams, R.

[1] *Sobolev Spaces.* Academic Press, New York, 1975.

Agmon, S.

[1] *Lectures on Elliptic Boundary Value Problem.* Van Nostrand, Princeton, 1965.

Agmon, S., Douglis, A., and Nirenberg, L.

[1] Estimates near the boundary for the solutions of elliptic differential equations satisfying general boundary values I and II. *Commun. Pure Appl. Math.* **12** (1959) 623–727 and **17** (1964) 35–92.

Akhiezer, N.I.

[1] *The Calculus of Variations.* Blaisdell, New York, 1962.

Albert, A.A.

[1] A quadratic form problem in the calculus of variations. *Bull. A.M.S.* **44** (1938) 250–253.

Almgren, F.J.

[1] *The Theory of Varifolds – a Variational Calculus in the Large for the k-Dimensional Area Integrand.* Mimeographed notes; Princeton Univ. 1965.

[2] Existence and regularity almost everywhere of solutions to elliptic variational problems among surfaces of varying topological type and singularity structure. *Ann. Math.* (2) **87** (1968) 321–391.

Anderson J.M., and Duchamp, T.

[1] On the existence of global variational principles. *Amer. J. Math.* **102** (1980) 781–868.

Antman, S.

[1] The influence of elasticity on analysis: modern developments. *Bull. A.M.S.* **9** (1983)

267–291.

[2] Geometrical and analytical questions in nonlinear elasticity. In: *Seminar on Nonlinear P.D.E.*, ed. by S.S. Chern. Springer, New York, 1984, pp. 1–30.

Atteia, M., and Dedieu, J.P.

[1] Minimization of energy in nonlinear elasticity. In: *Nonlinear Problems of Analysis in Geometry and Mechanics*, Proceedings, ed. by M. Atteia, D. Bancel, I. Gumowski. Pitman, Boston, 1981, pp. 73–79.

Attouch, H.

[1] *Variational Convergence of Functional and Operators*. Pitman, London, 1984.

Aubert, G.

[1] *Contribution aux problèmes du calcul des variations et application à l'élasticité non linéaire*. Thèse de doctorat d'état, Paris VI, 1986.

[2] A counterexample of a rank one convex function which is not polyconvex in the case $n = 2$. *Proc. R. Soc. Edinb.* **106A** (1987) 237–240.

Aubert, G., and Tahraoui, R.

[1] Sur la minimisation d'une fonctionnelle non convexe, non différentiable en dimension 1. *Boll. U.M.I.* **17** (1980) 244–258.

[2] Théorèmes d'existence pour des problèmes du calcul des variations. *J. Diff. Eq.* **33** (1979) 1–15.

[3] Sur la faible fermeture de certains ensembles de contraintes en élasticité non linéaire plan. *C.R. Acad. Sci., Paris* **290** (1980) 537–540.

[4] Théorèmes d'existence en optimisation non convexe. *Appl. Anal.* **18** (1984) 75–100.

Bachelot, A.

[1] Formes quadratiques A-compatibles dans les espaces de type L^p. Preprint Université de Bordeaux, 1984.

Ball, J. M.

[1] On the calculus of variations and sequentially weakly continuous maps. In: *Ordinary and Partial Differential Equations*, Proceedings, ed. by W.N. Everitt and B.D. Sleeman, Lecture Notes in Math., vol. 564. Springer, Berlin, 1976, pp. 13–23.

[2] Convexity conditions and existence theorems in nonlinear elasticity. *Arch. Ration. Mech. Anal.* **63** (1977) 337–403.

[3] Constitutive inequalities and existence theorems in elastostatics. In: *Nonlinear Analysis and Mechanics*, Proceedings, ed. by R.J. Knops, Res. Notes, vol. 17. Pitman, London, 1977, pp. 13–25.

[4] On the paper "Basic calculus of variations". *Pacific J. Math.* **116** (1985) 7–10.

[5] Does rank one convexity imply quasiconvexity. In: *Metastability and Incompletely Posed Problems*, ed. by S. Antman et al. Springer, New York, 1987, pp. 17–32.

Ball, J. M., Curie, J.C., and Olver, P.J.

[1] Null Lagrangians, weak continuity and variational problems of arbitrary order. *J. Funct. Anal.* **41** (1981) 135–174.

Ball, J. M., and James, R.D.

[1] Fine phase mixtures as minimizers of energy. (To appear).

Ball, J. M., Knops, R.J., and Marsden, J.E.

[1] Two examples in nonlinear elasticity. In: *Journées d'Analyse Non Linéaire*, Proceedings, ed. by P. Bénilan, and J. Robert. Lecture Notes in Math., vol. 665. Springer, Berlin, 1978, pp. 41–49.

Ball J. M., and Knowles, G.

[1] A numerical method for detecting singular minimizers. *Numer. Math.*, **51** (1987) 181–197.

Ball, J. M., and Mizel, V.J.

[1] One-dimensional variational problems whose minimizers do not satisfy the Euler-Lagrange equations. *Arch. Ration. Mech. Anal.* **90** (1985) 325–388.

Ball, J. M., and Murat, F.

[1] $W^{1,p}$ quasiconvexity and variational problems for multiple integrals. *J. Funct. Anal.* **58** (1984) 225–253.

Banyaga, A.

[1] Formes volume sur les variétés à bord. *Enseignement Math.* **20** (1974) 127–131.

Bauman, P., and Phillips, D.

[1] A nonconvex variational problem related to change of phase. (To appear).

Berkowitz, L.D.

[1] Lower semicontinuity of integral functionals. *Trans. A.M.S.* **192** (1974) 51–57.

[2] *Optimal Control Theory*. Springer, New York, 1974.

Berliocchi, H., and Lasry, J.M.

[1] Intégrandes normales et mesures paramétrées en calcul des variations. *C.R. Acad. Sci. Paris* **274** (1971) 839–842.

[2] Intégrandes normales et mesures paramétrées en calcul des variations. *Bull. S.M.F.* **101** (1973) 129–184.

Bliss, G.

[1] *Lectures on the Calculus of Variations*. University of Chicago Press, Chicago, 1951.

Boccardo, L., Murat, F., and Puel, J.P.

[1] Existence de solutions faibles pour des équations elliptiques quasi-linéaires à croissance quadratique. In: *Nonlinear PDE and Appl.*, Collège de France Seminar, vol. IV, ed. by H. Brézis, and J.L. Lions. Pitman, London, 1983.

[2] Résultats d'existence pour certains problèmes elliptiques quasilinéaires. *Ann. Sc. Norm. Sup. Pisa* **11** (1984) 213–235.

Bolza, O.

[1] *Lectures on the Calculus of Variations*. Chelsea Publ. Comp., New York, 1946.

Brézis, H.

[1] *Opérateurs maximaux monotones*. North Holland, Amsterdam, 1973.

[2] *Analyse fonctionnelle*. Masson, Paris, 1983.

Browder, F.E.

[1] Remarks on the direct method of the calculus of variations. *Arch. Ration. Mech. Anal.* **20** (1965) 251–258.

Busemann, H., Ewald, G., and Shepard, G.

[1] Convex bodies and convexity on Grassman cones. *Math. Ann.* **151** (1963) 1–41.

Busemann, H., and Shepard, G.

[1] Convexity on nonconvex sets. In: *Coll. on Convexity*, Proceedings, Copenhagen, 1965, pp. 20–33.

Buttazzo, G.

[1] Problemi di semicontinuità e rilassamento in calcolo delle variazioni. In: *Equazioni differenziali e calcolo delle variazioni*, Proceedings, ed. by L. Modica. ETS, Pisa, 1985, pp. 23–36.

[2] *Semicontinuity, Relaxation and Integral Representation Problems in the Calculus of Variations*. C.M.A.F., Lisbon, 1986.

Buttazzo, G., and Dal Maso, G.

[1] Integral representation and relaxation of local functionals. *Nonlinear Anal.* **9** (1985) 515–532.

[2] A characterization of nonlinear functionals on Sobolev spaces which admit an integral representation with a Carathéodory integrand. *J. Math. Pures Appl.*. (To appear).

Caccioppoli, R., and Scorza Dragoni, G.

[1] Necessità della condizione di Weierstrass per la semicontinuità di un integrale doppio sopra una data superficie. *Memorie Acc. d'Italia* **9** (1938) 251–268.

Caratheodory, C.

[1] Über die Variationsrechnung bei mehrfachen Integralen. *Acta Szeged Sect. Scient.*

Math. **4** (1929) 193–216.

[2] *Calculus of Variations and Partial Differential Equations of the First Order.* Holden Day, San Francisco, 1965.

Carbone, L., and Sbordone, C.

[1] Some properties of Γ-limits of integral functionals. *Ann. Mat. Pura Appl.* **122** (1979) 1–60.

Carr, J.

[1] Phase transitions via bifurcation from heteroclinic orbits. In: *Systems of nonlinear P.D.E.*, Proceedings, ed. by J.M. Ball. Reidel, Dordrecht, 1983, pp. 333–342.

Casadio Tarabusi, E.

[1] An algebraic characterization of quasiconvex functions. (To appear).

Cesari, L.

[1] Semicontinuità e convessità nel calcolo delle variazioni. *Ann. Sc. Norm. Sup. Pisa* **18** (1964) 389–423.

[2] A necessary and sufficient condition for lower semicontinuity. *Bull A.M.S.* **80** (1974) 467–472.

[3] Lower semicontinuity and lower closure theorems without seminormality conditions. *Ann. Mat. Pura Appl.* **98** (1974) 381–397.

[4] *Optimization Theory and Application.* Springer, New York, 1983.

Charrier, P., Dacorogna, B., Hanouzet, B., and Laborde, P.

[1] An existence theorem for slightly compressible material in nonlinear elasticity. *S.I.A.M. Math. Anal.* **19** (1988) 70–86.

Chipot, M., and Kinderlehrer, D.

[1] Equilibrium cnfigurations of crystals. (To appear).

Ciarlet, P.G.

[1] *Elasticité tridimensionnelle.* Masson, Paris, 1985; or *Lectures on Three-dimensional Elasticity.* Tata Inst., Springer, Berlin, 1973.

[2] *Mathematical Elasticity. vol. I: Three-dimensional Elasticity.* North-Holland, Amsterdam, 1987.

Ciarlet, P.G., and Geymonat, G.

[1] Sur les lois de comportement en elasticité non-linéaire compressible. *C.R. Acad. Sci. Paris* **295** (1982) 423–426.

Ciarlet, P.G., and Nečas, J.

[1] Unilateral problems in nonlinear three dimensional elasticity. *Arch. Ration. Mech. Anal.* **87** (1985) 319–338.

[2] Injectivity and self contact in nonlinear elasticity. *Arch. Ration. Mech. Anal.* **97** (1987) 171–188.

Ciarlet, P.G., and Quintela-Estevez, P.

[1] Polyconvexity and the von Karman Equations. (To appear).

Clarke, F.H.

[1] Admissible relaxation in variational and control problems. *Ann. Mat. Pura Appl.* **98** (1974) 381–397.

[2] *Optimization and Nonsmooth Analysis.* Wiley, New York, 1983.

Clarke, F.H., and Vinter, R.V.

[1] Regularity properties of solutions to the basic problem in the calculus of variations. *Trans. A.M.S.* **289** (1985) 73–98.

Coddington, E.A., and Levinson, N.

[1] *Theory of Ordinary Differential Equations.* Mac Graw Hill, New York, 1955.

Courant, R.

[1] *Calculus of Variations.* Courant Institute publications, New York, 1962.

Courant, R., and Hilbert, D.

[1] *Methods of Mathematical Physics.* Wiley, New York, 1966.

Dacorogna, B.

[1] A relaxation theorem and its applications to the equilibrium of gases. *Arch. Ration. Mech. Anal.* **77** (1981) 359–386.

[2] Minimal hypersurfaces in parametric form with non convex integrands. *Indiana Univ. Math. J.* **31** (1982) 531–552.

[3] Quasiconvexity and relaxation of non convex variational problems. *J. Funct. Anal* **46** (1982) 102–118.

[4] *Weak Continuity and Weak Lower Semicontinuity of Non Linear Functionals.* Lecture Notes in Math., vol. 922. Springer, Berlin, 1982.

[5] Regularization of non elliptic variational problems. In: *Systems of Nonlinear P.D.E.*, Proceedings, ed. by J.M. Ball. Reidel, Dordrecht, 1983, pp. 384–400.

[6] Relaxation for some dynamical problems. *Proc. R. Soc. Edinb.* **100A** (1985) 39–52.

[7] Remarques sur les notions de polyconvexité, quasi-convexité et convexité de rang 1, *J. Math. Pures Appl.* **64** (1985) 403–438.

[8] A characterization of polyconvex, quasiconvex and rank one convex envelopes. *Rend. Circolo Mat. Palermo* **15** (1987) 37–58.

[9] Convexity of certain integrals of the calculus of variations. *Proc. R. Soc. Edinb.* **107A** (1987) 15–26.

Dacorogna, B., and Fusco, N.

[1] Semi-continuité des fonctionnelles avec contraintes du type $\det F > 0$. *Boll. U.M.I.* **4-8** (1985) 179–189.

Dacorogna, B., and Marcellini, P.

[1] A counterexample in the vectorial calculus of variations. In: *Material Instabilities in Continuum Mechanics*, Proceedings, ed. by J.M. Ball. Oxford Science Public., Oxford, 1988, pp. 77–83.

Dacorogna, B., and Moser, J.

[1] On a partial differential equation involving Jacobian determinant. (To appear).

De Campos, L.T., and Oden, J.T.

[1] Non quasiconvex problems in nonlinear elastostatics. *Adv. Appl. Math.* **4** (1983) 380–401.

De Giorgi, E.

[1] Su una teoria generale della misura $(r - 1)$ dimensionale in uno spazio ad r dimensioni. *Ann. Mat. Pura Appl.* **36** (1954) 191–213.

[2] *Frontiere orientate di misura minima*, Sem. Mat. Scuola Norm. Sup. Pisa 1960-61. Editrice Tecnico Scientifica, Pisa, 1961.

[3] *Teoremi di semicontinuità nel calcolo delle variazioni.* Inst. Naz. di Alta Matem., Roma, 1968.

[4] Sulla convergenza di alcune successioni di integrali del tipo dell'area. *Rend Mat. Roma* **8** (1975) 227–294.

[5] Some continuity and relaxation problems. In: *E. De Giorgi Colloquium*, Proceedings, ed. by P. Kree. Pitman Res. Notes, vol. 125. London, 1985, pp. 1–11.

De Giorgi, E., Buttazzo, G., and Dal Maso, G.

[1] On the lower semicontinuity of certain integral functionals. *Alti Accad. Naz. Lincei Rend* **74** (1983) 274–282.

De Giorgi, E., and Spagnolo, S.

[1] Sulla convergenza degli integrali della energia per operatori ellitici del secondo ordine. *Boll. U.M.I.* **8** (1973) 391–411.

Demengel, F., and Temam, R.

[1] Convex functions of a measure and applications. *Indiana Univ. Math. J.* **33** (1984) 673–709.

Dunford, N., and Schwartz, J.T.

[1] *Linear Operators.* Wiley, New York, 1957.

Edelen, D.G.B.
[1] The null set of the Euler-Lagrange operator. *Arch. Ration. Mech. Anal.* **11** (1962) 117–121.

Eisen, G.
[1] A counterexample for some lower semicontinuity results. *Math. Z.* **162** (1978) 141–144.
[2] A selection lemma for sequences of measurable sets and lower semicontinuity of multiple integrals. *Manuscripta Math.* **27** (1979) 73–79.

Ekeland, I.
[1] Sur le contrôle optimal de systèmes gouvernés par des équations elliptiques. *J. Funct. Anal.* **9** (1972) 1–62.
[2] Discontinuités de champs hamiltoniens et existence de solutions optimales en calcul des variations. *Publ. Math. (IHES)* **47** (1977) 5–32.
[3] Nonconvex minimization problems. *Bull. A.M.S.* **1** (1979) 443–475.

Ekeland, I., and Temam, R.
[1] *Analyse convexe et problèmes variationnels.* Dunod, Paris, 1974; English translation: *Convex Analysis and Variational Problems.* North Holland, Amsterdam, 1976.

Ericksen, J.L.
[1] Nilpotent energies in liquid crystal theory. *Arch. Ration. Mech. Anal.* **10** (1962) 189–196.
[2] On the symmetry of deformable crystals. *Arch. Ration. Mech. Anal.* **72** (1979) 1–13.
[3] Some phase transitions in crystals. *Arch. Ration. Mech. Anal.* **73** (1980) 99–124.
[4] Twinning of crystals I. In: *Metastability and Incompletely Posed Problems*, ed. by S. Antman et al., Springer, New York, 1987, pp. 77–96.

Ericksen, J.L., and Toupin, R.A.
[1] Implications of Hadamard's condition for elastic stability with respect to uniqueness theorems. *Canad. J. Math.* **8** (1956) 432–436.

Evans, L.C.
[1] Quasiconvexity and partial regularity in the calculus of variations. *Arch. Ration. Mech. Anal.* **95** (1986) 227–252.

Federer, H.
[1] *Geometric Measure Theory.* Springer, Berlin, 1969.

Fenchel, W.
[1] On conjugate convex functions. *Canad. J. Math.* **1** (1949) 73–77.
[2] *Convex Cones, Sets and Functions.* Mimeographed notes, Princeton University, 1951.

Fichera, G.
[1] Semicontinuity of multiple integrals in ordinary form. *Arch. Ration. Mech. Anal.* **17** (1964) 339–352.

Fonseca, I.
[1] Variational methods for elastic crystals. *Arch. Ration. Mech. Appl.* **97** (1987) 189–220.
[2] The lower quasiconvex envelope of the stored energy function for an elastic crystal. *J. Math. Pures Appl.* **67** (1988) 175–195.

Fosdick, R.L., and Mac Sithigh, G.P.
[1] Minimization in incompressible nonlinear elasticity theory. *J. Elasticity* **16** (1986) 267–301.

Fusco, N.
[1] Quasiconvessità e semicontinuità per integrali multipli di ordine superiore. *Ricerche Mat.* **29** (1980) 307–323.
[2] Remarks on the relaxation of integrals of the calculus of variations. In: *Systems of Nonlinear P.D.E.*, ed. by J.M. Ball. Reidel, Dordrecht, 1983, pp. 401–408.

Fusco, N., and Hutchinson, J.

[1] $C^{1,\alpha}$ partial regularity of functions minimizing quasiconvex integrals. *Manuscripta Math.* **54** (1985) 121–143.

[2] Partial regularity for minimizers of certain functionals having non quadratic growth. (To appear).

Gamkrelidze, R.

[1] On some extremal problem in the theory of differential equations. *S.I.A.M. J. Control Optim.* **3** (1965) 106–128.

Gelfand, I.M., and Fomin, S.V.

[1] *Calculus of Variations*. Prentice-Hall, Englewood, 1963.

Giaquinta, M.

[1] *Multiple Integrals in the Calculus of Variations and Nonlinear Elliptic Systems*. Princeton University Press, Princeton, 1983.

Giaquinta, M., and Giusti, E.

[1] *Q*-minima. *Anal. Non Linéaire* **1** (1984) 79–107.

Giaquinta, M., and Modica, G.

[1] Partial regularity of minimizers of quasiconvex integrals. *Anal. Non Linéaire* **3** (1986) 185–208.

Gilbarg, D., and Trudinger, N.S.

[1] *Elliptic Partial Differential Equations of Second Order*. Springer, Berlin, 1977.

Giusti, E.

[1] *Minimal Surfaces and Functions of Bounded Variation*. Birkhäuser, Boston, 1984.

Goldstine, H.H.

[1] *A History of the Calculus of Variations from the 17th to the 19th Century*. Springer, Berlin, 1980.

Green, A.E.

[1] On some general formulae in finite elastostatics. *Arch. Ration. Mech. Anal.* **50** (1973) 73–80.

Gurtin, M.E.

[1] *An Introduction to Continuum Mechanics*. Academic Press, New York, 1981.

Gurtin, M.E., and Temam, R.

[1] On the antiplane shear problem in finite elasticity. *J. of Elasticity* **11** (1981) 197–206.

Hadamard, J.

[1] *Leçons sur la propagation des ondes et les équations de l'hydrodynamique*. Hermann, Paris, 1903.

[2] Sur quelques questions du calcul des variations. *Bull. Soc. Math. France* **33** (1905) 73–80.

Halmos, P.R.

[1] *Measure Theory*. Springer, New York, 1976.

Hanouzet, B.

[1] Applications bilinéaires compatibles avec un système à coefficients variables, continuité dans les espaces de Besov. *Commun. Partial Differ. Equations* **10** (1985) 433–465.

Hanouzet, B., and Joly, J.L.

[1] Bilinear maps compatible with a system. In: *Contributions to Nonlinear PDE*, ed. by C. Bardos et al., Research Notes, vol. 89. Pitman, London, 1983, pp. 208–217.

Hardy, G.H., Littlewood, J.E., and Polya, G.

[1] *Inequalities*. Cambridge University Press, Cambridge, 1961.

Hestenes, M.R.

[1] *Calculus of Variations and Optimal Control Theory*. Wiley, New York, 1966.

Hestenes, M.R., and Mac Shane, E.J.

[1] A theorem on quadratic forms and its applications in the calculus of variations. *Trans. A.M.S.* **47** (1949) 501–512.

Hildebrandt, S.
   [1] *Variationsrechnung mehrdimensionaler Integrale.* Preprint, Univ. Bonn, 1975.
   [2] *Variationsrechnung und hamiltonsche Mechanik.* Preprint, Univ. Bonn, 1977.
Hill, R.
   [1] Constitutive inequalities for isotropic elastic solids under finite strain. *Proc. Roy. Soc. London* **A314** (1970) 457–472.
Ioffe, A.D.
   [1] An existence theorem for a general Bolza problem. *S.I.A.M. J. Control Optim.* **14** (1976) 458–466.
   [2] Sur la semicontinuité des fonctionnelles intégrales. *C.R. Acad. Sci. Paris* **A284** (1977) 807–809.
   [3] On lower semicontinuity of integral functionals I and II. *S.I.A.M. J. Control Optim.* **15** (1977) 521–538, and 991–1000.
Ioffe, A.D., and Tihomirov, V.M.
   [1] Extensions of problems in the calculus of variations. *Trudy Moskovskovo Matem. Obsch.* **18** (1968) 188–246.
   [2] *Theory of Extremal Problems.* North Holland, Amsterdam, 1979.
John, F.
   [1] Uniqueness of nonlinear elastic equilibrium for prescribed boundary displacements and sufficiently small strains. *Commun. Pure Appl. Math.* **25** (1972) 617–634.
Kinderlehrer, D.
   [1] Twinning of crystals II. In: *Metastability and Incompletely Posed Problems*, ed. by S. Antman et al., Springer, New York, 1987, pp. 185–211.
Klotzler, R.
   [1] On the existence of optimal processes. *Banach Center Publications* vol. **1** Warszawa (1976) 125–130.
Knops, R., and Stuart, C.A.
   [1] Quasiconvexity and uniqueness of equilibrium solutions in nonlinear elasticity. *Arch. Ration. Mech. Anal.* **86** (1984) 233–249.
Knowles, J.K., and Sternberg, E.
   [1] On the ellipticity of the equations of nonlinear elastostatics for a special material. *J. Elasticity* **5** (1975) 341–361.
   [2] On the failure of ellipticity of the equations for finite elastic plane strain. *Arch. Ration. Mech. Anal.* **63** (1977) 321–336.
Kohn, R., and Strang, G.
   [1] Explicit relaxation of a variational problem in optimal design. *Bull. A.M.S.* **9** (1983) 211–214.
   [2] Optimal design and relaxation of variational problems I, II and III. *Commun. Pure Appl. Math.* **39** (1986) 113–137, 139–182, and 353–377.
Kohn, R., and Vogelius, M.
   [1] Relaxation of a variational method for impedance computed tomography. *Commun. Pure Appl. Math.* (To appear).
Ladyzhenskaya, O.A., and Uraltseva, N.N.
   [1] *Linear and Quasilinear Elliptic Equations.* Academic Press, New York, 1968.
Lavrentiev, M.
   [1] Sur quelques problèmes du calcul des variations. *Ann. Mat. Pura e Appl.* **4** (1926) 7–28.
Lebesgue, H.
   [1] Sur le problème de Dirichlet. *Rend. Circ. Mat. Palermo* **24** (1907) 371–402.
Le Dret, H.
   [1] Constitutive laws and existence questions in incompressible nonlinear elasticity. *J. Elasticity* **15** (1985) 369–387.

[2] Incompressible limit behaviour of slightly compressible nonlinear elastic materials. *Modélisation Mathématique et Analyse Numérique* **20** (1986) 315–340.

Lions, J.L., and Magenes, E.

[1] *Problèmes aux limites non homogènes et applications.* Dunod, Paris, 1968; English translation: *Non-homogeneous Boundary Value Problems and Applications.* Springer, Berlin, 1972.

Lions, P.L.

[1] *Generalized Solutions of Hamilton-Jacobi Equations.* Pitman, London, 1982.

Lurie, K.A., and Cherkaev, A.V.

[1] Optimal structural design and relaxed controls. *Opt. Control Appl. Math.* **4** (1983) 387–392.

Mac Shane, E.J.

[1] On the necessary condition of Weierstrass in the multiple integral problem of the calculus of variations. *Ann. of Math.* **32** (1931) 578–590.

[2] The condition of Legendre for double integral problems of the calculus of variations. Abstract 209, *Bull. A.M.S.* **45** (1939) 369.

[3] Generalized curves. *Duke Math. J.* **6** (1940) 513–536.

[4] Necessary conditions in the generalized curve problem of the calculus of variations. *Duke Math. J.* **7** (1940) 1–27.

[5] Relaxed controls and variational problems. *S.I.A.M. J. Control Optim.* **5** (1967) 438–485.

Mania, B.

[1] Sopra un esempio di Lavrentieff. *Boll. U.M.I.* **13** (1934) 147–153.

Marcellini, P.

[1] Some problems of semicontinuity. In: *Recent Methods in Nonlinear Analysis*, Proceedings, ed. by E. De Giorgi, E. Magenes, and U. Mosco. Pitagora, Bologna, 1979, pp. 205–222.

[2] Alcune osservazioni sull'esistenza del minimo di integrali del calcolo delle variazioni senza ipotesi di convessità. *Rend. Mat.* **13** (1980) 271–281.

[3] A relation between existence of minima for nonconvex integrals and uniqueness for non strictly convex integrals of the calculus of variations. In: *Mathematical Theories of Optimization*, Proceedings, ed. by J.P. Cecconi, and T. Zolezzi, Lecture Notes in Math., vol. 979. Springer, Berlin, 1983, pp. 216–231.

[4] Quasiconvex quadratic forms in two dimensions. *Appl. Math. Optim.* **11** (1984) 183–189.

[5] Approximation of quasiconvex functions and lower semicontinuity of multiple integrals. *Manuscripta Math.* **51** (1985) 1–28.

[6] On the definition and the lower semicontinuity of certain quasiconvex integrals. *Anal. Non Linéaire* **3** (1986) 391–409.

Marcellini, P., and Sbordone, C.

[1] Relaxation of non convex variational problems. *Atti Acad. Naz. Lincei* **63** (1977) 341–344.

[2] Semicontinuity problems in the calculus of variations. *Nonlinear Anal.* **4** (1980) 241–257.

[3] On the existence of minima of multiple integrals of the calculus of variations. *J. Math. Pures Appl.* **62** (1983) 1–10.

Marcus, M., and Mizel, V.J.

[1] Transformation by functions in Sobolev spaces and lower semicontinuity for parametric variational problems. *Bull. A.M.S.* **79** (1973) 790–795.

[2] Lower semicontinuity in parametric variational problems, the area formula and related results. *Amer. J. Math.* **99** (1975) 579–600.

Marques, M.D.P.M.
[1] Hyperelasticité et existence de fonctionnelle d'énergie. *J. Mécanique Théorique et Appliqée* **3** (1984) 339–347.

Mascolo, E., and Schianchi, R.
[1] Existence theorems for nonconvex problems. *J. Math. Pures Appl.* **62** (1983) 349–359.
[2] Nonconvex problems of the calculus of variations. *Nonlinear Anal.* **9** (1985) 371–380.

Meyers, N.G.
[1] Quasiconvexity and lower semicontinuity of multiple variational integrals of any order. *Trans. A.M.S.* **119** (1965) 125–149.

Miranda, M.
[1] Sul minimo dell'integrale del gradiente di una funzione. *Ann. Sc. Norm. Sup. Pisa* **19** (1965) 626–665.

Moreau, J.J.
[1] *Fonctionnelles convexes*. Séminaire sur les équations aux dérivées partielles, Collège de France, Paris, 1966-1967.

Morrey, C.B.
[1] Quasiconvexity and the semicontinuity of multiple integrals. *Pacific J. Math.* **2** (1952) 25–53.
[2] *Multiple Integrals in the Calculus of Variations*. Springer, Berlin, 1966.

Moser, J.
[1] On the volume elements on a manifold. *Trans. A.M.S.* **120** (1965) 286–294.

Murat, F.
[1] Compacité par compensation. *Ann. Sc. Norm. Sup. Pisa* **5** (1978) 489–507.
[2] Compacité par compensation II. In: *Recent Methods in Nonlinear Analysis*, Proceedings, ed. by E. De Giorgi, E. Magènes, and U. Mosco. Pitagora, Bologna, 1979, pp. 245–256.

Murat, F., and Tartar, L.
[1] Calcul des variations et homogénéisation. In: *Les méthodes de l'homogénéisation: théorie et applications en physique*. Dir. des études et recherches de l'EDF, Eyrolles, Paris, 1985, pp. 319–370.

Nečas, J.
[1] *Les méthodes directes en théorie des équations elliptiques*. Masson, Paris, 1967.
[2] Theory of locally monotone operators modeled on the finite displacement theory for hyperelasticity. *Beitrᗺge zur Analysis* **8** (1976) 103–114.

Nečas, J., and Hlaváček, I.
[1] *Mathematical Theory of Elastic and Elastoplastic Bodies: An Introduction*. Elsevier, Amsterdam, 1981.

Nirenberg, L.
[1] Remarks on strongly elliptic partial differential equations. *Commun. Pure Appl. Math.* **8** (1955) 648–674.

Oden, J.T.
[1] Existence theorems for a class of problems in nonlinear elasticity. *J. Math. Anal. Appl.* **69** (1979) 51–83.

Oden, J.T., and Reddy, J.N.
[1] *Variational Methods in Theoretical Mechanics*. Springer, Berlin, 1983.

Ogden, R.W.
[1] Large deformation isotropic elasticity, on the correlation of theory and experiment for compressible rubberlike solids. *Proc. Roy. Soc. London* **A 326** (1972) 565–584, and **A 328** (1972) 567–583.

Olech, C.

[1] Integrals of set valued functions and linear optimal control problems. In: *Colloque sur la théorie mathématique du contrôle optimal*, C.B.R.M. Vander Louvain, 1970, pp. 109–125.

[2] Weak lower semicontinuity of integral functionals. *J. Optim. Theory and Appl.* **19** (1976) 3–16.

[3] A characterization of L^1 weak lower semicontinuity of integral functionals. *Bull. Polish Acad. Sci. Math.* **25** (1977) 135–142.

Oppezzi, P.

[1] Convessità dell'integranda in un funzionale del calcolo delle variazioni. *Boll. U.M.I.* **1-B** (1982) 763–777.

Pars, L.

[1] *An Introduction to the Calculus of Variations.* Heinemann, London, 1962.

Podio-Giudugli, P., Vergara Caffarelli, G., and Virga, E.G.

[1] Discontinuous energy minimizers in nonlinear elastostatics: an example of J. Ball revisited. *J. Elasticity* **16** (1986) 75–96.

Raoult, A.

[1] Non polyconvexity of the stored energy function of a Saint-Venant–Kirchhoff material. *Aplikace Matematiky* **6** (1986) 417–419.

Raymond, J.P.

[1] Conditions nécessaires et suffisantes d'existence de solution en calcul des variations. *Anal. Non Linéaire* **4** (1987) 169–202.

Reid, W.R.

[1] A theorem on quadratic forms. *Bull. A.M.S.* **44** (1938) 437–440.

Reshetnyak, Y.

[1] General theorems on semicontinuity and on convergence with a functional. *Sibir. Math.* **8** (1967) 801–816.

[2] On the stability of conformal mappings in multidimensional spaces. *Sibir. Math.* **8** (1967) 91–114.

[3] Stability theorems for mappings with bounded excursion. *Sibir. Math.* **9** (1968) 667–684.

[4] Weak convergence and completely additive vector functions on a set. *Sibir. Math.* **9** (1968) 1039–1045.

Rockafellar, R.T.

[1] *Convex Analysis.* Princeton Univ. Press, Princeton, 1970.

[2] Integral functionals, normal integrands and measurable selections. In: *Nonlinear Operators and the Calculus of Variations*, ed. by J.P. Gossez et al., Lecture Notes in Math., vol. 543. Springer, Berlin, 1976, pp. 157–207.

Rudin, W.

[1] *Real and Complex Analysis.* Mac Graw-Hill, New York, 1966.

Rund, H.

[1] *The Hamilton-Jacobi Theory in the Calculus of Variations.* Van Nostrand, London, 1966.

Sbordone, C.

[1] Lower semicontinuity and regularity of minima of variational integrals. In: *Nonlinear PDE and Appl.*, Collège de France Seminar, vol. IV, ed by H. Brézis, and J.L. Lions, Pitman, London, 1983, pp. 194–213.

Serre, D.

[1] Formes quadratiques et calcul des variations. *J. Math. Pures Appl.* **62** (1983) 117–196.

Serrin, J.

[1] On a fundamental theorem of the calculus of variations. *Acta Math.* **102** (1959) 1–32.

[2] On the definition and properties of certain variational integrals. *Trans. A.M.S.* **101** (1961) 139–167.

Silverman, E.
[1] A sufficient condition for the lower semicontinuity of parametric integrals. *Trans. A.M.S.* **167** (1972) 465–470.
[2] Strong quasiconvexity. *Pac. J. Math.* **46** (1973) 549–554.

Sivaloganathan, J.
[1] A field theory approach to the stability of radial equilibria in nonlinear elasticity. (To appear).

Strang, G.
[1] *The Polyconvexification of $F(\nabla u)$*. Research report CMA-RO 9-83 of the Australian National Univ., Canberra, 1985.

Tahraoui, R.
[1] Sur le contrôle des valeurs propres. *C.R. Acad.Sci. Paris* **300**, 4 (1985) 101–104.
[2] Théorèmes d'existence en calcul des variations et applications à l'élasticité non linéaire. *C.R. Acad. Sci. Paris* **302**, 14 (1986) 495–498.
[3] Existence theorems in calculus of variations and applications to nonlinear elasticity. (To appear).

Tartar, L.
[1] *Topics in Nonlinear Analysis*. Preprint, Univ. of Wisconsin, Madison, 1975.
[2] Compensated compactness and applications to partial differential equations. In: *Nonlinear Analysis and Mechanics*, Proceedings, ed. by R.J. Knops, Res. Notes, vol. 39. Pitman, London, 1978, pp. 136–212.

Temam, R.
[1] A characterization of quasiconvex function. *Applied Math. and Opt.* **8** (1982) 287–291.

Temam, R., and Strang, G.
[1] Duality and relaxation in the variational problems of plasticity. *J. de Mécanique* **19**, 3 (1980) 493–527.

Terpstra, F.J.
[1] Die Darstellung biquadratischer Formen als Summen von Quadraten mit Anwendung auf die Variationsrechnung. *Math. Ann.* **116** (1938) 166–180.

Thompson, R.C., and Freede, J.L.
[1] Eigenvalues of sums of Hermitian matrices. *J. Research Nat. Bur. Standards B* **75 B** (1971) 115–120.

Tonelli, L.
[1] *Fondamenti di calcolo delle variazioni*. Zanichelli, Bologna, 1921.

Troutman, J.L.
[1] *Variational Calculus with Elementary Convexity*. Springer, New York, 1983.

Uhlig, F.
[1] A recurring theorem about pairs of quadratic forms and extensions: a survey. *Lin. Alg. and its Appl.* **25** (1979) 219–237.

Van Hove, L.
[1] Sur l'extension de la condition de Legendre du calcul des variations aux intégrales multiples à plusieurs fonctions inconnues. *Proc. Koninkl. Ned. Akad. Wetenschap* **50** (1947) 18–23.
[2] Sur le signe de la variation seconde des intégrales multiples à plusieurs fonctions inconnues. *Koninkl. Belg. Acad. Klasse der Wetenschappen*, Verhandelingen **24** (1949).

Vasilenko, G.N.
[1] Weakly continuous functionals of the calculus of variations in the spaces $W^{1,p}(U; R^m)$. *Soviet Math. Dokl.* **32** (1985) 706–709.

Warga, J.
[1] Relaxed variational problems. *J. Math. Anal. Appl.* **4** (1920) 111–128.
[2] *Optimal Control of Differential and Functional Equations.* Academic Press, New York, 1972.

Weyl, H.
[1] Geodesic fields in the calculus of variations for multiple integrals. *Ann. Math.* **36** (1935) 607–629.

Yosida, K.
[1] *Functional Analysis.* Springer, Berlin, 1966.

Young, L.C.
[1] Generalized curves and the existence of an attained absolute minimum in the calculus of variations. *C.R. Soc. Sci. Lett. Varsovie Classe III* **30** (1937) 212–234.
[2] Generalized surfaces in the calculus of variations I and II. *Ann. Math.* **43** (1942) 84–103, and 530–544.
[3] *Lectures on the Calculus of Variations and Optimal Control Theory.* W.B. Saunders, Philadelphia, 1969.

Zehnder, E.
[1] Note on smoothing symplectic and volume preserving diffeomorphisms. In: *Geometry and Topology*, Proceedings, ed. by J. Palis, and M. do Carmo, Lecture Notes in Math., vol. 597. Springer, Berlin, 1976, pp. 828–855.

Index

Applied Mathematical Sciences

cont. from page II

Springer-Verlag

Berlin Heidelberg New York London Paris Tokyo Hong Kong